Offshore Well Construction

published by
THE UNIVERSITY OF TEXAS
CONTINUING EDUCATION
PETROLEUM EXTENSION SERVICE
2005

Library of Congress Cataloging-in-Publication Data--to come

First Edition 2005
Printed in the United States of America

Catalog No. 1.13010
ISBN 0-88698-216-2

TABLE OF CONTENTS

CHAPTER 1
The Role of Well Construction in the Evaluation and Development of Oil and Gas Reserves

INTRODUCTION

The evaluation and development of oil and gas reserves is a complex process requiring the interaction of numerous different disciplines. Well construction forms a pivotal role in this process as it is responsible for constructing the conduit from the reservoir to the surface.

The process of exploring for oil and gas can be broken down into a number of successive operations, each more expensive and complex than the previous and each generating higher quality data. In addition, at the end of each operation, the data is reviewed and the process amended or terminated as required.

The main components are:

- Geological appraisal
- Geophysical prospecting
- Exploration drilling
- Appraisal drilling
- Development drilling

Geological Appraisal

The known geological data of a region is reviewed. Government bodies have an interest in the economic geology of its sovereign territory and usually enforce laws, which maintain a database of all geological activity within the territory. This means that data which may have been determined during exploration for a particular mineral resource is available to other explorationists. It is obvious that regions of the earth with easy access have been explored in greater detail.

The objective is to identify the types of rocks in which oil and gas may have accumulated. These are sedimentary rocks and the sequence of occurrence of the rocks can be related to other sequences in which oil and gas have already been found. Unfortunately, the patterns in deposition are unreliable and, although patterns in one area may be similar to those where oil and gas have been found, little confidence can be placed on them containing the correct structures to trap oil and gas and also actually containing oil or gas.

Onshore, a geological survey of surface features may be conducted to confirm the geological prognosis, or to fill in details which may be missing from existing surveys.

Offshore, this can be done with shallow drilling.

Geophysical Prospecting

Geophysical prospecting is the application of the principle of physics to the study of subsurface geology.

Geophysical prospecting enhances the geological information already known about a formation. The objective is to separate the basement rocks (those which were formed first and on which sedimentary basins may have subsequently formed) from the sedimentary rocks since oil and gas form in sedimentary rocks. Geophysical methods can be used to measure the thickness of sediments and to measure the shape of structures within the sediments.

Geophysical surveys can be divided into two broad categories:

1. Reconnaissance surveys to outline possible areas of interest where there are thick sediments and the possibility of structural traps
2. Detailed surveys to define well locations to test specific structures

The most commonly used geophysical methods are:

- Magnetic surveys which measure the anomalies in the earth's magnetic field produced by the magnetic properties of subsurface rocks
- Gravity surveys which measure the anomalies in the earth's gravitational field produced by the density of the subsurface rocks
- Seismic surveys that measure the time taken for sound waves to travel through subsurface rocks.

Magnetic and gravity surveys are generally reconnaissance methods.

Seismic surveys are generally detailed surveys.

The raw data from a seismic survey is electronically manipulated and output as a seismic section (fig. 1.1). This is then interpreted to determine the depth and type of rocks present in the subsurface and the structures. It contains no information of the fluid content of the rock.

Additional information on geophysical surveying techniques is attached for reference at the end of this section.

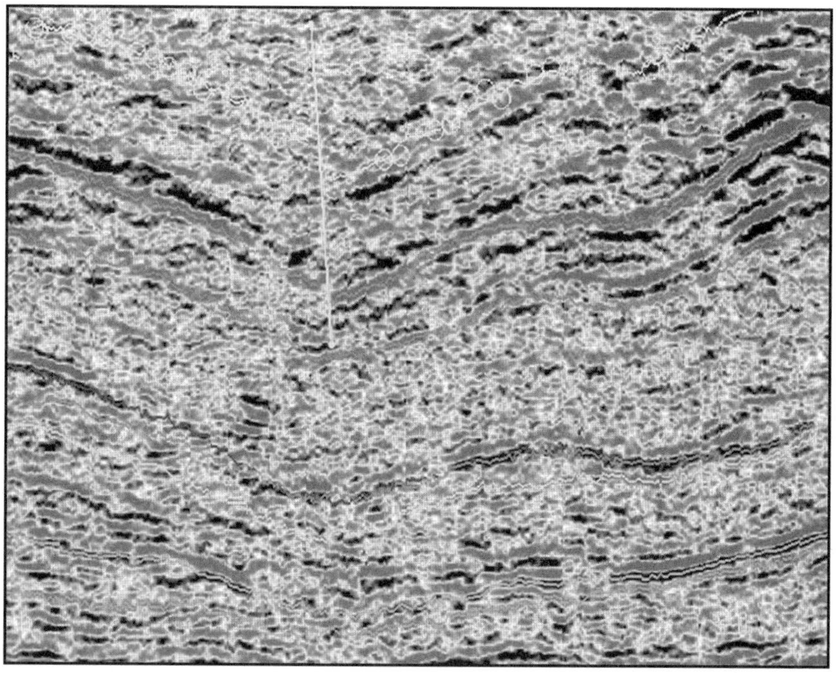

Figure 1.1 Typical seismic section

Exploration Well Drilling

Based on the interpretation of the geological and geophysical studies, the decision may be made to drill an exploration well. The location of the well is planned to intersect the features identified by the geophysical surveys. The cuttings from the well are analyzed by geologists at the well site to build a geological model of the area. As the well is drilled, electric logs are run before the well is cased. These logs measure the natural radiation and electrical potential of the sediments as well as resistivity and sonic travel time. The logs run depends on the geology of each section of the well; the sediments containing hydrocarbons are logged in greater detail.

The geological information and the log information are used to determine if there are hydrocarbon-bearing zones. If there are, the nature and quantity of hydrocarbon, flow properties, and pressure of the hydrocarbon-bearing zone must be assessed, as well as the depth at which the hydrocarbon exists, the thickness of the zone, and the presence of an aquifer.

Formation evaluation is the term used to cover this activity, although the techniques used vary widely.

Routinely, a repeat formation test (RFT) is conducted. A tool is lowered downhole and it is positioned against the side of the borehole. It can then measure the pore pressure (the pressure in the pores of the formation) at that depth and take a sample of the formation fluid. The tool is then released from the side of the borehole and repositioned to take another pressure reading. The pressures and depths can be correlated to examine the density of the fluid (and therefore the type of fluid) and the pore pressure profile within the formation. These results identify the zones that contain hydrocarbons but not the capacity nor the permeability of the formation. This is accomplished by flow tests termed drill stem tests (DST), (since the drill pipe is used as a flow conduit during the test). These are very sophisticated tests which allow sections of the formation to flow as though they were on production. During the test, the pressure at the bottom of the well, the flow rates at the surface, and the composition of the fluids produced is measured. These indicate the volume of hydrocarbons in the zone under test and the flow capacity or permeability of the zone. These tests may last 8 to 24 hours and the data (in exploration wells especially) is treated confidentially. These tests are very expensive and, when taken with the cost of drilling the well, represent a massive investment.

The data from the exploration well (even if it was dry) is reviewed and a decision taken to drill appraisal wells.

Appraisal Well Drilling

The object of appraisal well drilling is to delineate the boundaries of the reservoir. In general, if an exploration well has found economically interesting formations, appraisal wells are drilled in succession, in general, to the north, south, east, and west of the location, ideally to intersect the contacts between the oil and water and the oil and gas (if they are present). The exact location of the wells cannot be planned (or there would be no need for appraisal) and the data from each well is reviewed and the location of the next appraisal well changed accordingly. The logging and testing of the appraisal wells are basically of the same format as the exploration well.

Development Well Drilling

If the results of the appraisal wells are economically encouraging, a development program for the field is put into operation. This program will specify the number and location of development wells to be drilled to fully cover the field and allow production and injection from or into the formations. The development wells may or may not include the exploration well and the appraisal wells. As the wells are drilled, they are logged and tested, the data augmenting the geological model of the formation and modifying the flow model of the reservoir. The number of development wells dictates the size of the platform required and the amount of ancillary equipment (water injection facilities, etc.). The estimate of the reservoir size and the program of development well drilling allows the determination of the production profile for the field. This is significant from an engineering standpoint since it schedules the work involved in bringing the reservoir into production and the remedial work, or workovers, expected during the life of the field. It is also a financial schedule for the field since it indicates the cash flow associated with the production from the field. Taken together with the exploration costs, an estimate of the overall profitability of the field can be made and also the requirements for finance (either borrowing money or producing profit to pay back loans) during the life of the field. As part of this, the reserves must be calculated. These are not fixed figures, since the acquisition of data and reassessment of the reserves is related to the rate of drilling of the development wells and the data generated as the reservoir is produced.

LICENSING

In order to control the activities of companies engaged in the exploration and development of oil and gas reserves, governments will normally sell off or lease the right to explore for hydrocarbons on their sovereign territory. This licensing arrangement works in a multitude of different ways, dependent upon where in the world the operation is taking place.

LEGISLATION

Legislation varies from country to country, so it is always prudent to check the rules and regulations applicable to the particular area being worked in. In addition, it will be necessary to deal with a number of different governmental bodies.

As a general rule of thumb, there will always be a requirement for the following:

- Environmental impact assessment or statement
- Approval to locate a rig
- Approval to drill a well
- Approval to complete a well
- Approval to abandon or suspend a well
- Safety case and bridging documents

OPERATING COMPANY ORGANIZATION

The interaction of the exploration, drilling, petroleum engineering, and production personnel within an organization is a key factor in the efficient transfer of information and, hence, understanding of a specific project. Each oil company has its own organizational structure and ways of conducting business.

In some cases, the above groups form distinct departments within the organization, whereas in others, the structure evolves from a limited number of departments and therefore would involve a combination of groups, such as exploration and petroleum engineering or well services and production.

A typical structure includes geology/geophysics, well construction (which includes well engineering and drilling operations), petroleum engineering, well services and production (which comprises maintenance operations and planning). It can be seen that the range of disciplines involved in petroleum engineering is quite extensive and, in many situations, this broad range of capabilities is used to coordinate across the time span of the exploration, development, and production phases.

Exploration

The exploration department will be responsible for identifying structures for consideration for development and providing a substructure map of the prospect. The responsibility of exploration would be to further update, refine, and modify the substructure map and reservoir modelling in accordance with the increased amount of data which becomes available during the development program. The exploration department will further be required to provide guidance on the selection of final well locations in the development plan in conjunction with the reservoir engineers, within petroleum engineering, who will be assessing the recovery of oil or gas from the structure as a function of the final well locations.

Well Construction

The well construction department is responsible for the safe and efficient drilling of the well to defined targets and locations identified by exploration and petroleum engineering. They are further charged with the responsibility of ensuring that all evaluation work is conducted safely and in accordance with the requirements of the other departments. In this context, there will generally be two specific functions within well construction, namely—operations, which are responsible for the day-to-day supervision and planning of individual wells, and well engineering, which will be responsible for the adaptation and development of new or improved technology for inclusion in the drilling programs.

Petroleum Engineering

Petroleum engineering is a broad-based discipline which has a prolonged input to reservoir evaluation and development.

Petroleum Geology

Normally, there will be geological specialists within the department who will work closely with the petrophysicists and reservoir engineers to ensure that locations of individual wells and the evaluation process are carried out efficiently and yield the required information to improve the reservoir model developed by the company.

Petrophysics

A petrophysicist is responsible for recommending the wireline logs which will be run into individual wellbores and for the analysis of those logs to yield information relating to the reservoir structure and fluid composition. This function is therefore crucial to ensuring that the exploration and development wells yield the required information to provide detail within the geological structure model.

Reservoir Engineering

Reservoir engineering is a broad discipline and, as such, reservoir engineers will be responsible for the following areas of technology.

1. The properties and performance of reservoir fluids.
2. The response of the reservoir rock to the production process.

3. Assessment of the response of a reservoir to the production or depletion process.
4. Identifying and recommending the means by which oil recovery can be enhanced or improved, e.g., pressure maintenance or by the use of enhanced oil recovery.

In general terms, the reservoir engineer is charged with the responsibility of ensuring that the reservoir can be exploited as effectively as possible, and that the reservoir energy available within the fluid is fully utilized to maximize the potential recovery from the reservoir.

Production Technology

The production technologist, or engineer, is responsible for the wellbore and the completion equipment installed within it, and also with the consequences of production in terms of the reservoir fluids; e.g., the tendency for scale, wax, or asphaltene deposition. In the cycle of reservoir evaluation and development, production technologists will be heavily involved in the design and selection of equipment which will be installed inside the wellbore and will be required to withstand operating conditions and the fluids, but, in the longer term, development of the reservoir. The production technologist will be charged with maintaining the wells at their peak operating efficiency and ensuring that maximum recovery is achieved. This may necessitate the implementation of workovers to correct mechanical or reservoir problems which may arise as a result of continued production.

Operations

The operations group within petroleum engineering provides the necessary link between the operational groups within well construction, who will be responsible for the drilling of the exploration and development wells, and the evaluation and technical specialists within petroleum engineering for whom the well is being drilled to yield the necessary information for the reservoir modelling. The operations section therefore requires a detailed understanding of the role of well construction and also of the various disciplines within petroleum engineering to ensure they can provide the effective coordination necessary.

Economics

The role of economics is fundamental to the evaluation, development, and abandonment of reservoirs and wells. It is seen as being the means by which technical information can be transmitted into management terms to allow decisions to be made regarding future investment or abandonment of projects.

Well Services

The role of well services is to specify and prepare completion equipment for installation inside the wellbore and then to periodically conduct repair work within the wellbore to replace malfunctioning components.

Production

The production department is responsible for the ongoing and continuous production of fluids from the reservoir. Their responsibility is therefore to monitor and control production in such a way as to maximize the recovery of reserves from the reservoir. The planning of production rates and production plateaus are frequently based upon reservoir models generated by reservoir engineering within the petroleum engineering section and will be implemented by the production department. Since the production department is responsible for the development wells once they are in production, it is their responsibility to ensure the wells are maintained in peak operating capacity and, as such, they will be responsible for coordinating all maintenance work required within the platform and also around the individual wells.

APPENDIX
Geophysical Survey Types

MAGNETIC SURVEYS

The igneous and metamorphic rocks of the basement complex are magnetic in varying degrees and create anomalies in the earth's magnetic field. Sedimentary rocks are practically nonmagnetic and magnetic measurements on or above the surface of a sedimentary basin are, therefore, separated from the source of the anomalies by the thickness of the sediments. As the magnitude of an anomaly is related to the distance from its source, the method can be used to deduce the thickness of sediments overlying the basement. The major tectonic trends of the basement may also be revealed by magnetic surveys. Since lava flows and igneous intrusions usually have quite strong magnetic effects, their presence in sediments can be detected by this method.

The earth's magnetic field is very weak, varying from 60,000 gammas in a vertical direction at the magnetic poles to about half that intensity in a horizontal direction at the magnetic equator. The magnetic field between the poles of a small horseshoe magnet is about 1,000 times the strength of the earth's magnetic field. The magnitude of anomalies that are significant in oil exploration varies from a few gammas to a few hundred gammas.

The anomalies are measured by magnetometers suspended from aircraft flying at a specified altitude along specified flight lines which may be 1 mile apart or up to 20 miles apart, depending on the resolution of the survey. The instruments may measure anomalies to a few hundredths of a gamma. Continuous recordings are made of the magnetic field during the survey, and readings from a ground magnetometer ensure that there are no magnetic storms during the survey.

The results are corrected for variations in the earth's magnetic field and the effects of the sun, for errors in the survey procedure, and for known regional effects. Maps are constructed with contours to show the anomalies. In favorable circumstances, some indication of basement structure may be obtained, along with the separation of near surface and basement effects.

GRAVITY SURVEYS

Gravity surveys measure the effect on the earth's gravitational field of variations in the density of subsurface rocks. Basement rocks have, in general, a higher density than the overlying sediments and, where this is the case, anomalously high gravity values are recorded when the basement rocks approach the surface. Conversely, low gravity values are recorded over depressions in the basement surface. Gravity surveys can therefore be used to outline sedimentary basin development and show structural trends in the basin. Variations in the density also occur in the sedimentary rocks and, when older and denser rocks are brought near the surface in the cores of anticlines and other structures, anomalously high gravity values are recorded. The density of salt is usually lower than that of the surrounding rocks and anomalously low gravity values are frequently associated with salt structures such as salt domes.

The earth's gravity field varies from 983.221 gals at the poles to 978.048 gals at the equator. As anomalies of the order of 0.001 gal can be significant in oil exploration, the unit in gravity surveys is a milligal. Gravimeters are sensitive instruments that can measure changes in gravity of 0.01 milligal; i.e., one part in one-millionth of the earth's gravity field.

The instrument height must be known for each reading and the procedure is time consuming in rough terrain. Ship surveys house the instrument on a gyroscopically stabilized platform to minimize the movement of the ship.

The survey is conducted along a specified number of traverses, spaced from 0.5 to 1 mile apart. The readings are corrected for elevation, latitude, topography, and diurnal variations and are plotted on a map, contoured to show the variations in the magnetic field. The interpretation of the anomalies depends on knowledge of the shapes of the subsurface structures. This information is unknown during exploration and, therefore, the gravity survey data is usually used to provide leads for further geophysical exploration.

SEISMIC SURVEYS

The study of the form and occurrence of earthquake waves recorded by seismographs has been the principal source of knowledge of the constitution of the interior of the earth. Using a special type of seismograph, or geophone, seismic surveys explore the geological structure in the earth's sedimentary section by recording the ground movements produced by man-made explosions. The waves created by the explosions are reflected back to the earth's surface by the elastic discontinuities that occur at changes of rock types in the sediments. Seismic surveys are divided into two categories depending on the path taken by the waves in the sediments between the explosion and the geophones. They are termed the "reflection" and "refraction" methods. Seismic surveys provide more detailed information about the shape and depth of subsurface structures than any of the other geophysical methods. They are the methods most frequently used in the exploration for oil and gas.

Both seismic methods measure the time taken by the waves to travel from the explosion, or shot-point, to the geophones. It is therefore necessary to record both the time of the shot and the time of arrival of the waves at the geophones. The travel times are rarely longer than 6 seconds and are measured to one-thousandth of a second. The information is recorded on magnetic tape in the field and the tapes are subsequently processed in a data-processing center.

Seismic Reflection Method

Reflection surveying is similar in principle to echo-sounding at sea, where an acoustic signal is transmitted from a ship and reflected back to the surface by the sea bottom. The time taken by the signal to return to the surface is converted to the water depth from a knowledge of the velocity of the signal in water. The reflecting horizons in seismic surveys occur at changes in the circumstances of geological formations and these are not usually as clearly defined as the sea bottom. Also, there are generally many changes of formation within the sedimentary section so that a seismic reflection record may contain many reflections and is therefore more complicated than an echo-sounder record.

The reflected energy is recorded by groups of geophones laid on the ground at equally spaced intervals (50 or 100 metres apart) along a line. The number of groups used to record each shot may be 24, 48, or as high as 96, and hence the spread length with 50 m spacing between geophone groups varies from 1,200 m to 4,800 m. The ground movements resulting from the energy released by the shot cause the geophones to generate small electrical impulses which are taken by cable to a conveniently located recording station where the impulses from each of the groups are amplified and recorded in digital form on magnetic tape. A monitor record on photographic paper is also taken to check that a satisfactory record is obtained. The shot point can be located either at the center of the spread or at one end of the spread.

The geophones record all ground movements and a reflection record is complicated by the effects of extraneous movements from natural and man-made sources and from the shot itself. These movements, which are called "noise," tend to obscure the reflected energy, and field techniques are designed to minimize the noise on the record and enhance the reflections. The noise affecting a reflection record can be reduced by varying the number and the spacing of the geophones in a group by employing a pattern of shot holes instead of a single shot, and by the use of electrical filters in the amplifiers. However, the greatest improvement in signal to noise ratio is obtained by the use of multiple coverage of "common depth point" (CDP) shooting, as it is commonly called, and this technique is now almost universally employed on reflection surveys. In this technique, the shot point and geophone stations are moved along the line between shots and hence multiple records are obtained corresponding to reflections from the same subsurface points; i.e., CDPs. The number of stations by which the source and geophone stations are moved along the line between shots determines the multiplicity of cover. The records corresponding to each common depth point are added together during the processing of the data to enhance the reflections and cancel out the random noise.

Although the seismic reflection method can be applied to both marine and land surveying, the different problems associated with them require different operational techniques.

On land the energy source is normally a small explosive charge detonated in a shallow drilled hole, but falling weight and vibrating plate sources are also available. These have much less energy than a dynamite charge, but addition of signals from repeated drops or vibrations improves the signal to noise ratio to compensate for this. The geophone cable is connected in sections so that a "roll-along" technique may be used in which the last recording section can be moved to the front, thus moving the spread along the line.

The seismic reflection method has been adapted very successfully to the exploration of marine areas. The recording and shooting operations can be conducted from a single ship which houses the recording instruments and tows a neutrally buoyant cable containing the geophones.

In marine work, the energy sources used are generally nondynamite sources, such as compressed air or gasses exploding under water or an electric discharge under water. These sources are towed at the rear of the ship at a suitable depth beneath the surface of the sea. Under favorable weather conditions, surveys can proceed much faster at sea than on land because drilling of shot holes is eliminated and the geophone spread moves continuously along the line. As a consequence of this, the degree of multiplicity of subsurface coverage recorded on marine surveys is generally higher than that recorded on land surveys. As in all geophysical surveys, accurate position fixing is an important and integral part of the operation and, at sea, one of the radio navigational aids, such as the Decca system, is generally used in conjunction with the Satellite Navigational system.

Seismic data is recorded on magnetic tape in digital form and processing of the data is undertaken by computer. All of the standard processes that are applied to reflection records prior to the interpretation of the results can be handled by computer. The results are output as seismic sections through the sediments and skilled operators can then interpret these to determine structures.

Seismic Refraction Method

A portion of the seismic energy from a shot is refracted at the elastic discontinuities that occur within and at the base of the sedimentary section. When the formation below a discontinuity has a wave velocity higher than the overlying formations, the waves are refracted along the higher velocity formation and give rise to waves that return to the earth's surface. The waves are detected at the surface by geophones and recorded with equipment similar to that used for reflection surveys.

The distance between the shot and the return of the waves to the surface depends on the depth of the refracting formation. When the sedimentary section contains a number of refracting formations of increasing velocity at successively greater depths, waves from each formation are recorded in turn as the distance from the shot is increased. By selecting the appropriate shot to geophone spread distance and moving both shot and spread along a line, continuous subsurface coverage of the refracting horizon is obtained. The depth and shape of the refracting formation can be calculated from the travel time recorded along a line.

A refraction spread is laid out in line with the shot-point with the geophones spaced at equal intervals along the spread. The distance between the geophone stations is generally about 1,000 feet and a spread of 24 geophones covers a distance of nearly 5 miles. Refraction observations are made at distances of 15 miles or more when a deep formation is mapped and, at this distance, charges of up to 3 tons of explosive may be required to give adequate refracted energy.

Interpretation of Seismic Results

The object of a seismic survey is the location and detailing of structural traps in which oil may have accumulated. The initial program of seismic lines depends on the existing knowledge of the area and may vary from widely spaced reconnaissance lines over unexplored territory to a detailed survey to assist in the location of development wells. It is usual for lines to be added to the program either as the work develops or as a follow-up to the results of the first survey. In the case of marine seismic surveys, it is generally the latter, as the field work is often completed several weeks before the results can be processed and interpreted. When the field records have been processed, the travel times to the reflecting or refracting formations are plotted on maps and contours drawn through equal time values. A separate map is drawn for each formation. The time-contour, or isochron, maps show all the structural features, but depth-contour maps are more convenient for exploration purposes.

As the velocities that must be used in the conversion of time to depth maps can vary from 4,000 feet per second for unconsolidated near-surface sediments, to 6,000–13,000 feet per second for sandstones and shales, and 14,000–20,000 feet per second for limestones, it is essential to determine the appropriate velocity for each area. It is also essential to recognize any significant velocity variations in an area, as these can cause an appreciable change in the shape of the contours as they are converted to depth. The direct method of measuring velocities of seismic waves in the sedimentary section is to lower a geophone into a well and measure the travel times from shots near the surface to various depths in the well. However, velocity information is frequently required before the first well is drilled and can be obtained from a statistical analysis of reflection data. The development of common depth-point shooting methods has increased the amount of data available for velocity analysis. The velocities recorded on refraction lines are another source of information. When a well has been drilled, a continuous velocity log may be run in the hole. The log is recorded by an instrument which measures and integrates travel times over short portions of the well. The accuracy of the integration is checked at selected depths by comparison with the travel times measured directly with a geophone.

The picking of the times to the various reflection horizons to be mapped on a seismic section and the processing of the maps (contouring) can be done by computer, to allow the conversion of travel time into depth, and hence the production of isopach maps.

CHAPTER 2
Well Design Process

OVERVIEW

The well construction process can be broken down into five sequential phases of work, as follows:

1. Preliminary well design
2. Detailed well design
3. Prepare drilling program
4. Execute well program
5. Analyze and improve performance

Well design focuses primarily on the preliminary and detailed well design and the preparation of the drilling program.

PRELIMINARY WELL DESIGN

Preliminary well design is essentially a screening stage of the well design process. The major steps are shown below.

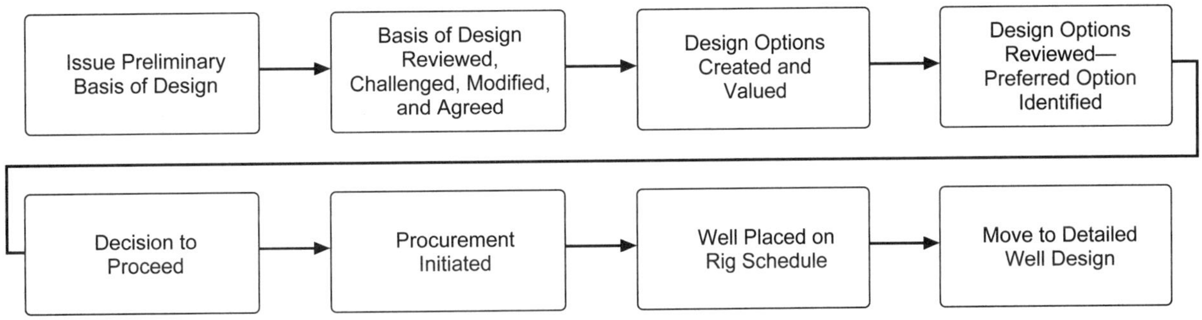

Issue Preliminary Basis of Design

Once the geological and geophysical studies have identified a potential well location, the subsurface team will work up a basis of design. This is the information that gets handed over to the well construction team and forms the basis of the well design. The basis of design will generally provide information on the following:

- Well name and number
- Well objectives
- Total depth
- Surface location
- Water depth
- Target location
- Target size and tolerance
- Target constraints
- Geological prognosis
- Seismic section
- Expected hydrocarbons
- Anticipated pore pressures
- Anticipated temperature profile
- Offset wells
- Geological hazards (shallow gas, faulting, H_2S, CO_2, lease line restrictions, flow lines, etc.)
- Additional constraints (drilled before a certain date, etc.)
- Evaluation program (details and justification of required wireline logs, coring, and testing)

Basis of Design Reviewed, Challenged, Modified, Agreed

The basis of design will be reviewed at a meeting held between well construction and the relevant subsurface groups. The aim of this meeting is to ensure a common understanding of the goals and objectives of the well and how they will be achieved.

If necessary, certain aspects of the basis of design will be challenged and, where necessary, modified. This normally relates to the evaluation program and criteria relating to the target size and tolerance.

Once any modifications are made, representatives of the subsurface and well construction teams sign off the basis of design.

Design Options Created and Valued

The drilling engineer will take the signed-off basis of design and generate a number of design options.

As a first step, the drilling engineer will review all of the available offset data and regional data. Typical offset data reviewed includes:

- Pore and fracture pressure plots
- Time depth curves
- Daily drilling reports
- Daily mud reports
- Final well reports
- Mud logging records
- Bit records
- Casing and cementing reports
- Survey records

This will give the drilling engineer an understanding of how previous wells were drilled, what problems were experienced and how they were solved, what casing program was used, what mud type and weights were used, any directional problems experienced, and how long the well took to drill, etc.

All of the offset data is normally compiled into an offset data pack for future reference.

The drilling engineer will take the offset data and the basis of design and work up a series of different design options. This will normally involve a number of different casing schematics or variations on well trajectories.

The selection of casing setting depths will be discussed in more detail in Chapter 3, Casing Design Section.

For each option, the drilling engineer will generate the following information:

- Provisional trajectory
- Casing schematic
- Provisional mud program, including mud types and weights
- Provisional cement program, including types of cements and slurry types
- Torque and drag assessment
- Budgetary time estimate
- Budgetary cost estimate
- Hazard assessment

Design Options Reviewed, Preferred Option Identified

The drilling engineer will present the various well design options at a peer review meeting. Present at this meeting will be the members of the subsurface team and various members of the well construction team.

The aim of the meeting is to ensure that all the requirements of the basis of design have been met by the various design options, that all hazards have been identified, and to agree on a preferred option to carry forward to detailed design.

If a radical new design is being proposed, then additional study work might be required to prove a particular aspect of the design or to eliminate or reduce a particular hazard; e.g., elimination of a casing string, using a surface stack from a semisubmersible, using unusual hole sizes, new mud systems, etc.

Decision to Proceed

Once the preferred option has been identified, the subsurface group inputs the budgetary cost estimate into an economics model to determine if the well meets the economic criteria laid down by the operating company.

It is also likely that, at this stage, a series of discussions will be held with the other partners in the well to confirm their acceptance of the proposed well design and economics.

Procurement Initiated

Once the decision to proceed has been received, then the well construction department initiates the procurement process. Procurement can be broken down into two main areas:

- Contracts
- Materials

Contracts

Contracts are required to cover all of the services needed to drill a well. Typical contracts are required to cover the following:

- Site survey
- Drilling rig
- Rig moving
- Mud logging
- Wireline logging
- Directional drilling and surveying
- ROV
- Helicopters
- Supply boats
- Supply base facilities
- Drilling tools (jars, accelerators, etc.)
- Fishing tools
- Cementing
- Drilling fluids

How the contracts are tendered and awarded depend upon the particular operating company practices and any applicable legislation. For example, in the European Union (EU), all contracts must be prequalified according to a specific set of rules.

Contracting Strategy

Typical contracts that are used for drilling rigs include the following:

- Daywork drilling
- Incentive-based daywork drilling
- Lump sum
- Footage
- Limited turnkey drilling
- Integrated project management

Each of these has various merits, as discussed below. Which type of contract is awarded depends very much on the particular operator's preference and capabilities.

Daywork Drilling

- Conventional operating arrangement
- Operator assumes ALL the risk of drilling the well
- Drilling contractor and other third-party service providers compensated for work done on a daily basis
- All consumables charged to the operator's account
- Most clear cut

Incentive-Based Daywork Drilling

- Risk shared by operator and contractors
- Basic metrics are set
- Performance measured against the predicted drilling curve
- If the well is drilled ahead of schedule, a proportionate bonus is paid
- If the well takes longer than predicted, a penalty is applied to the contractor compensations
- The bonus and penalty are usually capped

Lump-Sum Operating Rate Daywork Drilling

- Operator with limited resources depends upon another operator or contractor to supply a majority of the required services and materials
- The service company is compensated on a daily lump-sum basis and reimbursed for the consumables
- Feasible for small operators—limited budget
- Generally, the least common contracting strategy

Footage Drilling

- Tends to apply over certain sections of a well
- Win-Win for both operator and drilling contractor if applied correctly
- Needs to be fully evaluated
- Operator does lose some control—needs to be evaluated
- More applicable on long wells with long hole sections

Limited Turnkey Drilling

- Shared risk agreement between operator and turnkey contractor
- Operator supplies some third-party services
- Operator supplies required tubulars and other tangible items
- Turnkey contractor supplies the rig and performs agreed tasks on a lump-sum basis
- Can be good where sporadic wells are planned—operator does not have to overstaff the project
- Good for remote locations or foreign locations (local turnkey contractor = better local understanding)

Integrated Project Management

- Usually includes a large, integrated service company
- Service company is a primary participant in the well planning and engineering
- Operator relies on the service contractor to determine material and service requirements and provide these items
- Integrated service company is usually compensated on a daily lump-sum basis plus operational charges
- This strategy allows the operator to shift day-to-day well operations and logistical planning responsibilities to the service contractor
- Integrated service provider can hire the third-party services

Materials

The term *materials* typically covers the following types of equipment:

- Casing
- Tubing
- Wellheads
- Christmas trees
- Drilling mud
- Cement and additives
- Casing accessories
- Drill bits

Well Placed On Rig Schedule

Once the decision to proceed has been received, a preferred spud date is determined based on the amount of time to complete the detailed well design, lead time to procure material, constraints identified in the basis of design, or any other constraints that might exist, e.g., monsoon season, wait-on-weather issues, etc., and the well is placed on the rig schedule.

DETAILED WELL DESIGN

The major steps involved in detailed well design are shown below.

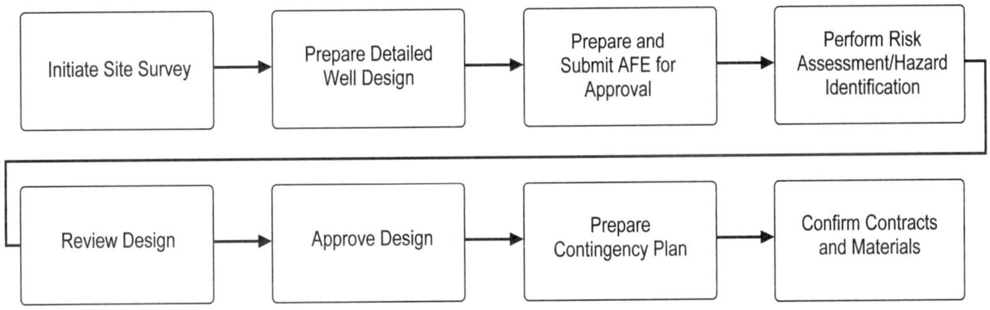

Initiate Site Survey

If a site survey has not been performed, then this will be initiated.

For offshore locations, the site survey is used to determine the following information:

- Water depth
- Seabed conditions (location of debris, anchor-holding assessment, etc.)
- Shallow geology
- Presence of shallow gas
- Soil strength (jackup leg penetration and conductor load capability)

In addition, if required, environmental data on wind, wave, and currents will also be collated and their impact on the well design assessed.

For onshore locations, the site survey is used to determine the following information:

- Site location
- Road access
- Site preparation
- Shallow geology
- Presence of shallow gas
- Local hazards (flooding, mud slides, etc.)
- Weather impact

For all locations, the support requirements are also evaluated at this point and any impact on the well design assessed. The following are typical areas that are evaluated:

- Transportation of personnel and supplies to the location
- Emergency response
- Medical facilities
- Local infrastructure

Prepare Detailed Well Design

Detailed drilling design entails taking the preliminary well design and developing it further to the point at which the drilling program can be prepared.

Detailed well design includes, but is not limited to, a detailed engineering study and design of the following areas of the well:

- Pore and fracture pressure profiles
- Temperature profiles (HPHT wells)
- Casing design
- Casing and components
- Drilling fluids
- Hydraulics and hole cleaning
- Cementing design
- Trajectory and surveying
- Torque and drag
- Drill string design
- Well abandonment
- Completion design
- Well cost and duration
- Contingency planning

Obviously, the amount of time spent on each area is a function of the complexity of the well being planned.

As a number of these issues are interrelated, it is essential that a system of change control be used to ensure that the effect of changing a parameter is carried throughout the complete design. For example, changing mud weight can affect casing design, hydraulics, hole cleaning, etc.

Prepare and Submit AFE

An authorization for expenditure (AFE) is required for any well construction operation. The AFE requires that all partners sign off on the well (and in some countries to be approved by government) before the well is spudded.

The AFE provides an estimation of the well duration and cost together with a detailed breakdown of the major components that make up the total cost.

The well duration is an estimate of how long the well will take to drill and complete. The timings are normally based on historical well times, often with additional contingency for weather.

The cost is a combination of the following cost types:

Services

- Rig
- Mud engineering
- Cementing
- Directional drilling
- Fishing
- Wireline logging
- Rental tools
- Etc.

Consumables

- Mud
- Cement and additives
- Casing and tubing
- Welhead and Christmas tree
- Bits and nozzles
- Fuel

Logistics

- Helicopters
- Supply boats
- Transport
- Supply base
- Telecomms and IT

Support

- Supervision
- Well planning
- Operator overhead

Perform Risk Analysis

The hazard assessment performed as part of the preliminary well design is reviewed and updated as required based on the detailed design. Any additional hazards identified are recorded, and appropriate safeguards developed.

For any high risk or high consequence activities, a risk assessment will be performed and documented. Quantitative techniques may be applied as appropriate.

The final hazard assessment and any risk analysis is normally reviewed and approved by senior well construction management.

The purpose of performing hazard assessment and risk analysis is as follows:

- To ensure that all well construction hazards and their effects on personnel, environment, and property are identified and assessed.
- To ensure that there are adequate safeguards in place to reduce risks to as low as reasonably practical (ALARP).

Peer Review Design

The detailed design is normally subjected to a series of peer reviews at various stages, depending upon the well complexity, and will invariably be signed off at a senior management level.

It is now becoming quite common for rig site personnel to be involved in some of these peer reviews (see later section on Technical Limit).

Approve Design

Once any actions evolving from the peer reviews have been closed out, the final well design will be approved and signed off.

Prepare Contingency Plans

Contingency planning, based on simple "what if" scenarios, is performed to ensure that

- sufficient material and equipment are available.
- procedures are developed to mitigate the likelihood of the contingency occurring (a large number of contingency options are developed from the hazard assessment).

- procedures are developed so that personnel know what to do in the event of an emergency.
- the well design is robust enough to cope with sudden changes of plan.

Typical "what if" scenarios reviewed are as follows:

- Well control event
- The pipe gets stuck
- The casing won't run to bottom
- Too much angle is being built
- The target will be missed
- Pore pressure is higher than predicted
- Losses occur
- Hole instability occurs
- Geology is not coming in as predicted

The purpose of contingency planning is to ensure that unforeseen events do not result in a poorly planned response that results in injury to personnel or damage to the environment or equipment.

Confirm Contracts and Materials

The drilling engineer ensures that all contracts are in place for all the required services and that materials procurement is being carried out according to the final design specifications.

PREPARE WELL PROGRAM

The major steps involved in the preparation of the well program are as follows:

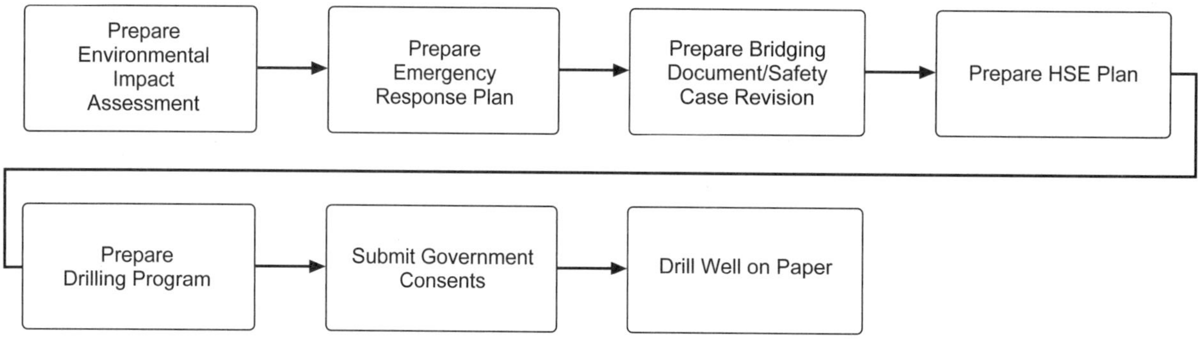

Prepare Environmental Impact Assessment

Environmental impact assessments are now required for most operations worldwide. Once completed, they are submitted to government bodies for approval, which can sometimes take months, especially if the well is to be drilled in an environmentally sensitive area. This activity is often performed in parallel to the detailed well design phase.

Prepare Emergency Response Plan

An emergency response plan is required to bridge between the drilling contractor's and operator's emergency response and oil spill contingency plans.

Prepare Bridging Document

A bridging document or safety case revision is required to bridge between the rig safety case and the operator's management system.

Prepare HSE Plan

A number of operators require that there is an HSE plan developed for each well or series of wells.

Prepare Drilling Program

This document essentially provides guidance on how the well is to be drilled, ensuring that any constraints in the well design are adhered to. The contents of a typical drilling program are shown in Appendix 1.

The drilling program will be signed off prior to distribution.

Prepare Consent Documentation

A number of documents are normally required to be submitted to government agencies prior to spudding the well. This is done to gain consent or permission from the government to perform various activities. Typical approvals or consents required are:

- Consent to Drill
- Consent to Move a Rig
- Consent to Locate a Rig

Drill Well On Paper

A drilling the well on paper exercise is normally held prior to spudding the well. Both office and rig site personnel, from the operator, drilling contractor, and additional service providers, attend. The exercise has three main objectives.

1. A dry run of the well, aimed at identifying any problems ahead of time
2. Explaining why the well is being drilled the way it is
3. Obtaining ideas from the rig site personnel as to performance improvements that could be made.

EXECUTE WELL PROGRAM

As the well is being drilled, the progress is monitored and reported, often against a time-depth curve or other performance measures, such as days per 1,000 ft, percent of nonproductive time, etc.

Progress is also monitored against the well design parameters and, if required, additional design verification is made. For example, if a formation comes in deeper than predicted, or a leak-off test is lower than anticipated.

Although the bulk of these variations should have been addressed in the contingency planning, it is still necessary to complete an as-built design and verify that it meets the various acceptance criteria laid down by the basis of design, the operating companies' internal policies, and any government legislation.

ANALYZE AND IMPROVE PERFORMANCE

As the well is drilled, the performance is analyzed and performance improvement recommendations made. This is normally done on a continual basis and is explained in further detail in the Technical Limit section later on.

APPENDIX

Sample Well/Drilling Program Format

A typical well program should contain the following information. The amount of detail within each category will obviously vary depending on the well type (conventional, ERD, deepwater, HPHT).

1. Well Information
   a. General Information
   b. Well Objectives
   c. Geological Prognosis
   d. Pore and Fracture Pressure Plot
   e. Well Montage
   f. Directional Plot
   g. Risks/Hazards/Potential Problems
   h. Time-Depth Curve
   i. BOP Configuration
2. Drilling Procedure
   a. Rig Move/Prespud
   b. Individual Hole Section Details
      - Objective
      - Potential Problems
      - Offset Data Summary
      - BHA
      - Bit and Hydraulics
      - Drilling Operations/Practices
      - Drilling Fluids
      - Casing and Cementing
      - Wellhead
      - Contingency Procedures
3. Directional Drilling and Surveying Program
4. Wireline Logging Program
5. Coring Program
6. Well-Testing Program
7. Completion Program
8. Abandonment Program
9. Emergency Procedures
   a. Weather
   b. Well Control
   c. Other
10. Appendixes
   a. Seabed Survey
   b. Structural Maps
   c. Bit Records
   d. Offset Data
   e. Offset Logs
   f. Drilling Fluids Program
   g. Cement Program
   h. Service Providers and Contact Details
   i. Hazard and Risk Assessment
11. References and Drawings
   a. Technical Literature
   b. Equipment Specifications
   c. BOP and Wellhead

CHAPTER 3
Casing Design

INTRODUCTION

Casing design is about achieving the total depth of the well safely, with the most cost effective number of casing or liner strings.

Purpose of Installing Casing

In order to allow the drilling and completing of a well, it is necessary to line the drilled open hole with steel pipe or casing. Once in place, this pipe is cemented, supporting the casing and sealing the annulus in order to

- strengthen the hole.
- isolate unstable, flowing, underbalanced, and overbalanced formations.
- prevent the contamination of freshwater reservoirs.
- provide a pressure-control system.
- confine and contain drilling, completion, produced fluids, and solids.
- act as a conduit for associated operations (drilling, wireline, completion, and further casing or tubing strings) with known dimensions (IDs, etc.).
- support wellhead and additional casing strings.
- support the BOP and Christmas tree.

There are primarily six types of casing installed in an onshore or offshore well:

- Stove pipe, marine conductor, foundation pile
- Conductor string
- Surface casing
- Intermediate casing
- Production casing
- Liners

Stove Pipe, Marine Conductor, Foundation Pile

Stove pipe is used for onshore locations and is either driven or cemented into a predrilled hole. The pipe protects the immediate soil at the base of the rig from erosion caused by the drilling fluid.

Marine conductor is a feature of offshore drilling operations where the BOP stack is above the water. It provides structural strength and guides drilling and casing strings into the hole. It is usually driven or cemented in a predrilled hole. The string helps isolate shallow unconsolidated formations and protects the base of the structure from erosion by the drilling fluid.

Foundation pile is usually jetted in or cemented into a predrilled hole from a floating drilling unit—where the BOP stack is on the seafloor. Again, the string isolates unconsolidated formations and supports the guide base for the BOP stack, Christmas tree, or flowbase, and guides drilling and casing strings into the hole.

Conductor String

This string is used to support unconsolidated formations, protect freshwater sands from contamination, and case off any shallow gas deposits. The string is usually cemented to the surface onshore and to the seabed offshore. This is the first string onto which the BOP is installed. If surface BOPs are used (i.e., jackups) the conductor string also supports the wellhead, the Christmas tree, and subsequent casing strings.

Surface Casing

Provides blowout protection for deeper drilling, structural support for the wellhead or subsequent casing strings, and is often used to isolate troublesome formations. The string is either cemented to the surface or to the inside of the conductor string.

Intermediate Casing

Again, this string provides blowout protection for deeper drilling and isolates troublesome formations that could impair well safety and/or hamper drilling operations. An intermediate casing string is commonly set when a well is likely to encounter an influx and/or loss of circulation in the open hole, thus providing blowout protection by upgrading the strength of the well. The cement height is determined by the design requirement to seal off any hydrocarbon or flowing salt zones. The top of the cement does not need to be inside the surface string.

Production Casing

This is the name applied to the casing that has the production tubing run within it and could potentially be exposed to reservoir fluids. It can either extend to the surface as an integral string or be a combination of a production liner (7") and the previously set production casing (9⅝"). The purpose of production casing is to isolate the producing zones, allow for reservoir control, act as a conduit for safe transmittal of fluids, gas, or condensate to the surface, and prevent the influx of unwanted fluids.

Liners

A liner will be suspended a short distance above the previous casing shoe and will be cemented along its whole length to insure a good seal isolating the annulus. Often a liner top packer can be set as a precautionary second barrier. HPHT wells that incorporate a long liner may only cement the shoe and squeeze the liner lap. Liners permit deeper drilling, separate productive zones from reservoir formations, and can thus be installed for testing purposes.

Drilling liners are set:

- to provide a deeper shoe.
- to isolate unstable formations.
- to achieve a drilling casing at a reduced cost.
- due to rig limitations.

Production liners are set:

- to complete the well at a reduced cost.
- to allow for a larger production conduit providing a range of choices for the tubing.
- due to rig limitations.

CASING PROPERTIES

Casing is usually specified by the following properties:

- Outside diameter and wall thickness
- Weight per unit length
- Grade of steel
- Type of connection
- Length of joint

Outside Diameter and Wall Thickness

The outside diameter refers to the pipe body and not to the coupling. Coupling diameter is important as it determines the minimum hole size that the casing can be run into.

Wall thickness determines the inside diameter of the pipe and hence the maximum bit size that can be run through the pipe.

The permitted tolerance on outside diameter and wall thickness is given in API Spec 5CT. As a general rule:

Casing outside diameter >= 4½"	Tolerance +1.00%, –0.50%
Casing outside diameter < 4½"	Tolerance ±0.031%
Wall thickness	Tolerance –12.5%

Weight per Unit Length

The nominal weight of casing is used primarily to identify casing during ordering. Nominal weights are not exact and are based on the calculated theoretical weight of a 20-foot length of threaded and coupled pipe.

Grades of Steel

The mechanical and physical properties of casing are dependent upon the chemical composition of the steel and the heat treatment it receives during manufacture.

API defines nine grades of steel for casing.

H40 J55 K55 C75 L80 N80 C95 P110 Q125

The number in the designation gives the API minimum yield strength in thousands of psi. Hence, L80 casing has a yield strength of 80,000 psi.

The letter in the designation gives an indication of the type of steel and the treatment it received during manufacture.

A more detailed section on material selection can be found later.

Types of Connection

There are a host of connection types available on the market today. Selection of a suitable connection should be based upon the intended application, the required performance, and cost.

The table below can act as a rough guide as to whether API or Premium threads should be used.

Production Tubing			
Liquids	API Threads	<5,000 psi>	Premium Threads
Gas	API Threads	<3,500 psi>	Premium Threads
Production Casing			
Liquids	API Threads	<7,500 psi>	Premium Threads
Gas	API Threads	<5,000 psi>	Premium Threads

Surface and Intermediate Casing

If the pressure differential across the connection is ≥7,500 psi, a premium thread is the preferred option. An API thread with an enhanced coupling design can be used, although its sealing qualities are not as reliable. Leak-resistance values for API connections can be found in API bulletin 5C2.

Connection Properties

The connection collapse, burst, and tensile properties should be compared with the pipe body properties. Whichever are the lowest should be used in all casing design connections.

In addition, some connections have a very low capacity in compression when compared to their tensile strengths. If compression or compression/bending is a critical load, query the manufacturer on their coupling capacity under these conditions (i.e., Vam SC has only 25% cap in compression versus tension).

Six generic connection types are available. These are shown below with some general characteristics.

API 8-Round, STC or LTC

- Good availability and price
- Liquid sealability up to about 210°F
- Sealing is a combination of connection geometry and thread dope.
- Poor gas tightness
- Gauges and expertise are widely available for rework and refurbishment.
- Prone to galling and cross-threading due to out of roundness, especially in larger ODs
- High assembly circumferential (hoop) stress in coupling
- Tensile efficiency 70–75% depending on thread type
- Leak resistance must be verified per API Bulletin 5C3.

API BTC

- Good availability and price
- Poor gas tightness
- Liquid sealability up to about 210°F
- Sealing is a combination of connection geometry and thread dope.
- Tin plating improves leak resistance.
- Gauges and expertise are widely available for rework and refurbishment.
- Prone to galling and to cross-threading due to out of roundness, especially in larger ODs
- High assembly circumferential (hoop) stress in coupling
- Tensile efficiency is generally 85–95% of pipe body
- Leak resistance of BTC must be verified per API Bulletin 5C3

Metal-to-Metal Seal, Threaded, and Coupled

- Availability to depend on propriety type, e.g., Vam, Fox, NS-CC, etc.
- Good gas tightness, generally
- Special clearance couplings manufactured from some higher grade material are available to improve hole clearance.
- Susceptible to handling damage if not treated with care. Pins must be bored concentric to seals for effective gas sealing.
- Particularly suited to use on cold-worked high alloys that cannot be upset
- Generally good makeup characteristics due to reduced thread interference compared to API connections
- Gauges and expertise are available, depending on type, for rework and refurbishment and can readily be recut.
- Assembly circumferential (hoop) stress in coupling can be controlled by reduced thread interference since sealing in the thread is not a requirement.
- Tensile efficiency is generally at least equal to BTC and in many instances equal to or exceeds pipe body.

Metal-to-Metal Seal, Upset, and Integral (or Coupled)

- Poor availability of couplings and limited upset recuts for pipe refurbishment
- Costly, especially upsetting

- Good gas tightness
- Usually exhibiting very good repeated make/break capabilities
- Susceptible to handling damage if not treated with care
- Pins must be bored concentric to seals for effective gas sealing.
- Tensile efficiency at least equal to or greater than pipe body

Metal-to-Metal Seal, Formed, and Integral (Flush)

- Hole clearance characteristics excellent, flush pipe OD
- Reasonable availability, easy to refurbish/recut, no requirement for couplings
- Good gas tightness
- Pins must be bored concentric to seals for effective gas sealing.
- Tensile efficiency = 50%–75% of pipe body depending on type of connections
- Connections may be weaker than the pipe body for internal pressure rating.

Weld on, Upset, and Integral

- Very costly (connector, weld, and NDT)
- Elimination of mill end with weld on box
- Coarse threads to resist cross-threading or galling
- Continuous threaded product resists disengagement under severe bending.
- Grades limited to weldable (line pipe) or H-40, K/J-55
- Tensile efficiency generally greater than pipe body

Length of Joint

Casing joints are not manufactured in exact lengths. API has specified three ranges in which pipe lengths must lie.

Range	Length (ft)	Average Length (ft)
1	16–25	22
2	25–34	31
3	>34	42

THE CASING DESIGN OPERATION

There are two phases of casing design.

1. The first takes place during the preliminary well design and involves the casing scheme selection and casing setting depth determination.
2. The second takes place during the detailed well design and involves a determination of the loads that the casing will be exposed to during the life of the well, and the selection of tubulars with suitable mechanical and physical properties that can withstand the predicted loads.

PRELIMINARY DESIGN

Casing Setting Depth Determination

The initial selection of casing setting depths is based on the anticipated pore pressure and fracture gradients. The drilling engineer is responsible for ensuring that, as far as possible, all the relevant offset data has been considered in the estimation of pore pressure and fracture gradients and that, for directional wells, the effect of hole angle on offset fracture gradient data has been considered.

The total depth of the well, and hence the setting depth of the production casing or liner, is driven by logging, testing, and completion requirements. The shoe must be set deep enough to give an adequate sump for logging, perforating, and testing on production activities.

The initial estimate of determining casing setting depths is best determined graphically, as follows—plotting pore pressure and fracture gradient, expressed in equivalent density against depth.

1. Draw the mean pore pressure gradient curve along with lithology, if available. Note any intervals which are potential problem areas such as differential sticking, lost circulation, or high-pressure gas zones.
2. Draw the mud weight curve. The mud weight curve should include a trip margin of around 200 to 400 psi.
3. Draw the predicted fracture gradient curve. Draw a fracture gradient design curve, which parallels the predicted fracture gradient curve with a reduction of 0.3 to 0.5 ppg for kicks and ECD during cementing.
4. Plot offset mud weights and LOTs to provide a check of the pore pressure predictions or highlight the need for further investigation.

A typical plot is shown below in Figure 3.1. The initial casing setting depths can be determined as follows.

1. Working from the bottom up, enter the mud weight curve at Point A.
2. Move up to Point B, which determines the initial estimated setting depth for the production casing.
3. Move across to Point C, which identifies the mud weight requirement for that depth.
4. Move up to Point D, which determines the initial estimated setting depth for the intermediate casing.
5. Move across to Point E to identify the mud weight required at that depth. For the example shown, Point E is the normal pressure range and no further casing is required to withstand the associated mud weight. However, conductor and surface casing are required and the setting depth for these casings is discussed later.

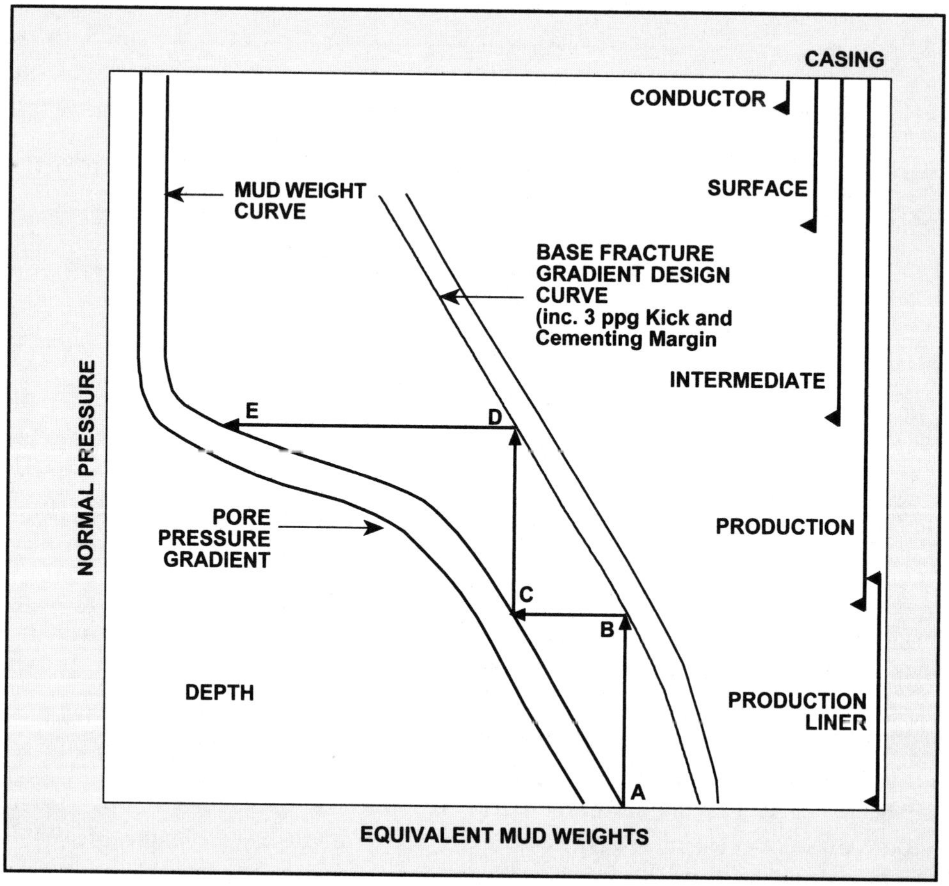

Figure 3.1 A typical plot

Other factors that may impact casing depth selection, in addition to pore pressure and fracture pressure, are:

- shallow gas zones.
- lost circulation zones.
- formation stability which is sensitive to exposure time or mud weight.
- directional well profile. It is important to line out the well trajectory before setting casing and attempt to achieve a consistent survey ahead of a tangent section. Also, long open-hole sections may require casing to reduce the occurrence of stuck pipe and the level of torque.
- sidetracking requirements as specified in the basis of design; e.g., 13⅜" casing might be set high to allow 9⅝" casing to be cut and pulled for a sidetrack in a 12¼" hole.
- freshwater sands (drinking water).
- hole cleaning, particularly if a long section of 17½" hole is required.
- salt sections.
- high-pressure zones.
- lithology—casing shoes should, where practicable, be set in competent impermeable formations.
- uncertainty in depth estimating due to seismic uncertainty.

All of the above need to be considered and the initial casing setting depths adjusted accordingly.

Kick Tolerance

Once the initial casing setting depths have been selected, the kick tolerance associated with those depths should be calculated. Starting from TD up to the surface, the kick tolerance and preferred setting depth for each casing string should be calculated.

Kick tolerance is the maximum kick size that can be taken into the wellbore and circulated to the shoe without breaking down the formation. It is dependent on the mud weight in use, the open-hole weak point (normally assumed to be the previous casing shoe), the formation pressure, the size and density of the influx, and the hole geometry.

There are two methods for calculating kick tolerance. The first calculates a kick intensity and the second a kick volume.

Note that both methods neglect any temperature effects and assume an ideal gas.

Kick Intensity

Kick intensity is a measure of how much the mud weight can be raised for a given kick volume. In other words, if you drill into an overpressured zone, by how much can you raise the mud and still circulate out the kick.

For casing design purposes, the kick volume is assumed to be 25 bbl and the minimum acceptable kick intensity is 0.5 ppg. If the kick intensity is below this value, then further approval should be sought.

Kick intensity is calculated using the following equation.

$$KI = \frac{MAASP - (MW \times 0.052 \times Hi)}{0.052 \times TVD} \quad \text{(ppg)}$$

where

KI = kick intensity (ppg)
$MAASP$ = maximum allowable annular surface pressure (psi)
MW = mud weight in the hole (ppg)
Hi = height of influx (ft)
TVD = true vertical depth of well (ft).

Example

12½" hole TD	13,123 ft
BHA	697 ft × 8" DC
DP	5"
Mud weight	13.2 ppg
Previous casing shoe	8,842 ft
LOT at shoe	14.3 ppg EMW

$$MAASP = 8{,}842 \text{ ft} \times (14.3 \text{ ppg} - 13.2 \text{ ppg}) \times 0.052$$
$$= 506 \text{ psi}$$
$$Hi = 25 \text{ bbl}/[(12.25^2 - 8^2) \times 0.0009714]$$
$$= 300 \text{ ft}$$

(Note that this is the height of the influx around the drill collars. As this volume passes up around the drill pipe, the height will reduce due to a larger annular capacity between DP and open hole.)

$$KI = [506 \text{ psi} - (13.2 \text{ ppg} \times 0.052 \times 300 \text{ ft})]/(0.052 \times 13{,}123 \text{ ft})$$
$$= 0.44 \text{ ppg.}$$

Kick Volume

The kick volume determines the maximum kick size that can be taken into the wellbore and circulated to the shoe.

The following process can be used to calculate kick volume.

1. **Estimate the safety margin to be applied to the leak-off pressure at the open-hole weak point.**

   When the influx is displaced from the hole, there will be additional pressures acting in the wellbore. The following are possible causes of such additional pressure during circulation:
   - Annulus friction
   - Choke operator error
   - Choke line losses (if not compensated for)

   The total safety margin to be applied to the leak-off pressure will be the sum of these additional pressures. The maximum allowable static weak-point pressure can therefore be determined. (This is the maximum allowable weak-point pressure before circulation is initiated.)

   The drilling engineer must use his/her judgement to determine the most appropriate safety factor to be applied to the leak-off pressure at the open-hole weak point. This safety factor should be based on operating area experience.

2. **Calculate the maximum allowable static weak-point pressure (P_{max}).**

   The maximum allowable pressure is given by:

$$P_{max} = P_{lo} - (\text{safety margin}) \text{ (psi)}$$

   where

   P_{max} = maximum allowable weak-point pressure (psi)
   P_{lo} = leak-off pressure at the open-hole weak point (psi)

3. **Calculate the maximum allowable height of influx in the open-hole section.**

   The maximum height of influx that can be taken in the open hole without exceeding P_{max} at Dwp is given by:

$$H = \frac{P_{max} - Pf - (TD - Dwp)\, MW\, 0.052}{0.052\, MW - gg} \text{ (ft)}$$

   where

   H = height of influx (ft)
   P_{max} = maximum allowable pressure at the open-hole weak point (psi)

MW = mud weight in the hole (ppg)
gg = gas gradient psi/ft
TD = bit depth (ft)
Pf = formation pressure at TD (psi)
Dwp = depth of shoe or weak point.

4. **Calculate the volume that this height corresponds to at initial shut-in conditions.**

At initial shut-in conditions, this can be converted to an influx volume as follows:

$$V_1 = H \times C1 \text{ (bbl)}$$

where
V_1 = kick tolerance for initial influx (bbl)
$C1$ = annular capacity at the BHA (bbl/ft).

$C1$ must be determined bearing in mind the hole dimensions in relation to the height of the influx, H. For example, if H is greater than the height of the BHA, both the capacity of the drill pipe/open-hole annulus and the drill collar/open-hole annulus must be used to calculate the kick tolerance, V.

5. **Calculate the volume that this height corresponds to when the top of the influx has been circulated to the open-hole weak point.**

This height corresponds to a volume at the open-hole weak point given by:

$$V_{wp} = H \times C2 \text{ (bbl)}$$

where
$C2$ = annular capacity below the open-hole weak point (bbl/ft).

$C2$ must be determined bearing in mind the hole dimensions immediately below the open-hole weak point in relation to the height of the influx, H.

6. **Calculate what V_{wp} (as calculated in (5) would be at initial shut-in conditions).**

Using Boyle's law to convert this volume to its original volume at initial shut-in conditions (V_2):

$P1 \times V_1 = P2 \times V_2$ or in this case:

$$Pf \times V_2 = P_{max} \times V_{wp}$$

$$V_2 = \frac{P_{max} \times V_{wp}}{Pf}.$$

7. **The kick tolerance is the lower of V_1 and V_2.**

As a general rule, a kick tolerance of less than 100 bbl should be justified by a review of type of well, rig equipment for kick detection, operator/driller experience, area experience, and geology.

The proposed kick tolerance for casing seat design should be noted for each casing, approved by the drilling superintendent, and recorded on the drilling program.

Sensitivity studies should be considered to identify the contingency provisions required to accommodate pore or fracture pressures during drilling which differ from those assumed during design.

Example

12½" hole TD	13,123 ft
BHA	697 ft × 8" DC
DP	5"
Mud weight	13.2 ppg
Previous casing shoe	8,842 ft
LOT at shoe	14.3 ppg EMW

1. Assume a 150-psi safety margin
2. $P_{max} = (14.3 \text{ ppg} \times 8{,}842 \text{ ft} \times 0.052) - 150 \text{ psi}$
   $= 6{,}424 \text{ psi}$
3. $Pf = 13{,}123 \text{ ft} \times 13.2 \text{ ppg} \times 0.052$
   $= 9{,}008 \text{ psi}$

Note that this assumes the well is on balance.

$$H = \frac{6{,}424 \text{ psi} - 9{,}008 \text{ psi} + [(13{,}123 \text{ ft} - 8{,}842 \text{ ft}) \times 13.2 \text{ ppg} \times 0.052]}{(0.052 \times 13.2) - 0.1}$$

$= 604 \text{ ft}$

4. $V_1 = 604 \text{ ft} \times [(12.25^2 - 8^2) \times 0.0009714]$
   $= 50.5 \text{ bbls}$
5. $V_{wp} = 604 \text{ ft} \times [(12.25^2 - 5^2) \times 0.0009714]$
   $= 73.4 \text{ bbls}$
6. $V_2 = 6{,}424 \text{ psi} \times 73.4 \text{ bbls}/9{,}008 \text{ psi}$
   $= 52.3 \text{ bbls}$
7. Kick volume, in this case, is $V_1 = 50.5$ bbls.

Surface and Conductor Setting Depth Design

The minimum setting depth for surface and conductor casings is the depth at which the bottomhole pressure created by the *circulating drilling fluid* (*ECD*) is exceeded by the fracture value of the formation.

The ECD can be significantly affected in large diameter holes by high ROP and poor hole cleaning.

Water depth can play a significant part and results in the depth that the ECD is exceeded by the fracture value being pushed that much deeper, and can result in additional casing strings being run.

In deepwater areas, bending and axial loading are primary considerations in the design of the conductor casing. Due to the complexity of the interaction of the various parameters that affect bending and axial loading, a computer program is needed to accurately model the loads and the behavior of the conductor.

In addition, the load bearing capacity (which directly relates to the soil strength below the mudline) must be determined.

DETAILED DESIGN

The detailed design stage is about determining the loads each casing string will be exposed to during the life of the well, and the selection of tubulars with suitable mechanical and physical properties that can withstand the predicted loads.

The major steps in detailed design for each casing string are shown below.

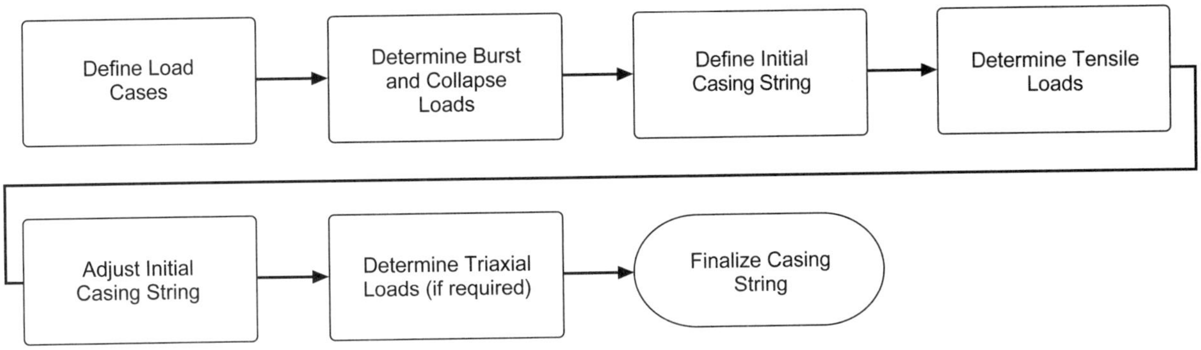

Design Load Cases

The anticipated design load cases should be calculated in the order in which they occur. This helps identify all the loads to which a casing string may be exposed.

Note that not all of these load cases will be applicable to every casing string. For example, production loads do not have to be considered for an intermediate casing string on an exploration well.

Installation Loads

Typical installation loads include

- casing running.
- cementing.
- conventional cementing, stab-in, etc.
- plug bump.

Drilling Loads

Typical drilling loads include

- pressure testing after waiting on cement.
- maximum mud weight.
- well control.
- lost circulation.

Production Loads

Typical production loads include

- pressure testing with completion or kill weight fluid.
- functioning DST tools.
- near surface tubing leak.
- collapse due to plugged perforations.
- special production operations (stimulation, gas lift, injection).

Design Factors

To account for factors that are improperly addressed or not directly accounted for, the properties of the casing are downrated by a design factor before being compared with the calculated design loads.

Typical design factors used are as follows

Collapse	1.0
Burst	1.1
Tension	1.3
Triaxial	1.25

Local legislation should be checked to ensure that other more stringent design factors are not stipulated.

Collapse Design

For all casing strings, a collapse load occurs when the external pressure is greater than the internal pressure.

Collapse design focuses on the internal and external pressure profiles.

Generally speaking, the collapse load will be highest at the casing shoe.

Collapse Installation Loads

The worst case collapse load during installation occurs during cementing.

For conventional cement jobs, the worst case occurs with the cement column on the outside of the casing.

Example

9⅝" Production casing set at 11,450 ft
Mud weight 11 ppg
Top of cement at 7,000 ft
Cement weight 16 ppg
Top of spacer at 5,500 ft
Spacer weight 13 ppg

Internal Pressure Profile

At surface = 0 psi
At casing shoe = 11,450 ft × 11 ppg × 0.052
= 6,549 psi

External Pressure Profile

At surface = 0 psi
At top of spacer = 5,500 ft × 11 ppg × 0.052
= 3,146 psi
At top of cement = 3,146 psi + [(7,000 ft – 5,000 ft) × 13 ppg × 0.052]
= 4,160 psi
At casing shoe = 4,160 psi + [(11,450 ft – 7,000 ft) × 16 ppg × 0.052]
= 7,862 psi
Net collapse load at casing shoe = 7,862 psi – 6,549 psi
= 1,313 psi

The net collapse load is best represented graphically, as shown below, by plotting the internal and external pressure profiles and the net collapse load (fig. 3.2).

For stab-in cement jobs, the possibility of bridging needs to be taken into account (fig. 3.3). If bridging occurs, then the external hydrostatic pressure will be increased by the circulating pressure, with a subsequent increase in the net collapse load.

Example

20" Surface casing set at 1,500 ft
Mud weight 9.5 ppg
Cement to surface
Cement weight 16 ppg
Bridge occurs with 1,000 psi surface pressure

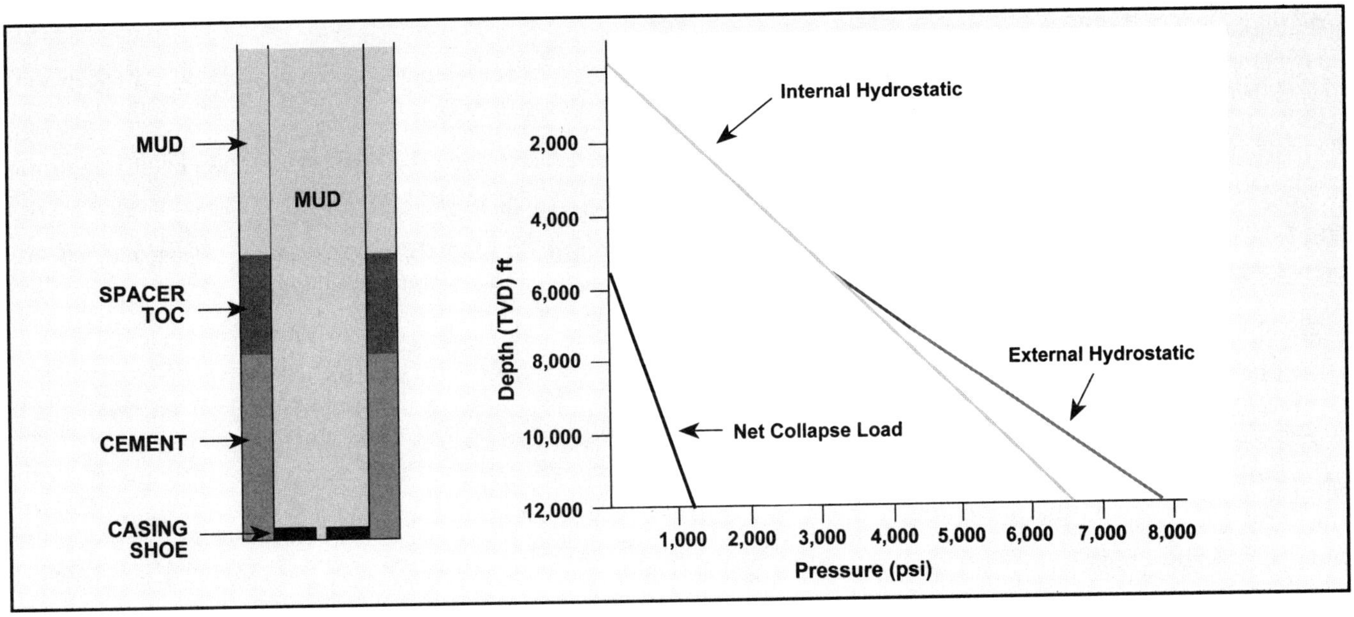

Figure 3.2 Collapse–installation

Internal Pressure Profile

At surface = 0 psi
At casing shoe = 1,500 ft × 9.5 ppg × 0.052
= 741 psi

External Pressure Profile

At surface = 1,000 psi
At casing shoe = 1,000 psi + (1,500 ft × 16 ppg × 0.052)
= 2,248 psi
Net collapse load at casing shoe = 2,248 psi – 741 psi
= 1,507 psi

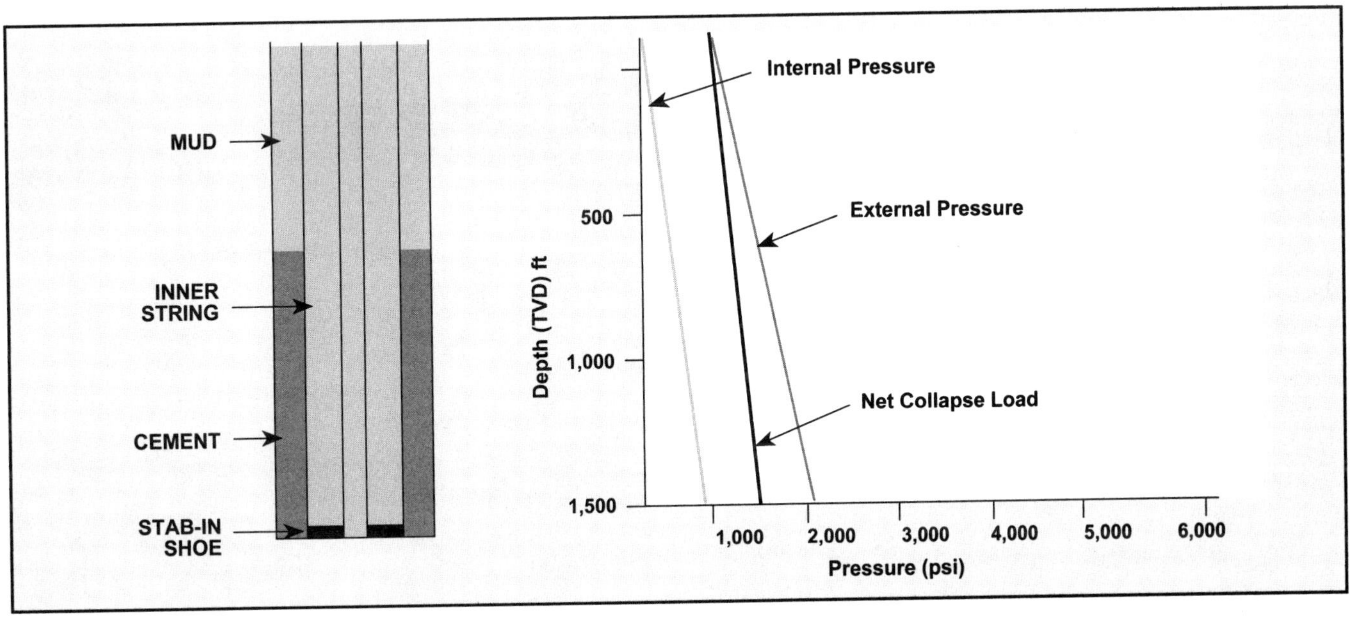

Figure 3.3 Collapse–installation (stab-in)

Collapse Drilling Loads

The worst case collapse load during drilling occurs if lost circulation is encountered and the internal hydrostatic pressure decreases (fig. 3.4).

By convention, the external fluid is deemed to be the mud that was in place when the casing was run. This is due to the uncertainty of complete cement isolation around the casing caused by channelling or washouts.

The level that the internal fluid drops to can be anything from hundreds of feet to total evacuation of the casing, and depends on the internal mud weight in use and the pore pressure of the loss zone.

Example

13⅜" Intermediate casing set at 9,750 ft
External mud weight 11 ppg
Internal mud weight 11.2 ppg
Drilling ahead 12¼" hole at 13,360 ft
Experienced total losses and fluid level dropped to 2,528 ft

Internal Pressure Profile

At surface	=	0 psi
At 2,528 ft	=	0 psi
At casing shoe	=	(9,750 ft – 2,528 ft) × 11.2 ppg × 0.052
	=	4,206 psi

External Pressure Profile

At surface	=	0 psi
At casing shoe	=	9,750 ft × 11 ppg × 0.052
	=	5,577 psi
Net collapse load at casing shoe	=	5,577 psi – 4,206 psi
	=	1,371 psi

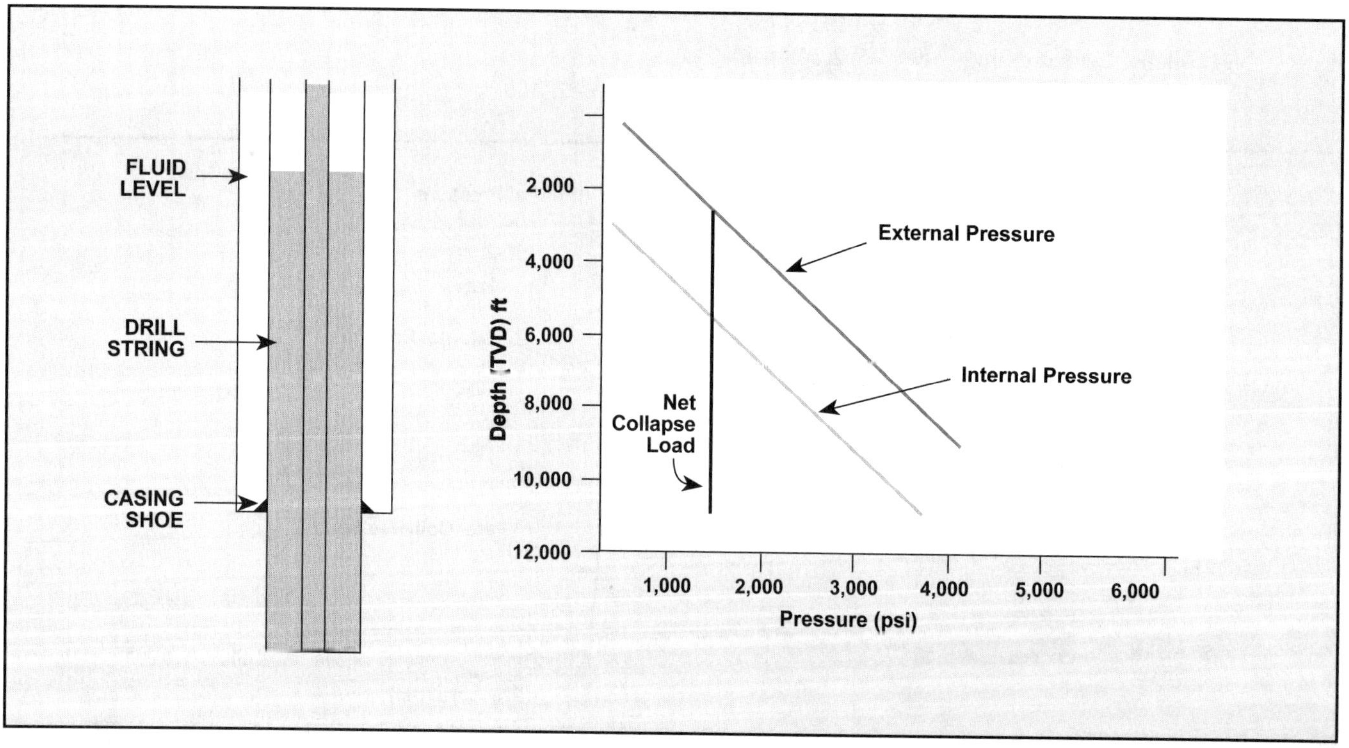

Figure 3.4 Collapse–drilling

Collapse Production Loads

Collapse loads that production casing and liners will be exposed to need to be considered for the entire life of the well (fig. 3.5). This depends upon what the well will be used for, but consideration should be given to the following as applicable:

- DST operations
- Stimulation techniques
- Gas Lift
- Drawdown

By convention, the external fluid is deemed to be the mud that was in place when the casing was run. After a period of time (typically one year), this can be relaxed as follows:

Position In Well	External Fluid
Uncemented Casing/Casing Annulus	Mud that casing was run in
Cemented Casing/Casing Annulus	Mud that casing was run in
Cemented Casing/Open-Hole Annulus	Pore Pressure

The internal hydrostatic pressure will vary, depending upon the position of the production packer, and the collapse considerations should be separated above the packer or below the packer.

Example

9⅝" Production casing set at 15,700 ft
Production packer set at 12,000 ft
Completion fluid weight 9.2 ppg
Mud weight behind casing 11 ppg
Gas gradient 0.1 psi/ft
Perforations have been plugged and the well has been drawn down to 0 psi at the surface.

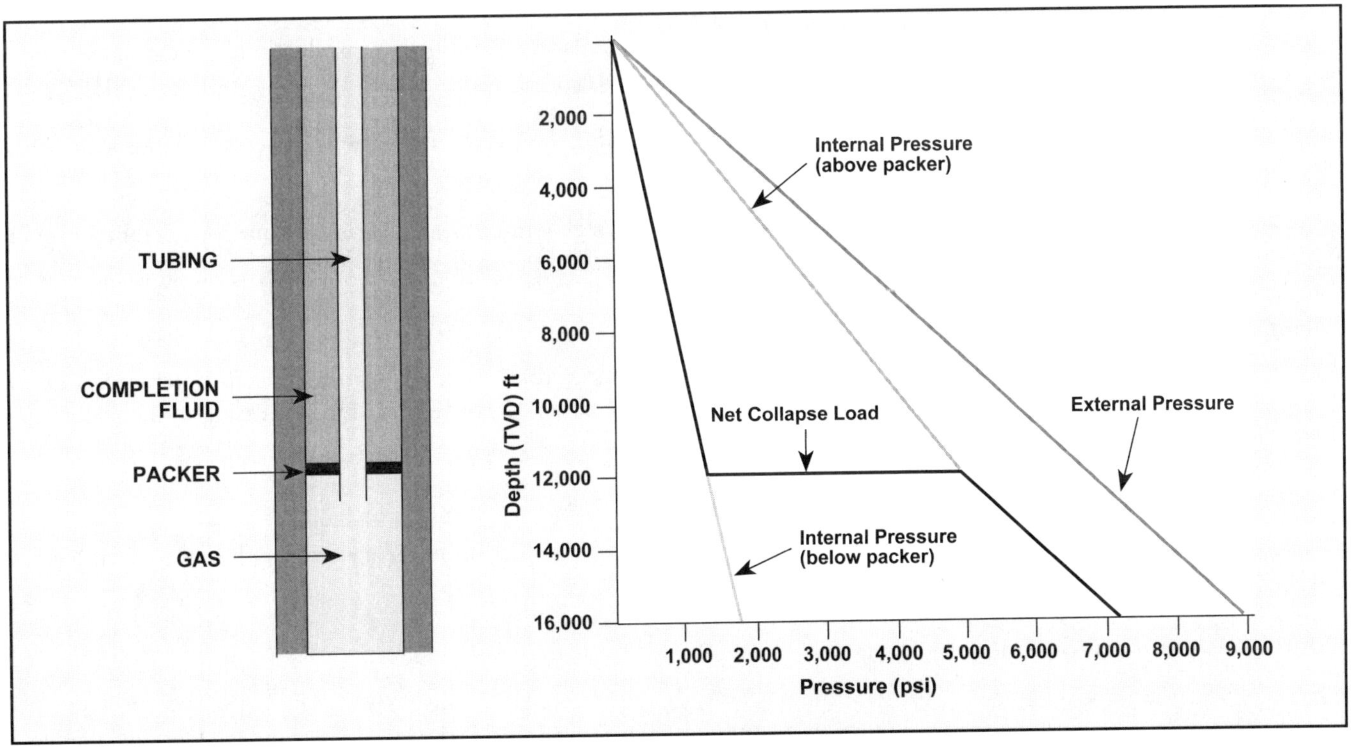

Figure 3.5 Collapse–production

Above Packer

Internal Pressure Profile

At surface	=	0 psi
At packer	=	12,000 ft × 9.2 ppg × 0.052
	=	5,741 psi

External Pressure Profile

At surface	=	0 psi
At packer	=	12,000 ft × 11 ppg × 0.052
	=	6,864 psi
Net collapse load at packer	=	6,864 psi – 5,741 psi
	=	1,123 psi

Below Packer

Internal Pressure Profile

At packer	=	12,000 ft × 0.1 psi/ft
	=	1,200 psi
At casing shoe	=	15,700 ft × 0.1 psi/ft
	=	1,570 psi

External Pressure Profile

At packer	=	12,000 ft × 11 ppg × 0.052
	=	6,864 psi
At casing shoe	=	15,700 ft × 11 ppg × 0.052
	=	8,980 psi
Net collapse load at packer	=	6,864 psi – 1,200 psi
	=	5,664 psi
Net collapse load at casing shoe	=	8,890 psi – 1,570 psi
	=	7,320 psi

Selecting a Casing that Meets the Collapse Loads

From the net collapse loads generated by looking at the various installation, drilling, and production load cases, the worst case collapse load that the casing will have to tolerate can be determined.

From casing tables, it is then possible to choose a casing or series of casings that match the worst case collapse load. This stage is often ignored until the burst loads have been calculated.

The casing collapse figures that are quoted in most casing tables are generated from a series of equations detailed in API Bulletin 5C3 and are a function of the casing OD, the wall thickness, and the casing yield strength.

Biaxial Loading

The collapse figures determined using API Bulletin 5C3 are for casings that are under zero axial load. In practice, due to the weight of the casing or to the combined action of internal and external pressures, it is rare for casing to be under zero axial load.

The effect of axial load is to decrease the collapse strength of casing.

For casing design purposes, unless the collapse strength is critical, the reduction in collapse strength due to axial load is normally ignored.

Other Considerations for Collapse Design

Casing Wear

If casing wear is anticipated to be an issue (i.e., high sidewall forces, extended drill-string-to-casing contact, localized doglegs, abrasive or damaged hard banding in use, etc.), then this needs to be accounted for.

As the collapse strength of casing is related to the wall thickness, then if casing wear estimates that 20% of the wall thickness will be worn away during drilling, then the residual collapse strength of the casing will be 80% of that of new casing.

More detail on casing wear can be found later.

Burst Design

For all casing strings, a burst load occurs when the internal pressure is greater than the external pressure (fig. 3.6).

As with collapse, burst design focuses on the internal and external fluids and the hydrostatic pressures that they exert.

Burst Installation Loads

The worst case burst load during installation occurs during cementing.

Two cases need to be considered here.

- During the displacement immediately prior to the spacer exiting the shoe
- Bumping the plug

Example. Cement Displacement

13⅜" Intermediate casing set at 9,750 ft
Mud weight 11 ppg
5,000 ft of 16 ppg cement
2,000 ft of 13 ppg spacer

Internal Pressure Profile

At surface	=	0 psi
At top of cement	=	(9,750 ft – 7,000 ft) × 11 ppg × 0.052
	=	1,573 psi
At top of spacer	=	1,573 + (5,000 ft × 16 × 0.052)
	=	5,733 psi
At casing shoe	=	5,733 psi + (2,000 ft × 13 ppg × 0.052)
	=	7,085 psi

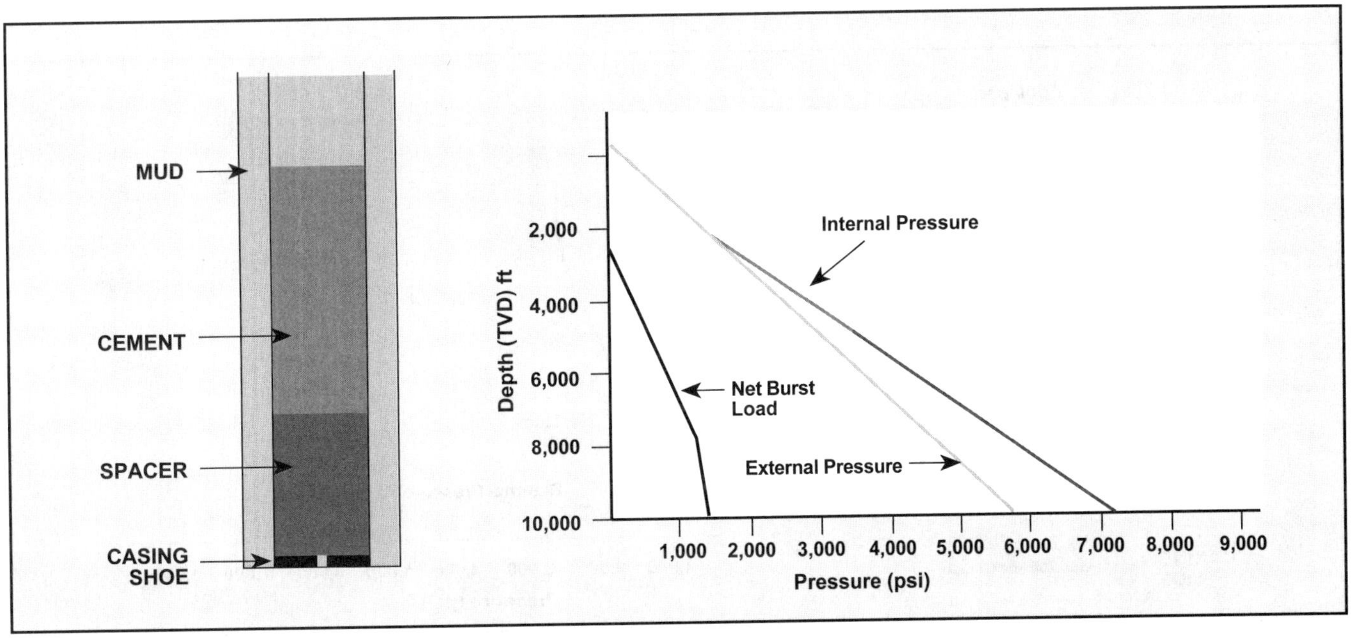

Figure 3.6 Burst—installation—cement displacement

External Pressure Profile

At surface	=	0 psi
At casing shoe	=	9,750 ft × 11 ppg × 0.052
	=	5,577 psi
Net burst load at surface	=	0 psi
Net burst load at casing shoe	=	7,085 psi – 5,577 psi
	=	1,508 psi

The net burst load is best represented graphically, as shown below, by plotting the internal and external pressure profiles and the net collapse load (fig. 3.7).

Example. Plug Bump

13½" Intermediate casing set at 9,750 ft
Mud weight 11 ppg
Top of spacer at 1,750 ft
Spacer weight 13 ppg
Top of cement at 3,000 ft
Cement weight 16 ppg
Plug bump pressure 2,500 psi

Internal Pressure Profile

At surface	=	2,500 psi
At casing shoe	=	2,500 psi + (9,750 ft × 11 ppg × 0.052)
	=	8,077 psi

External Pressure Profile

At surface	=	0 psi
At top of spacer	=	1,750 ft × 11 ppg × 0.052
	=	1,001 psi
At top of cement	=	1,001 psi + [(3,000 ft – 1,750 ft) × 13 ppg × 0.052]
	=	1,846 psi
At casing shoe	=	1,846 psi + [(9,750 ft – 3,000 ft) × 16 ppg × 0.052
	=	7,462 psi

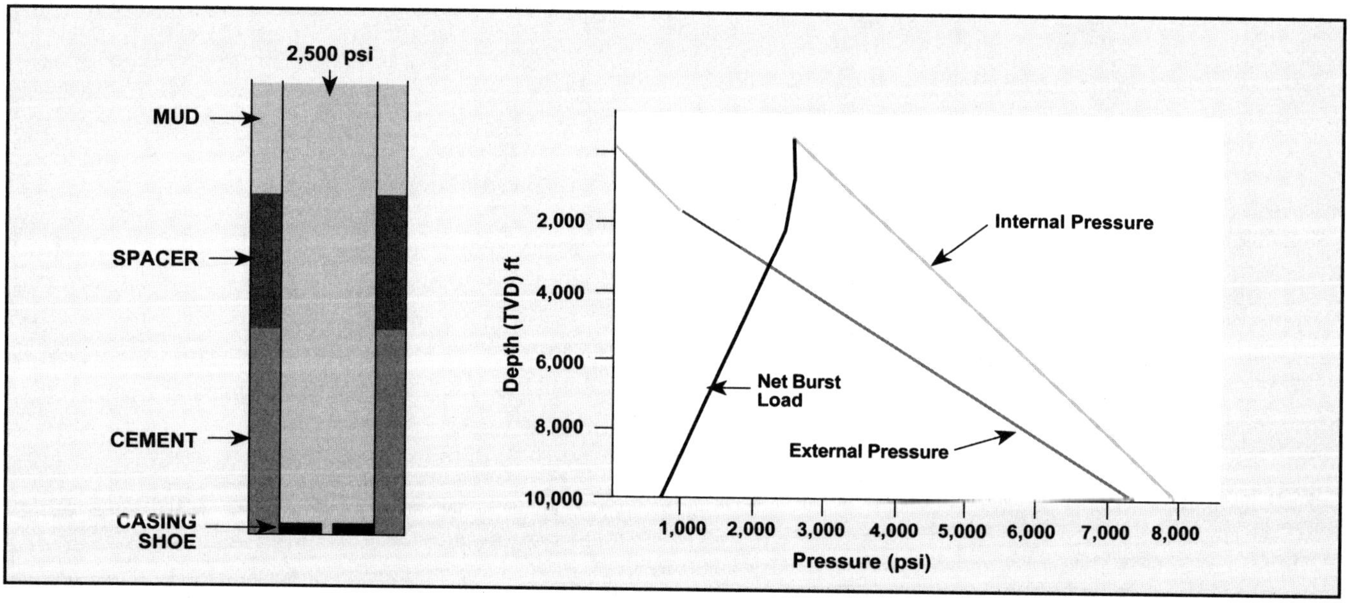

Figure 3.7 Burst—installation—plug bump

Net burst load at surface	=	2,500 psi
Net burst load at casing shoe	=	8,077 psi – 7,462 psi
	=	615 psi

Burst Drilling Loads

The worst case burst drilling loads occur either during pressure testing or during a well-control event (fig. 3.8).

Example. Pressure Test

13⅜" Intermediate casing set at 9,750 ft
Mud weight 11.5 ppg
Top of cement at 3,000 ft
Previous casing shoe at 1,500 ft
Normally pressured
Pressure test 3,000 psi

Internal Pressure Profile

At surface	=	3,000 psi
At casing shoe	=	3,000 psi + (9,750 ft × 11.5 ppg × 0.052)
	=	8,831 psi

External Pressure Profile

At surface	=	0 psi
At top of cement	=	3,000 ft × 8.33 ppg × 0.052
	=	1,299 psi
At casing shoe	=	1,299 psi + [(9,750 ft – 3,000 ft) × 8.6 ppg × 0.052
	=	4,318 psi
Net burst load at surface	=	3,000 psi
Net burst load at casing shoe	=	8,831 psi – 4,318 psi
	=	4,513 psi

NOTE: The external pressure profile after cement has set is a matter of great debate, and requires careful consideration.

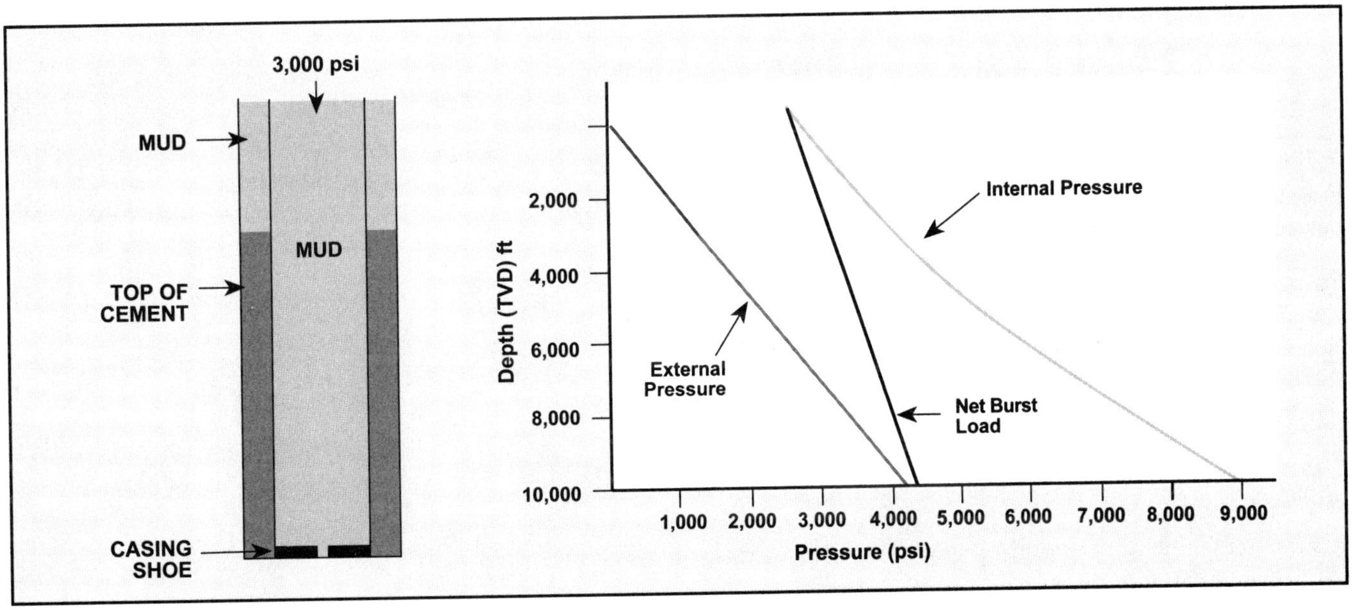

Figure 3.8 Burst—drilling

In the example on the previous page, it has been assumed that for casing in contact with formation via cement, the external pressure is equal to the expected pore pressure. For casing exposed to formation via an uncemented section (top of cement below previous casing shoe), the external pressure is equal to a column of mud-mix fluid with zero surface pressure (i.e., the mud has wholly settled out—in the above example, a water-based mud was assumed).

Burst Production Loads

Burst loads that need to be considered during production include pressure testing with a completion or kill weight fluid, or a near-surface tubing leak.

The internal load is the hydrostatic pressure of the fluid plus any applied pressure.

The external load will be as for the burst drilling load.

Example. Near-Surface Tubing Leak

9⅝" Production casing set at 15,700 ft
Production packer set at 12,000 ft
Completion fluid weight 9.2 ppg
Mud weight behind casing 11 ppg
Gas gradient 0.1 psi/ft
Top of cement 8,000 ft
Pore pressure 9.0 ppg EMW
SITHP 5,778 psi

Above Packer

Internal Pressure Profile

At surface	=	5,778 psi
At packer	=	5,778 psi + (12,000 ft × 9.2 ppg × 0.052)
	=	11,519 psi

External Pressure Profile

At surface	=	0 psi
At TOC	=	8,000 ft × 8.33 ppg × 0.052
	=	3,465 psi
At packer	=	3,465 psi + [(12,000 ft – 8,000 ft) × 9.0 ppg × 0.052]
	=	5,337 psi
Net burst load at packer	=	11,519 psi – 5,337 psi
	=	6,182 psi

Selecting a Casing that Meets the Burst Loads

From the net burst loads generated by looking at the various installation, drilling, and production load cases, the worst case burst load that the casing will have to tolerate can be determined.

From casing tables, it is then possible to choose a casing or series of casings that match the worst case burst and worst case collapse loads.

Remember that the burst or internal yield of the casing needs to be downrated by the design factor.

Example. 13⅜" Intermediate Casing

From the previous examples, the worst case collapse (during drilling) and the worst case burst (during drilling) are plotted on the same graph.

Assuming that the following 13" casing is available in stock

13⅜" 68 lb/ft K55
13⅜" 72 lb/ft N80

The collapse ratings for both of these casings are determined and plotted on the graph. From this it can be seen that either of these casings is acceptable from a collapse standpoint.

The burst ratings for both of these casings are determined, downrated by the design factor of 1.1 and then plotted on the graph. From this, it can be seen that the 68 lb/ft K55 casing can only be used for the top 1,500 ft.

In practice, it is more likely that a full string of 72 lb/ft N80 casing would be run. See Figure 3.9.

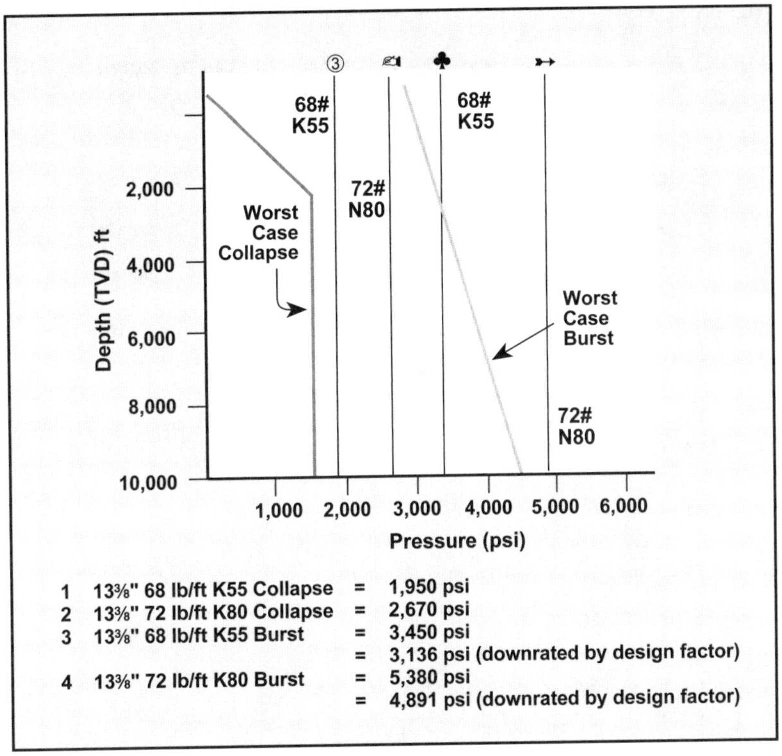

Figure 3.9 Selecting casing for collapse and burst

Other Considerations for Burst Design

Casing Wear

As for collapse, if casing wear is anticipated to be an issue, then this needs to be accounted for.

To allow for casing wear, the API burst rating should be multiplied by a casing wear factor (CWF) as shown in the following table. These values of CWF have been derived from finite element analysis of different sizes, weights, and grades of casing subjected to different wear grooves.

Percent Casing Wear	CWF
0	0
10	0.9
20	0.8
30	0.7
40	0.6

More detail on casing wear can be found later.

Tensile Design

Using the casing selected that meets the collapse and burst loads, it is then necessary to confirm that this casing will also meet the requirements of the tensile design.

Tensile Installation Loads

This stage involves assessing the suitability of the selected casing for withstanding running loads, cementing loads, and any pressure testing.

It is assumed that the casing is fixed at the surface but free to move at the shoe.

The loads that need to be considered are as follows.

Weight in Air

The weight of casing in air is simply the nominal casing weight multiplied by the true vertical depth of the casing.

$$F_{air} = W \times TVD$$

where

W = nominal casing weight (lb/ft)
TVD = TVD below point of interest to casing shoe (ft).

Buoyancy

Buoyancy can be calculated using the pressure area method and is normally the hydrostatic pressure multiplied by the casing cross-sectional area.

Care needs to be taken if tapered strings of casing are used as the buoyancy force will vary depending upon the depth and outside and inside diameters.

Buoyancy is always subtracted.

When the same fluid is on the inside and outside of the casing (i.e., when casing is being run), the following equation can be used.

$$F_{buoy} = P_e \times (A_o - A_i)$$

where

P_e = hydrostatic pressure at the bottom of the casing (psi)
A_o = area of the outside diameter (in.^{2})
A_i = area of the inside diameter (in.^{2}).

When different fluids are on the inside and outside of the casing (i.e., during cementing), the following equation can be used.

$$F_{buoy} = (P_e \times A_o) - (P_i \times A_i)$$

where

P_e = external hydrostatic pressure at the bottom of the casing
A_o = area of the outside diameter (in.^{2})
P_i = internal hydrostatic pressure at the bottom of the casing
A_i = area of the inside diameter (in.^{2}).

Bending

Whenever pipe is forced around a bend or curve in the well, a bending force will occur. The bending force is a tensile load occurring on the outer wall and compressive loads on the inner wall of the casing.

Bending loads are calculated using the following formula

$$F_{bend} = 64 \times DLS \times OD \times W$$

where

DLS = dogleg severity (°/100 ft)
OD = casing outside diameter (in.).

In nominally vertical wells, the DLS can be assumed to be 1°/100 ft. For deviated wells, the bending load only applies to the point where curvature exists; i.e., in build sections.

Drag

Drag is the result of sliding resistance between the wellbore and the pipe. It occurs in deviated wells and in sticky tight holes. It is not easy to compute manually and is best left to computer simulations.

Shock

Shock loading is the load resulting from movement of the casing as it is run in the hole or when the slips are set or when the casing encounters a ledge downhole.

Shock loads are calculated using the following formula

$$F_{shock} = 1{,}780 \times V \times A_s$$

where

V = instantaneous velocity (ft/sec)

$A_s = A_o - A_i$ (in.²).

Care should be taken that the instantaneous velocity used in this calculation is not exceeded during rig operations.

Pressure Testing

The purpose of a casing test is to verify that the casing string can withstand the maximum anticipated burst loads. Therefore, it should exceed the greatest predicted loads during both drilling and production operations.

$$F_{ptest} = P_{ptest} \times A_i$$

where

P_{ptest} = plug bump pressure or applied pressure test (psi).

Example

Running 13⅜" 72 lb/ft Intermediate casing to 9,750 ft
Inside diameter 12.347"
Mud weight 11 ppg
Instantaneous velocity 5 ft/sec

$$\begin{aligned}
F_{air} &= 9{,}750 \text{ ft} \times 72 \text{ lb/ft} \\
&= 702{,}000 \text{ lb} \\
F_{buoy} &= 9{,}750 \text{ ft} \times 11 \text{ ppg} \times 0.052 \times [(13.375^2 - 12.347^2) \times \pi/4] \\
&= 115{,}821 \text{ lb} \\
F_{bend} &= 64 \times 1°/100 \text{ ft} \times 13.375" \times 72 \text{ lb/ft} \\
&= 61{,}632 \text{ lb} \\
F_{shock} &= 1{,}780 \times 5 \text{ ft/sec} \times [(13.375^2 - 12.347^2) \times \pi/4] \\
&= 184{,}832 \text{ lb} \\
F_{total} &= F_{air} - F_{buoy} + F_{bend} + F_{shock} \\
&= 702{,}000 - 115{,}821 + 61{,}632 + 184{,}832 \text{ lb} \\
&= 832{,}643 \text{ lb}
\end{aligned}$$

Example

Cementing 13⅜" 72 lb/ft Intermediate casing at 9,750 ft
Inside diameter 12.347"
Mud weight 11 ppg
Top of lead slurry at 3,000 ft
Weight of lead slurry 11.6 ppg
Top of tail slurry at 9,000 ft
Weight of tail slurry 15.8 ppg
Plug bump pressure 3,000 psi

$$\begin{aligned}
F_{air} &= 9{,}750 \text{ ft} \times 72 \text{ lb/ft} \\
&= 702{,}000 \text{ lb} \\
\text{External Hydrostatic} &= (3{,}000 \text{ ft} \times 11 \text{ ppg} \times 0.052) + (6{,}000 \text{ ft} \times 11.6 \text{ ppg} \times 0.052) + (750 \text{ ft} \times 15.8 \text{ ppg} \times 0.052) \\
&= 5{,}951 \text{ psi}
\end{aligned}$$

$$
\begin{aligned}
\text{Internal Hydrostatic} &= 9{,}750 \text{ ft} \times 11 \text{ ppg} \times 0.052 \\
&= 5{,}577 \text{ psi} \\
F_{buoy} &= (5{,}951 \text{ psi} \times 13.375^2 \times \pi/4) - (5{,}577 \text{ psi} \times 12.347^2 \times \pi/4) \\
&= 168{,}425 \text{ lb} \\
F_{bend} &= 64 \times 1°/100 \text{ ft} \times 13.375" \times 72 \text{ lb/ft} \\
&= 61{,}632 \text{ lb} \\
F_{ptest} &= 3{,}000 \text{ psi} \times 12.347^2 \times \pi/4 \\
&= 359{,}198 \text{ lb} \\
F_{total} &= F_{air} - F_{buoy} + F_{bend} + F_{ptest} \\
&= 702{,}000 - 168{,}425 + 61{,}632 + 359{,}198 \text{ lb} \\
&= 954{,}405 \text{ lb}
\end{aligned}
$$

Tensile Drilling and Production Loads

This stage involves assessing the suitability of the selected casing for withstanding the loads that could be imposed on the casing after the cement has set.

It is assumed that the casing is fixed at the surface and fixed at the top of the cement in the annulus.

The loads that need to be considered are as follows.

Landing Casing

Any additional tension applied to the casing after waiting on cement. This is normally restricted to jackup or platform operations where tension is applied prior to setting slips in the wellhead.

Buckling

The potential for buckling exists if any of the following occur:

- Internal mud density increases
- Internal surface pressure increases
- Annular fluid is removed or annulus mud density is reduced
- Casing is landed with less than full hanging weight
- Temperature of the casing increases

If any of the above occur, this results in ballooning of the uncemented portion of the casing string, which could result in buckling.

Buckling itself does not mean that the casing has failed. However, as buckling develops into a helical form, this can promote increased dogleg severity and restrict tool drift length and can promote casing wear.

In general, for all wells deeper than 10,000 ft, casing designs must be evaluated for the impact of buckling.

The effects of buckling can be reduced by raising the top of the cement (reducing the amount of uncemented pipe), or (if practical), increasing the amount of tension applied prior to setting the slips.

Confirming that the Selected Casing Meets the Tensile Loads

From the tensile loads generated by looking at the various installations, drilling and production load cases, the worst case tensile load that the casing will have to tolerate can be determined.

The tensile capacity of the casing is downrated by the design factor and, if this exceeds the worst case tensile load, then the selected casing is acceptable for service.

If the tensile capacity after downrating by the design factor is less than the worst case tensile load, then the selected casing is not acceptable for service. The next weight and/or grade up that still meets the collapse and burst loads is then selected, and the tensile loads recalculated (changing the weight of the casing affects the internal diameter of the casing and these two properties affect all of the loads that make up the tensile load). This process is repeated until a suitable casing is selected.

Example

13⅜" 72 lb/ft N80 Intermediate casing set at 9,750 ft

From previous examples, this casing meets the collapse and burst worst case loadings.

From previous calculations, worst case tensile load occurs at plug bump.

$$F_{total} = 954{,}405 \text{ lb}$$

From casing tables, the tensile rating of the pipe body and connections are obtained and downrated by the design factor of 1.3

	Tensile Capacity	Downrated Capacity
Pipe body	1,661,000 lb	1,277,000 lb
STC Connection	1,040,000 lb	800,000 lb
BTC Connection	1,693,000 lb	1,302,000 lb

From this it can be seen that the pipe body and the buttress (BTC) connection have an acceptable tensile rating and are suitable for this application. However, the short round (STC) connection is not acceptable as its downrated tensile capacity is below F_{total}.

It is important, when looking at tensile capacity, that the weaker of the pipe body and connection is used.

Triaxial Design

The collapse, burst, and tensile loads calculated so far have all assumed that the stresses are in a single or uniaxial direction. In practice, service loads generate triaxial stresses.

The three principal stresses for casing are axial (σ_a), radial (σ_r) and tangential (σ_t) as shown in Figure 3.10.

The recommended theory for calculating triaxial stress is known as the Von Mises theory. This theory consists of defining an equivalent stress (σ_{VME}) and then relating that stress to the minimum specified yield stress (σ_y) of the casing.

The calculation to determine triaxial stress is best conducted using a suitable casing design program.

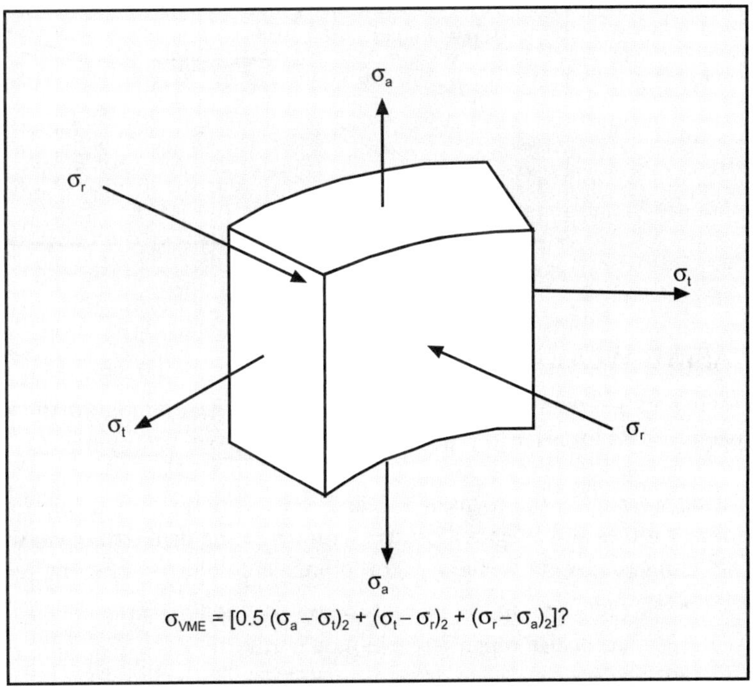

Figure 3.10 Von Mises theory

Triaxial design should be performed whenever any of the following conditions apply

- expected pore pressure >12,000 psi.
- temperature >250°F.
- H_2S service.
- OD/t <15.

According to the Von Mises theory, an axial tensile stress can increase the tangential stress capacity and vice versa. This is shown in the following diagram (fig. 3.11).

The equivalent stress (σ_{VME}) should be calculated at the top and bottom of each casing interval of weight and grade, at the top of the cement, at a particular depth where there is a specific change in internal or external pressure, or a specific hole geometry (DLS, washout, etc.).

Specialist casing design software is required to perform a triaxial casing design.

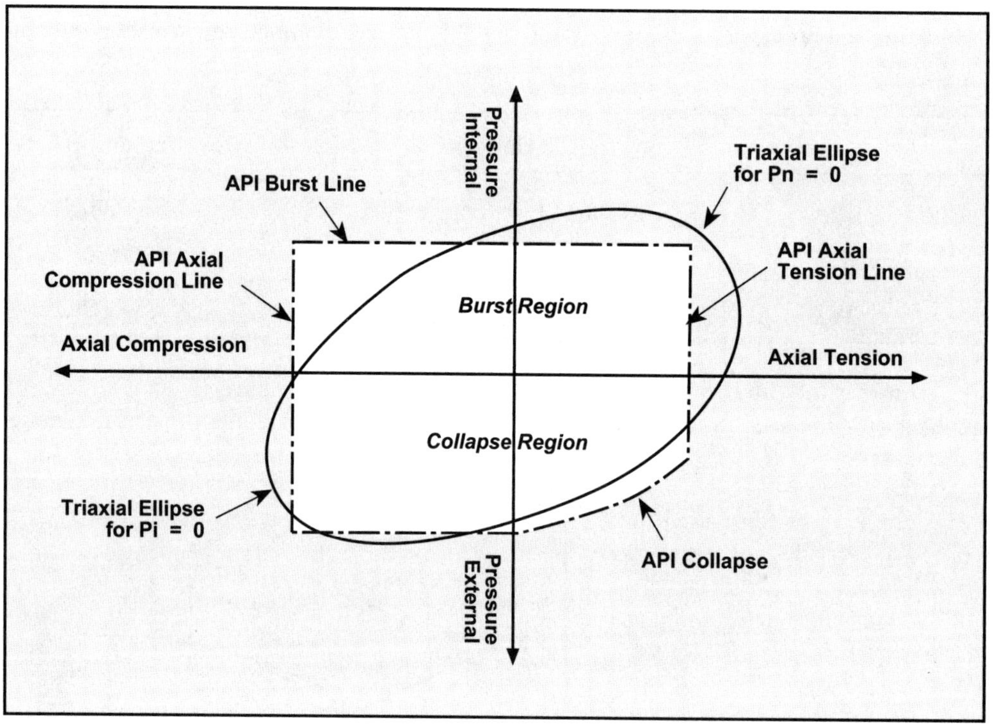

Figure 3.11 Von Mises theory

CASING WEAR

Having an understanding of the cause of casing wear will allow the well planner to optimize the well design and drill pipe specifications in order to reduce the occurrence to a minimum.

Casing wear takes the form of a wear groove generated by a rotating drill string that is forced into the casing wall. High sidewall forces and extended contact with a rotating drill string will wear the section down. Areas that are commonly identified with casing wear include kickoff points and doglegs. The implications of casing wear can be recognized as:

- reduction in the pressure integrity due to wear groove(s), reducing the burst/collapse values.
- expensive repairs to drill pipe hardfacing.
- friction (surface torque) can be high.
- wear groove may act as a starting point for future corrosion.

The types of casing wear are as follows:

Adhesive Wear

The transfer of material from a low-strength body to a high-strength body by solid-phase welding. However, the weld bond is weak and often the material 'falls' off the tool joint and is incorporated into the drilling fluid system as a flake.

Abrasive Wear—Machining

Often stems from pieces of exposed tungsten carbide on tool joints removing material from the casing wall. Again, debris is generated that is incorporated into the fluid system.

Abrasive Wear—Grinding/Polishing

Hard particles produce a standoff between the casing wall and the tool joint. The casing is cyclically loaded by the particles due to the drill pipe rotation causing fatigue and the surface to become brittle.

These three types of casing wear are given wear factors to highlight their severity:

Wear Mechanism	Debris Produced	Wear Factor
Adhesive	Cuttings	400–1800
Abrasive—Machining	Flakes	20–50
Abrasive—Grinding	Powder	0.1–10

Casing Wear Prediction

Casing wear mechanisms can be identified prior to the start of a well, and it is the control of the mechanism that reduces the amount of casing wear.

Contact Pressure and Load

Initially this is very high due to the small contact area resulting in high pressures and a large amount of wear. As wear continues, it will expose a larger surface area and will help to distribute and reduce the wear pressure. Doglegs and build and drop-off sections deserve special attention though. Transocean sets acceptable sidewall force limits as 2,200 for water-based mud systems and 2,500 for oil-base mud systems.

Well Design

Anticipate and consider 'what if' scenarios—likelihood of additional drilling, stuck pipe, and fishing operations for example.

Doglegs

Doglegs are unavoidable in many wells and, when designing a well, it is important to understand their effect on casing wear. While drilling the dogleg, it is important to maintain drilling parameters as constant as possible to control the dogleg severity. Smoothing out the dogleg profile can also have an impact on wear reduction and is strongly advised. Implementing a deep kickoff will lower the amount of sidewall forces exerted on the casing.

Control of Casing Wear

Certain areas have been identified that require consideration during well planning to minimize casing wear. In the field, an MFCT log can be run before and after the predicted wear to measure the extent of the damage.

Casing Material

It cannot be assumed that the higher the strength for resisting collapse and burst, the higher the resistance to wear. Materials must be looked at closely if significant casing wear is expected.

Wear Bushing, Crossovers, Centralizers, and Cementing

Regular inspection of the wear bushing is important based purely on the number of times it is exposed to tool joints travelling in and out of the hole.

When crossing over between two different weights and grades of casing, this section needs to be supported due to the different stiffness of the two sections. Support can either come from cement or adequate centralization; otherwise, a local dogleg might develop. The internal wall of the crossover is shouldered and will experience wear from drill pipe tool joints. The shoulder should be chamfered as much as possible.

It is prudent to locate the planned top of cement away from a zone of high sidewall forces, as immediately above the cemented pipe a local dogleg is likely to develop.

Centralizers should be placed across the pipe body as opposed to locating them at the coupling. The additional stand-off provided by the centralizer may result in a local dogleg establishing itself above the coupling.

Drill Pipe Hardbanding

Today the vast majority of hardbanding is a 'wear resistant alloy', such as Armacor M and Arnco 200 XT, that exhibits a uniform surface hardness as opposed to needing the addition of tungsten carbide. Nevertheless, it is still important to check what is received to make sure it is 'user friendly'.

Drill Pipe Protectors

These are placed on the body of the drill pipe a few feet above the tool joint to provide standoff between the tool joint and the casing. They are an elastomer and are manufactured to be ½" larger than the OD of the tool joint. During their use, they require regular calipering for wear and subsequent renewal. An alternative is to incorporate rotating sleeve subs within the drill string.

Mud Types

Water-Based Muds

Severe adhesive wear with friction. Caused by a lack of solid barriers in the mud.

Oil-Based Muds

Reduced friction. Potential for casing wear dependent upon drill pipe hardbanding and sidewall contact pressures.

Unweighted/Weighted Muds

Incorporating a weighting material will provide solids to the mud system that will act as particulate matter between the tool joint and the casing, providing a layer between the rolling surfaces. Particle size and hardness are relevant. A larger particle will provide greater standoff, and a softer material will perform better. For example, barite performs better than additions of hematite and quartz.

Sand/Silt

Sand particles are too large to 'roll' in the space between tool joint and casing. Unweighted systems experience so much adhesive wear as to make the abrasive influence of sand unnoticed. In a weighted system, the effect of sand is again unnoticed due to the diluting effect of the weighting product.

Lubricants

Their effect is dependent on surface conditions of the casing/tool joint and the amount of solids in the system. Lubricants produce a film that lies on the casing and the tool joint providing a surface of low resistance. However, as the solids content of a drilling system may increase, these solids will penetrate the film to such a degree as to prevent any film/film contact, rendering the lubricant ineffective. If lubricants are included, solids control will be a high priority.

MATERIAL SELECTION

Two types of service need to be considered:

- strings that are exposed to brines and muds.
- strings that are exposed to reservoir fluids.

Strings Exposed to Brines and Muds

Casing material is commonly composed of carbon or low-alloy steels; there is a wide range available as shown in API 5CT. Depending on geographical location, the buyer may have access to non-API standard tubulars that claim to have a better performance.

When dealing with completion brines and drilling muds, corrosion resistance is often not critical to the design. However, if exposure to hydrogen sulphide is anticipated, casing that is manufactured for sour service will have to be selected.

API-controlled grades of steel for high-pressure sour wells can lead to very heavy wall casing designs. There is growing interest to use specialized high-strength grades from non-API vendors.

Strings Exposed to Reservoir Fluids

Primarily this section concerns itself with production tubing; however, production casing/liners need to be considered for possible exposure due to a leak or tubing failure.

A rough guide exists for materials required, based on the partial pressures of carbon dioxide and hydrogen sulphide in the gas phase.

1. CO_2 <3 psia and H_2S <0.05 psia
2. CO_2 <3 psia and H_2S >0.05 psia
3. CO_2 >3 psia and H_2S <0.05 psia
4. CO_2 >3 psia and H_2S >0.05 psia

To determine the partial pressure in the gas phase, multiply the mole fraction by the bottomhole pressure.

For example:

10,000 psi bottomhole pressure with 50 ppm H_2S = 0.5 psia
(50/1,000,000 × 10,000)

10,000 psi bottomhole pressure with 3% CO_2 = 300 psia
(3/100 × 10,000)

1. CO_2 <3 psia and H_2S <0.05 psia

Carbon or low alloy steels. If carbon steels are selected, a rigorous look at their service life and corrosion potential is needed.

2. CO_2 <3 psia and H_2S >0.05 psia

Carbon or alloy steels. However, the materials need to meet the standards of NACE MR-0175-91. Grades that are for sour service are found in API 5CT and NACE MR-0175-91.

3. CO_2 >3 psia and H_2S <0.05 psia

It is not recommended to use carbon or low alloy steels because they will have insufficient corrosion resistance. Inhibitors and plastic-lined pipe is one option; however, the job starts to get a little complicated.

A material that has been proven in the field is 13% chrome steel tubulars. Although a number of limitations do exist, above grade C-95 the material becomes brittle if in sour service and above 125°C there is an increased risk of pitting or general corrosion.

If the grade or temperature is affecting the design, then a more exotic tubular is required, like duplex stainless steel, for example.

4. CO_2 >3 psia and H_2S >0.05 psia.

This environment is indicative of a corrosive well and a sour well. Corrosion resistant alloys and chloride/sulphide stress cracking resistant metals are needed for the design. 13% chrome steel tubulars are not adequate and consideration can be given for 22% or 25% Cr, but, with an H_2S partial pressure above 1.5 psia, austenitic stainless steels need to be used.

As much information on the operating conditions is required in order to complete the final selection. Such as:

- Design life
- Fluid types
- Partial pressures of CO_2 and H_2S
- Operating and design pressures
- Operating temperature
- Flow rate/regime
- Possibility of sand production
- Water chemistry
- Any changes likely during the life of the field/tubulars
- Is gas lift likely?

Considerations for Corrosion Resistant Alloys

Chlorides and Bromides

Unexpectedly high levels for the casing design are likely to cause pitting and chloride stress cracking. An increase in temperature will heighten this. It may be necessary to restrict the temperature and/or chloride/bromide levels to complete the design.

Mechanical Properties

Tensile Properties

The yield strength can be anything up to 10% less in the transverse direction with materials that depend, for their strength, on cold working. If a specific application is required, then the exact amount it is reduced by will need to be determined.

Yield strengths that are quoted in manuals are for an ambient temperature; hence, for any increase in the temperature, a decrease in the yield strength will occur. This is particularly important when selecting corrosion resistant alloys, and the amount of reduction for the designed temperature will need to be known.

The highest grade recommended when using duplex and highly alloyed austenitic steels is 125k psi. Above this and there is a chance for there to be a reduction in the corrosion resistance.

Toughness

This varies with temperature, and tubulars can be ductile or brittle. Brittle behavior needs to be avoided because it can be responsible for sudden failures. A control measure that can be used is the Charpy Impact Test, which is relatively inexpensive.

Erosion—Corrosion

The fluid velocity can advance tubular corrosion, and the design of the system needs to remain below the critical velocity. The critical velocity will naturally vary between the different alloys.

Corrosion Mechanisms

Carbon Dioxide

Carbon dioxide combines with water to form carbonic acid and will attack carbon steels where the iron carbonate layer is missing; hence, localized pitting develops.

The rate of corrosion is controlled by a number of factors including CO_2 partial pressure, pH, temperature, fluid velocity, and other chemicals.

Hydrogen Sulphide

Sources include: Well fluids
Bacterial activity
Breakdown products of other chemical additives

Hydrogen sulphide dissolves in water and can form a protective layer of iron sulphide scale. Pitting occurs where the scale is not present.

Hydrogen is a natural by-product of water corrosion and the hydrogen molecules are too large to diffuse into the metal. In the presence of hydrogen sulphide, the hydrogen atoms are prevented from combining into molecules, leaving the atoms to diffuse into the metal.

The hydrogen atoms tend to concentrate in areas that are already under stress. Their accumulation increases the stress levels and reduces the strength of the material. Stressed areas are stressed even further and cracking results which can occur rapidly.

Oxygen

Dissolved oxygen attacks iron converting it to an oxide and/or hydroxide. The rate of corrosion is controlled by the ability of the oxygen to diffuse to the area to support the corrosion process.

This can be a problem when using water-based drilling mud and in a water injection system. Scavengers are added to the mud and degassers or stripping units are associated with water injection.

Halide Ions

Cause localized pitting or crevice corrosion on materials used for downhole equipment. Halide ions are also responsible for causing stress corrosion cracking, in particular materials such as austenitic steels. Cracking will normally occur above 50°C and the result is brittle failure well below the material's yield strength.

Erosion—Corrosion

Damage is increased because of the velocity of a passing fluid. Velocity limits to minimize this form of erosion are related to API RP 14E.

Galvanic Corrosion

Corrosion takes place when two metals are coupled in an erosive environment. One metal will act as an anode and corrode faster, the other the cathode. The controls on this sort of erosion are the conductivity of the corrosive medium, surface area of the two metals, and the difference in equilibrium potentials of the two metals.

Localized Corrosion

Corrosion pitting where certain areas of the metal act as an anode. Process is enhanced by dissolved oxygen and is strongly influenced by temperature. Pitting is particularly damaging, as it is more penetrative than general corrosion.

Crevice corrosion is localized and confined to a gap between two materials. Within the gap the environment can be quite different to that outside, enhancing the corrosive process. Higher temperatures are conducive to this process.

PORE PRESSURE AND FRACTURE GRADIENT PREDICTION

Pore Pressure Prediction

Pore pressure can be predicted before, during, and after drilling by applying various methods, some directly and others indirectly or by correlation. The 'before drilling' is important as a basis of design for the drilling program. The 'during drilling' determination will serve to confirm the design while the 'after drilling' determination will provide data for future well designs in the area giving a greater level of confidence.

Before drilling estimation:

Lithology correlation—knowledge regarding abnormally pressured formations can be applied to occurrences of the same lithology if predicted along the proposed wellpath.
Shallow seismic (bright spots)—used to detect shallow gas and can infer seabed characteristics; e.g., firm for jacking up.
Seismic depth correlation—used to identify salt structures and predict depths of rocks whose signatures were determined on previous offset wells.
Seismic interval velocity plots—if a shot point corresponds with the well location, can be used as a sonic log.

During drilling:

Corrected d exponent dc—in over-pressured zones, compaction trends are reduced and formations are more porous/permeable than expected. Hence, drilling rates increase as there is less resistance (less drill chip hold-down force).
MWD data—similarly, as compaction trends decrease in over-pressured zones, the MWD will identify higher porosities than predicted against a normal trend.
Kicks—indicate pore pressure greater than hydrostatic.
Drill stem tests—measure pore pressure directly.

Changes in pore pressure indications:

Gas-cut mud—indicates increasing pore pressure.
Flow line temperature changes
Change in drilling rate
Chloride change—increase as pore pressure increases (influx of formation water).
Torque and drag—decrease as pore pressure increases.
Volume, shape, and size of cuttings—increase in pore pressure yields 'pressure cavings' in increasing quantities.

After drilling estimation:

Well logging—any logs that allow porosity or resistivity to be determined; e.g., sonic, density. Readings should be taken in shale to avoid washouts. Again, deviations from trend lines of normally pressured formations will indicate the onset of an abnormally pressured zone.
Bottomhole pressure surveys
Note:

1. Trend lines and pressure correlation can vary from area to area and even with formations of differing geological age within the same lithological column.
2. Analysis assumes consistent shale properties, which is unlikely.
3. In hard rock areas (carbonate sequences), compaction trends are interrupted since carbonates do not compact in the same manner as clays.

Fracture Gradient Prediction

Fracture gradient prediction is now largely based on the work carried out by Hubbert and Willis in 1957. This work predicted fracture pressure for vertical fractures based on a relationship between the overburden pressure gradient, pore pressure gradient, and the matrix stress coefficient (ratio of horizontal to vertical stress). They assumed that fracture pressure was equal to pore pressure plus the horizontal component of the vertical stress in the rock.
Later workers have developed this basic relationship into the more usual forms today.

Mathews and Kelly 1967: $F = P/D + Ki(\sigma/D)$

where

F = fracture gradient, psi/ft
P = formation pore pressure, psi
σ = matrix stress at point of interest, psi
D = depth at point of interest, ft tvd
Ki = matrix stress coefficient, no units.

Eaton 1976: $F = ((S/D - Pf/D).(y/(1\text{-}y)) + (Pf/D)$

where

Pf = formation pressure, psi
D = depth at point of interest, ft tvd
S = overburden gradient, psi
y = Poisson's ratio (shear strain/axial strain).

Offset well data may yield leak-off test data that can be plotted against vertical depth and lithology, and correlated against the proposed well plan to give additional fracture gradient reference points.

CHAPTER 4
Drilling and Completion Fluids

FUNCTIONS OF A DRILLING FLUID

The primary functions of a drilling fluid are:

- Well control
- Maintain hole stability
- Hole cleaning
- Transmit hydraulic horsepower to the bit
- Formation evaluation

These functions are achieved by careful selection of the drilling fluid and maintenance of its properties.

Additional functions of a drilling fluid are:

- Suspend cuttings and weighting agent while the fluid is static; e.g., connections
- Release entrained cuttings at surface
- Cool and lubricate the bit and drill string
- Create a thin, impermeable filter cake to reduce fluid invasion
- Support tubulars through buoyancy effect
- Prevent and control corrosion of drill string, etc.

TYPES OF DRILLING FLUID

There are three main types of drilling fluid, distinguished by their base fluid formulation.

Air/Gas

Used for drilling hard, dry formations or to combat lost circulation. Rarely used offshore, except for underbalanced or coiled-tubing drilling.

Water-Based Mud

The main types of water-based mud are

- Nondispersed
- Dispersed
- Calcium treated
- Polymer
- Low solids
- Salt water

Nondispersed Muds

Generally includes lightly treated, low-weight muds, and spud muds. No thinners added. Usually top hole and shallow well applications.

Dispersed Muds

With increasing depths and mud weights, mud formulations require dispersant additives (lignosulphonates, lignites, tannins) to cancel the interparticle attractive forces that create viscosity in water-based mud. This effectively extends the use of the mud system until it has to be replaced.

Calcium Treated Muds

Typically, this group would include the gyp-ligno and lime-based muds. Here, a source of calcium (gypsum, lime) is added in excess to ensure a constant supply of calcium ions, which is effective in slowing down the shale hydration process. These muds tend to be relatively cheap to run, and dump and dilute policies are the norm for ultimate solids control.

Polymer Muds

These muds utilize long chain, high molecular weight polymers, which can encapsulate drilled solids to prevent dispersion or coat them for inhibition. They also provide viscosity and fluid loss properties. Common examples would be partially hydrolyzed polyacrylate (PHPA), carboxymethylcellulose (CMC), and polyanionic cellulose (PAC). They are intolerant of calcium contamination and will not withstand temperatures in excess of 300°F.

Low-Solids Muds

These are generally polymer-based muds which are designed to have a maximum of no more that 6–10% solids content by volume.

Saltwater Muds

These would include the salt saturated and seawater polymer systems where polymeric additives provide viscosity and fluid loss properties.

General Comments on Water-Based Muds

Freshwater-based mud (generally used for land wells). Note that prehydrating bentonite into a 50 ppb slurry with fresh water is often a first stage in building muds offshore.

Seawater-based mud—this often involves cutting back prehydrated bentonite from 50 ppb to 20 ppb with seawater to provide a base fluid with some initial clay content to provide viscosity and support for filter cake formation. Additional polymers are added to control fluid loss and enhance viscosity while barites is used to adjust fluid density. Once drilling commences, additional fluid would be built simply with seawater and polymers. Drilled solids would replace the bentonite component.

Brine-based mud—used to create the most inhibited water-based muds; i.e., prevention of shale swelling. The brine base can comprise sodium, potassium or calcium chloride and, as before, additional polymers are added to control fluid loss and enhance viscosity while barites are used to adjust fluid density. Additional mud would be built in the same manner maintaining the relevant salt concentration. Examples include:

- A salt NaCl (sodium chloride) saturated base would be used for drilling through a massive salt zone.
- A potassium chloride (KCl) brine base would most often be used for drilling large hole sections (17½") through reactive shales, the potassium ion being the "active" ingredient. It is preferentially absorbed into the vacant sites within the lattice structure of the clay particles instead of water molecules thus slowing down the hydration process.

When drilling reservoir sections, it is more usual now for barites to be replaced as a weighting agent by either acid soluble weighting material (calcium carbonate) or brine initially formulated to the correct density. This is particularly relevant when completing with sand screens where minimal solids contamination is required.

Oil-Based Mud

The main types of oil-based mud are:

- Diesel-based
- Invert emulsion
- Oil-based—all oil
- Synthetic

Diesel-Based Muds

These comprise diesel oil as the base fluid mixed with an emulsified brine and are still used in some areas of the world despite the high aromatic hydrocarbon content and HSE concerns (adverse skin reactions, carcinogenic). The aromatic content (carcinogenic component) of diesel is about 30% by volume.

Invert Oil-Based Mud

These are essentially mineral oil-based formulations, which include emulsified calcium chloride brine occupying from 5 to 50% of the liquid phase. The aromatic content of the base oil is less than 10%.

Oil-based Muds (All Oil)

These are formulated using 100% oil as the base fluid and are usually regarded as coring or reservoir drilling fluids.

Synthetic-Based Muds

These are formulated as with the invert muds but the base fluid used contains no aromatics; e.g., esters, ethers, PAOs (polyalphaolefins), paraffins.

General Comments on Oil-Based Muds

Essentially, all these fluids except the all oil fluid comprise the same basic components. The base fluid (diesel, mineral oil, paraffin, ester, etc.) is mixed with emulsifiers and calcium chloride brine to create a water-in-oil emulsion followed by a filtrate reducer and organophilic clay. Barites can then be added to adjust the density.

Usually several emulsifiers are added, one being a low end rheology modifier.

The lime is present to aid one of the emulsifiers used while also creating an initial, alkaline buffer against acid gas contamination.

An organophilic ("oil loving") clay is added to create the required viscosity prior to any barites addition. This is an amine treated clay that can yield in the oil/water emulsion.

The calcium chloride is present to balance the water phase salinity with that of the interstitial water in the formations being drilled, particularly the shales, to prevent destabilizing the hole.

The all oil fluid is simply an oil-based mud in which there is no water component. Additives comprise lime, acid soluble bridging material, and emulsifiers, which are selected to minimize any potential formation damage.

During well-control problems, when the mud system is being treated with large amounts of barites, it is usual to add an oil wetting chemical and a thinner, as well as extra emulsifier, to ensure that all the barites remain oil wetted and supported by the fluid.

The usual polymeric additives for a water-based mud are ineffective in an oil-based mud.

DRILLING FLUID SELECTION

The selected drilling fluid is often a compromise of the available choices. The following criteria need to be addressed before determining what mud system to use.

- Well type
  - Offset data, if available, should be used to identify any problems experienced with previous mud systems
- Environmental
  - Local legislation may prohibit certain mud types or, in the case of oil-based mud, may require the use of cuttings containment
- Well-control requirements
  - Mud system must be capable of being weighted up to the maximum required to control formation pressure

- Hole stability
  - Either chemical (reactive shales requiring an inhibited system or water-soluble formations such as salt or anhydrite) or mechanical (stress induced requiring control by mud weight)
- Temperature/chemical stability of the mud
  - System must be chemically stable at the maximum expected bottomhole temperature
- Drilling performance
  - System must provide optimum rheology to help maximize penetration rate
  - System should minimize formation damage in reservoir sections
- Cost
  - Needs to be balanced with expected benefits and performance
- Product availability
  - In remote areas, may limit choice of system

TABLE 4.1
DRILLING FLUID ADDITIVES

Function	Description	Product
Alkalinity, pH control	Control the degree of acidity or alkalinity of a fluid	Caustic soda Soda ash Bicarbonate of soda
Bactericides	Products used to prevent degradation of natural organic additives such as starch and zanthan gum	Glutaraldehyde Quaternary ammonium
Calcium reducers	Used to reduce calcium in seawater and treat calcium contamination from cement, gypsum, anhydrite	Soda ash Bicarb SAPP (sodium acid pyrophosphate)
Corrosion inhibitors	Control corrosion by creating a protective film	Amine based
Defoamers	Reduce foaming tendencies	
Emulsifiers	Create a heterogeneous mixture of two insoluble liquids	OBM Fatty acids Amine based WBM Detergents Soaps Surfactants
Filtrate reducers	Reduce the fluid loss into the formation	Starch CMC Bentonite
Flocculants	Clarify fluids and create temporary increase in viscosity	Lime Acrylamide-based polymers
Foaming agents	Typically surfactants	Typically surfactants
Lost circulation materials (LCMs)	Plug loss zones. Can be fibrous, granular, flake materials, or cross-linked polymer	Crushed walnut Mica
Lubricants	Reduce the coefficient of friction between fluid and pipe wall	Glycols Oils Surfactants

Table 4.1—Drilling Fluid Additives, cont.

Function	Description	Product
Pipe-freeing agents	Spotted across stuck pipe zone. Increase lubricity and chemically attack filter cake	Detergents Soaps Oils
Shale control inhibitors	Slow down hydration mechanism in shales	Soluble calcium and potassium sources Glycol
Surface active agents	Surfactants. Reduce interfacial tension between two surfaces (water/oil, water/solid, etc.)	Emulsifiers Flocculants Wetting agents
Temperature stability agents	Increase rheological and filtration stability under high temperature conditions	Acrylic polymers Sulphonated polymers Lignite Lignosulphonate
Thinners, dispersants	Modify the relationship between viscosity and solids content. Reduce gels. Reduce attraction of clay particles	Tannins Lignite Lignosulphonate
Viscosifiers	Increase viscosity for hole cleaning/suspension	Bentonite CMC PAC XC polymer
Weighting agents	Increase density of a fluid to control formation pressure, increase high angle hole stability	Barites Iron oxide Calcium carbonate

CONTAMINATION

Sources of Contamination

A contaminant is any undesirable component that causes a detrimental effect to the drilling fluid.

Typical contaminants are:

- Drilled solids
- Evaporite salts—interfere with emulsification in *oil-based muds* (*OBM*). Result —mud flips due to water wet barites.
- Water flows—well control and water wetting of barites in OBM
- Acid gases—CO_2, H_2S
- Hydrocarbons—well-control problem
- Temperature—thermal degradation of polymers in *water-based mud* (*WBM*)
- Cement
- Seawater—surface line leaks
- Bacteria—biopolymers, starches prone to bacterial degradation if not treated
- Spotting fluid (diesel/base oil)—stuck pipe with WBM
- Bicarbonate—excessive treatment of cement contamination
- Carbonate—excessive treatment of Ca contamination

The most common source of contamination found in all types of drilling fluid is caused by entrained drilled solids.

In WBM, this eventually results in a dump and dilute policy being adopted to maintain fluid properties. Typically, in a 17½" hole section, the sand traps would be dumped periodically, the logic being that, with no agitation, gravity settling should occur allowing the accumulation of fines which have passed through the shaker screens. The new mud added would effectively dilute the solids remaining in the system and help to bring the fluid properties (viscosity, fluid loss, density) back into specification.

OBMs are more tolerant of solids contamination. However, once overloaded, the system must be changed out since dumping is not an option. Most often, extra volume required to make up for losses on cuttings and filling new hole serves to dilute the solids problem.

Solids Control

The quality of the solids control equipment onboard can extend the useful life of a mud system but, eventually, the re-cycling of the solids causes particle degradation to such an extent that they become colloidal-sized and untreatable by mechanical means. A typical solids control package would include the following equipment:

Gumbo box—in some areas of the world (Gulf of Mexico), top hole sections produce "gumbo" clay cuttings—clay aggregates that, once discharged from the riser, would block any flow-line conduit. Typically, the box comprises a very coarse inclined grid that permits rapid conductance of fluid while the large clay aggregates are discharged off the end.

Shale shakers—it is worth noting that the shale shakers screen 100% of the flow. As such, they are critical in controlling the solids content as long as attention is paid to the screens (no holes and optimum mesh size). Rigs are now more commonly equipped with four or more linear motion, primary shakers (e.g., Thule VSM 100) capable of handling the 1,200+ gpm flow rates typical of 17½" hole sections. With 200 mesh screens, shale shakers can remove particles down to 150 microns.

Desilters—downstream, there are usually 4" cone desilters in various grouping sizes. A 4" cone can handle 50 gpm and so 2 × 16 cones theoretically could handle a 1,600 gpm flow. Normally, sufficient cones are provided to enable processing of 150% of the maximum flow expected. In practice, desilters are only used with WBM due to the high fluid content in the discharge stream. The underflow (waste stream) can be screened over a 200 mesh (75 micron) screen to recover the fluid (mud cleaner for OBM use) but it should be noted that

- each pass through a centrifugal pump will degrade the solids further.
- hydrocyclones are ineffective at removing clay solids but do work well when drilling sands—the sand grains do not degrade when passed through a centrifugal pump.
- with the improvement in the primary solids control equipment (shakers), it is now possible to match the cut point of hydrocyclones with shaker screens much earlier while drilling, making the former almost redundant. Once 200 mesh screens are on the shakers, there is no advantage in using the desilter.

The final item of equipment is the centrifuge, usually one or two in series. For weighted muds, i.e., in excess of about 12 ppg, it is normal to use two in series—the first utilizing a low bowl speed (<1,800 rpm) to recover the more dense barites and the latter then able to operate at much higher speeds (>2,500 rpm) to remove the low-gravity solids. However, it should be noted that a typical centrifuge will only process fluid at 30–50 gpm. With pump rates of 900–1,200 gpm and high penetration rates, they will clearly have minimal impact in the solids control process. Extracting solids at the first pass (shale shaker) is the single most effective action.

In areas where dumping of WBM is prohibited, there is the option to flocculate the solids and then recover them by centrifugation. This creates a solid cake for formal disposal while rendering the original fluid portion fit to be reused as a clear base fluid or discharged.

Drilled Solids Classification

Drilled solids are classified by particle size.

Drilled Solid	Particle Size
Coarse	Greater than 2,000 micron
Intermediate	Between 250 and 2,000 micron
Medium	Between 74 and 250 micron
Fine	Between 44 and 74 micron
Ultra–Fine	Between 2 and 44 micron
Colloidal	Less than 2 micron

DRILLING FLUID PROPERTIES

Density

Density or mud weight is the most critical property of any drilling or completion fluid since this provides primary well control. The drilling fluid density must be adjusted so that the hydrostatic pressure it exerts is sufficient to counterbalance formation pressure (except when deliberately drilling underbalanced) and allow usually for a safety margin of 200 psi. However, if the overbalance is too high, differential sticking, formation damage (excessive fluid invasion), and hydraulic fracturing (creating fluid losses) can occur.

Typical weighting agents include the minerals barites (SG 4.2), dolomite (SG 2.8), and the individual salts constituting a particular brine formulation.

Weighting agent	Maximum weight (ppg)	Maximum weight (SG)
Barites	19.5	2.34
Dolomite	11.5	1.38
Potassium Chloride	9.7	1.16
Sodium Chloride	10.0	1.20
Sodium Formate	11.1	1.33
Calcium Chloride	11.8	1.42
Potassium Formate	13.3	1.60
Calcium Bromide	15.4	1.85
Caesium Formate	19.7	2.30
Zinc Bromide	20.5	2.46

Note: In HPHT wells, temperature must be taken into account and the "true-weight" (pressurized) mud balance used.

Increasing Density

Mud density is normally increased by the addition of more weighting agent.

Brines tend to be blended; i.e., mixing brines of different density. This avoids carrying excessive sacked quantities of salt on inventory at the rig site and labour intensive rig activities. However, care should be taken that the blend mixed remains in solution at the temperatures expected; i.e., ambient, riser, and downhole.

Reducing Density

A reduction of mud density can be achieved by dilution or by mechanical removal of the weighting agent or contaminant solids:

- Mechanical removal is done by using efficient solid control equipment (e.g., centrifuge). It is generally the preferred method.
- Dilution is also an efficient way to reduce the drilling fluid density but it can more easily upset the other drilling fluid properties. With brines, any dilution using simply water must utilize fresh water, not seawater, thus avoiding potential problems.

Funnel Viscosity

The funnel viscosity, determined by using a Marsh Funnel according to API recommended procedures, should be used only as a guideline for determining flow properties of low density drilling fluids. As the density increases, the funnel viscosity becomes less and less reliable. Nevertheless, trends can easily be established and a drastic change in the funnel viscosity can indicate drilling fluid contamination.

Plastic Viscosity

Plastic viscosity (*PV*) is defined as the "resistance to flow" due to mechanical frictions. PV depends primarily on the solids content and the shape and size of these solids.

Increasing PV

Solids particles such as bentonite and barites, etc., are required to enable the drilling fluid to perform satisfactorily, but excessive drilled solids are undesirable; their presence is the primary cause of any increase of PV or YP.

If these drilled solids are allowed to remain in the mud, they are gradually ground into smaller particles by shearing action through the bit and pumps, thus increasing PV by creating more particles surface area.

Reducing PV

The concentration of solids in the mud must be reduced in order to lower PV. This can most easily be achieved by two methods:

- Use of efficient solids control equipment (often the preferred choice)
- Whole mud/base fluid dilution

Chemical treatment can be used under certain circumstances to reduce PV; i.e., flocculation to remove solids contamination. However, it is not widely applicable to most drilling fluids in common usage for either technical or economic reasons.

Yield Point

The *yield point* (*YP*) is the resistance to flow caused by electrochemical forces rather than mechanical friction. These forces are the result of particle attraction between negative and positive charges located on their surfaces. Thus, the yield point is a measure of these attractive forces under conditions of flow.

In unweighted drilling fluids, the YP is maintained at a level required for adequate hole cleaning.

In weighted drilling fluids, a moderately high YP is required to support the heavy weighting agent particles in suspension.

Increasing YP

Occurs naturally when flocculation results after the introduction of specific soluble contaminants; e.g., salt, anhydride, and gypsum encountered while drilling.

Also occurs naturally through solids contamination: an increase in drilled solids prompting an increase in interparticle attractions.

Through chemical treatment: additions of chemical viscosifiers (i.e., polymers and clays) are often made to maintain yield point specifications.

Reducing the Yield Point

The YP can be reduced by either chemical or mechanical treatment:

- *Chemical treatment*—Dispersion, deflocculation, or thinning will neutralize the attraction forces.
- *Mechanical treatment*—Use of efficient solid control equipment is the preferred method. It can also be done by dilution but, unless the solid concentration is high, dilution can upset the other drilling fluid properties.

Gel Strength

The gel strength is a measure of the mud ability to form a gel when stationary. This ability prevents cuttings and weighting agent material such as barites from settling down when stationary.

Gel strength is due to the same attraction forces as the yield point, but it relates to a condition of the drilling fluid at rest. It is a function of both the concentration and size of the solids in the drilling fluid.

High gel strengths can be undesirable because trapped gas and cuttings are not easily released at the surface.

The breaking of the gel strength developed each time the mud pumps are started creates a pressure surge as flow is initiated. This can be sufficient to fracture the formation and induce losses if the mud gradient is close to the fracture gradient.

Increasing and reducing the gel strength is achieved in the same way as for the YP.

Note: The 10-minute gel test is a good indication of gel strength.

Filtration

The drilling fluid engineer conducts filtration tests on site.

- Standard API filtration conducted at ambient temperature and 100 psi pressure.
- *High-pressure high-temperature* (*HPHT*) conducted at 500 psi differential pressure across the mud sample (600 psi on top of the fluid and 100 psi back-pressure held within the receiver). The temperature of the test is usually dictated by expected bottomhole temperatures, the test normally being conducted at 25°F above the maximum expected. Typically, tests are conducted within the 200–300°F range for most North Sea work.

These tests give indications about two important parameters:

Filter Cake

The nature of the filter cake is important. The ideal filter cake is thin (maximize the hole diameter and reduce differential sticking) and impermeable (prevent filtrate from the mud entering the formation). Generally measured in 1⁄32 inch or millimetres. A good quality filter cake should be of the order of 2⁄32 inch in thickness.

Fluid Loss

This parameter provides an indication of filtrate invasion into the formation. For reactive shales being drilled with a WBM, it could provide an indication of wellbore stability. When drilling a reservoir, it could provide an indication of the scale of filtrate invasion and hence potential formation damage. Generally quoted in millimetres/30 minutes. Filtrate values of less than 2 ml for WBM using the API test and the equivalent with an OBM using the HPHT test could be termed good quality filtrate properties.

Note: The API test is the primary filtrate test for water-based muds. It is never conducted on oil-based muds. The HPHT test is always conducted on oil-based muds.

Both the above filtration tests determine the volume of filtrate and describe the character of the filter cake. An API test conducted at room temperature and 100 psi differential on an OBM would yield no filtrate due to the emulsion strength in the fluid. For most applications a controlled fluid loss is required, suggesting controlled invasion by mud filtrate into the surrounding rock. A thin, pliant filter cake is desirable as this both minimizes differential sticking and indicates controlled filter loss.

Solids

High-Gravity and Low-Gravity Solids

A sample of mud is placed in a retort cell (typically 10, 20, or 50 mls) and heated to about 600°C (dull red heat). This drives off the water and oil which are both collected in a calibrated receiver and measured directly, the oil layer lying above the water. The solid residues in the cell include barites (HGS), drilled solids (LGS), and salt from the aqueous phase. By knowing the volume and density of mud retorted and the volume of water and oil collected, it is possible to calculate the concentrations of both HGS and LGS in the original sample.

Sand Content

A sample of mud is decanted into a tube to a measured mark. The contents are then flushed through a 200-mesh (approx. 75 micron) sieve. The retained solids on the sieve are then back-flushed into the tube and allowed to settle. The volume of solids is read directly off calibrations on the tube to give a percentage solids figure. Since the screen will only retain sand-sized particles, the volume of solids is assumed to be sand. Normally, this figure is less than 0.5% but can rise to >3%, especially if the solids control equipment is overloaded; i.e., fast drilling in 17½" hole through sand. Apart from its unwelcome contribution to the overall solids content in the mud (impact on mud rheology), a high sand content can cause abrasion problems with pump liners and MWD tools and should be avoided at all times.

Clay Content

The *methylene blue test* (*MBT*) is performed on water-based muds. It is based on the capacity for reactive clays to absorb methylene blue dye until all the reactive sites are filled with dye. The test focuses on being able to detect excess dye. A sample of mud (1 ml) is diluted with water and boiled for 10 minutes along with 0.5 ml of 5 N sulphuric acid and 15 ml of 3% hydrogen peroxide. These are added to remove organic materials present (e.g., polymers) that are also able to absorb the dye and would thus interfere with the results. Methylene blue dye is then added in 0.5 ml increments and the solution spot tested on a filter paper. The presence of a spreading, blue halo indicates the presence of excess dye.

$$\text{MBT capacity} = (\text{ml of methylene blue})/(\text{ml of mud})$$

$$\text{Bentonite ppb} = 5 \times \text{MBT capacity}$$

For a new mud system, the MBT value is less than 5 ppb equivalent bentonite. When this figure reaches 20-25, the mud system is beginning to be overloaded with clays and a dump and dilute policy is normally advocated.

Chemical Analysis

Water-based Mud

For water-based mud, chemical analysis by titration entails determining the alkalinity, chloride, Ca, and total (Ca and Mg) hardness and potassium concentration for a KCl-based mud.

While a pH meter can be used to determine the pH directly, the alkalinity titration determines the character of the alkalinity. This is based on the presence and concentration of the hydroxide, carbonate, and bicarbonate ions.

Calcium ions should be present but not in excess of around 500 mg/lt for polymer-based muds. If they exceed 1,000 mg/lt, they can begin to affect the mud constituents; i.e., cause precipitation of polymers.

The potassium level should be monitored to ensure that there is sufficient presence to continue to inhibit clay swelling, i.e., typically of the order 30–45 ppb potassium chloride (*note:* usually provided as an 80 ppb KCl premix with 5% glycol if required).

Oil-based Mud

For oil-based mud, chemical analysis is restricted to the water phase and the water-in-oil emulsion needs to be broken down first. Analyses include alkalinity, chloride, and calcium.

The alkalinity in this instance determines the excess lime in the mud necessary to enable some of the emulsifiers to work and also to ensure an initial buffer against acid gas contamination.

The chloride level of the water phase is important and should be balanced against the formation water salinity. If it is too high, the mud will draw water from the formation and render it brittle (mechanical collapse), too low and water will pass from mud to formation via the process of osmosis and result in hydration (clay swelling). Usual water phase salinities are of the order 150,000–200,000mg/lt chloride for North Sea shales.

The calcium determination enables the sodium chloride salt contribution to be determined by difference. Calcium chloride is usually the principal salt additive for OBMs—it is more soluble than NaCl and thus is able to balance higher formation salinities.

pH

The pH value is a property of water-based mud and is a measure of the acidity of the mud.

A typical WBM formulation would operate in the pH 9–10 range. Maintaining this slightly alkaline environment also serves to reduce corrosion by eliminating any acid gases generated (CO_2 and H_2S).

Note: Certain WBM, e.g., silicates, operate in a much higher pH range, typically 11–12.

Electrical Stability

This test measures the ability of an OBM to transmit a current. Generally, the stronger the water-in-oil emulsion, then the greater the applied voltage required to transmit between two electrodes. Water contamination will result in decreasing electrical stability values being recorded. Typical values are in the order 400–600V.

TREND ANALYSIS

Various critical fluid properties should be monitored 2–3 times daily during the drilling phase. The objective is to establish trend lines for these properties to enable early detection of any problems. It must be remembered that any variation in a mud property is an indication that something has changed.

The following Tables 4.2 and 4.3 show the possible cause of various mud property trends in both water- and oil-based mud systems.

Note that a full mud check will be required to accurately determine what the problem is. The trends help point in the right direction.

TABLE 4.2
WATER-BASED MUD TREND ANALYSIS

Mud Property	Trend Change	Possible Cause
Mud Weight	Increase	Drill solids increase Heavy spot from barite sag Overtreatment during weight up
	Decrease	Formation fluid influx Light spot from barite sag Excessive water additions
Funnel Viscosity	Increase	Reactive shale drilled Drill solids increase Low water content Calcium contamination from cement Anhydrite formation drilled
	Decrease	Formation water influx Excessive water content
Plastic Viscosity	Increase	Unconsolidated sand drilled Drill solids increase Low water content
	Decrease	Formation water influx Excessive water additions Solids content decrease
Yield Point	Increase	Reactive shale drilled Anhydrite formation drilled Low water content Calcium concentration from cement
	Decrease	Formation water influx Excessive water additions Decrease in low gravity solids Additions of chemical thinners
Gel Strength	Increase	Reactive shale drilled Low water content Calcium contamination from cement Anhydrite formation drilled
	Decrease	Formation water influx Excessive water additions Additions of chemical thinners

Table 4.2—Water-based Mud Trend Analysis, cont.

Mud Property	Trend Change	Possible Cause
API/HPHT Fluid Loss	Increase	Low gravity solids increase Flocculation from cement, chloride, calcium contamination Low gel content
	Decrease	Mud treatment taking effect
pH	Increase	Addition of pH control additives Calcium contamination
	Decrease	Addition of mud products Anhydrite formation drilled
Chlorides	Increase	Salt formation drilled Pressure transition shale drilled Formation water influx
	Decrease	Water additions
Total Hardness	Increase	Salt or calcium formation drilled Formation water influx
	Decrease	Fresh water addition Chemical addition
MBT	Increase	Reactive shale is drilled Addition of bentonite
	Decrease	Water additions Solids removal equipment

TABLE 4.3
OIL-BASED MUD TREND ANALYSIS

Mud Property	Trend Change	Possible Cause
Mud Weight	Increase	Drill solids increase Heavy spot from barite sag Overtreatment during weight up
	Decrease	Formation fluid influx Light spot from barite sag Excessive base oil additions

Table 4.3—Oil-based Mud Trend Analysis, cont.

Mud Property	Trend Change	Possible Cause
Plastic Viscosity	Increase	Addition of water Addition of calcium carbonate Addition of primary emulsifier Low gravity solids increase
	Decrease	Addition of base oil Decrease in low gravity solids
Yield Point	Increase	Increase in organophilic clay Addition of emulsified water or synthetic polymer
	Decrease	Addition of base oil or degellant Decrease in organophilic clay
Gel Strength	Increase	Addition of organophilic clay Addition of water
	Decrease	Large base oil additions Increase in mud temperature
Oil Water Ratio	Change	Large addition of water or water influx Large additions of base oil High bottomhole temperature
Electrical Stability	Increase	Increase in emulsifier concentration Adding wetting agent or base oil
	Decrease	Decrease in emulsifier concentration Newly prepared OBM has low electrical stability but increases with time
Water Phase Salinity	Increase	Water % of oil water ratio decreasing Addition of calcium chloride
	Decrease	Water % of oil water ratio increasing from water addition or formation water influx
HPHT Fluid Loss	Increase	Addition of base oil Decrease in emulsifier concentration Water present in filtrate
	Decrease	Increase in primary emulsifier concentration
Excess Lime	Increase	Addition of lime anhydrite formation drilled
	Decrease	CO_2 or H_2S kick Additions of base oil or water

FORMATION DAMAGE

Formation damage can be caused by many mechanisms. Although some of these may be due to well conditions, the majority are caused by contamination of the formation by foreign substances, not only during the drilling and completion phase, but also during production and well servicing.

Formation damage results in the reduction in productivity or injectivity of a reservoir.

Causes of formation damage (mud related):

- Interaction between invading mud filtrate and reservoir fluid
- Interaction between invading mud filtrate and reservoir rock constituents
- Interaction between mud solids and reservoir rock
    - ❖ Particle plugging in reservoir pores
    - ❖ Interstitial clays swelling
    - ❖ Mobilization, by filtrate invasion, of interstitial fines which then plug pores
- Inorganic scale formation due to incompatible mud filtrate and reservoir fluid
- Reduction in relative permeability of reservoir in near wellbore area—flushed by filtrate
- Emulsion blocking—created by mixing of filtrate/reservoir fluid

Additional causes:

- Pores blocked with wax or asphaltenes
- Gas breakthrough and water coning reduces oil permeability
- Stress-induced permeability changes in near wellbore area
- Sand production—screen blocking

Prevention is better than cure

The prevention of formation damage should be a priority and the following steps considered:

- Ensure compatibility of mud filtrate or completion fluid with reservoir fluid beforehand through lab testing. Prevent scale, emulsion blocking, precipitation. Maintain tight control over fluid loss during operations.
- Prevent excessive lost circulation in the reservoir section.
- Ensure that mud filtrate/brine is sufficiently inhibited to prevent any interstitial clays from swelling.
- Ensure that any fluid invasion does not alter the water wet native state of the reservoir that is more conducive to higher oil permeability.
- To prevent solids plugging, tight control on filter loss during drilling is essential and solids-free clear brines during completion. This highlights both the importance of cleanliness and the need for brine filtration during completion phases.
- Avoidance of pressure surging in the reservoir to prevent losses/fluid invasion.
- Liner cement recipes should be formulated with sufficient fluid loss control.
- Prior to completion activities, ensure that the wellbore is clean; e.g., total removal of any residual OBM when the completion is being run in solids-free brine. The use of detergent and scouring pills is commonplace. Mechanical means using casing scrapers and wire brushes is also recommended along with riser/BOP jetting while functioning the rams.
- The use of easily removable (acid soluble) plugging materials for control of losses should be mandatory; e.g., calcium carbonate, oil soluble resin, sized salt.
- Perforating should ideally be conducted underbalance and the wells flowed as soon after as possible to clean out any debris.
- Control of sand production can be affected using screens and/or gravel packs instead of the conventional liner and cement job. Care is required not to damage the shroud around the screens while running them in.

CHAPTER 5
Cementing

OBJECTIVES

Primary Cementing

- Isolation of casing shoe
- Isolation of production zones—prevent cross-flow between intervals at different pressures
- Protection of water zones—prevent drilling fluid contamination of aquifers
- Isolation of problem interval – extreme losses, well control, side tracking
- Protection of casing—from corrosive formation fluids; e.g., H_2S, CO_2
- Casing support—e.g., support for conductor (load bearing—bending moments derived from supporting BOP/tree/riser and potential snag loads from fishing activities), prevent thermal buckling

Secondary or Remedial Cementing

Additional cementing done at a later stage; e.g., sealing off perforations, top up job on conductor, repair casing leaks, squeeze casing shoe, setting plugs, etc.

PLANNING

Planning for a cement job consists of evaluating a number of features, including:

- Assessment of hole conditions (hole cleaning, size, washouts, temperature)
- Mud properties
- Slurry design
- Slurry placement
- Additional equipment (float equipment, centralizers, ECPs)

COMMON CEMENTING PROBLEMS

Common problems that affect all cement jobs include:

- Poor hole condition (doglegs, borehole stability, washouts, hole fill, cuttings beds, etc.)
- Poor mud condition (high gel strengths and yield point, high fluid loss, thick filter cake, high solids content, lost circulation material, mud/cement incompatibility)
- Poor centralization (cement not placed uniformly around the casing, leaving mud in place)
- Lost circulation
- Abnormal pressure
- Subnormal pressure
- High temperature

CEMENT TYPES

API defines 9 different classes of cement (A to H) depending on the ratio of the four fundamental chemical components (C_3S, C_2S, C_3A, C_4AF where C = calcium, S = silicate, A = aluminate, and F = fluoride).

API Class	Mix water gal/sx	Slurry wt. ppg	Depth ft	BHST °F
A (Portland)	5.2	15.6	0–6,000	80–170
B (Portland)	5.2	15.6	0–6,000	80–170
C (High Early)	6.3	14.8	0–6,000	80–170
D (Retarded)	4.3	16.4	6,000–10,000	170–230
E (Retarded)	4.3	16.4	6,000–10,000	170–290
F (Retarded)	4.3	16.4	10,000–16,000	230–320
G (Basic California)	5.0	15.8	0–8,000	80–200
H (Basic Gulf Coast)	4.3	16.4	0–8,000	80–200

Notes:

Class A & B—Shallow depth use. Composition 50% C_3S, 25% C_2S, 10% C_3A, 10% C_4AF.
Class C—Produces high early strength due to high C_3S content.
Class D, E, and F—Retarded cements due to coarse grind or inclusion of organic retarders (lignosulphonates).
Class G and H—General purpose, compatible with most additives and able to be used over a wide range of temperature and pressure. H coarser than G—better retarding in deeper wells.
Class G is the most common type of cement used.

Other common cement variants without API specification include:

Pozmix cement—50% Portland, 50% pozzolan (ground volcanic ash), and 2% bentonite.
Gypsum cement—Mixture of Portland cement and gypsum. Used for remedial work.
Diesel oil cement—'Gunk squeeze'. Mixture of basic cement with base oil used to seal off loss zones. Will set if water present.
Silica flour—At temperatures above 230°F, cement will initially strengthen and then later weaken due to the subsequent formation of Dicalcium Silicate Hydrate (C_2SH). By adding 30–40% silica flour to the cement, CSH forms in preference to C_2SH thus extending the temperature range of the mix.

CEMENT PROPERTIES

Yield

The yield of the cement, in cubic feet per sack, is the volume that will be occupied by the cement, mix water, and additives when the slurry is mixed.

It will vary depending upon the cement class.

Slurry Density

A standard mix comprising 5 gallons of water and 94 lb (1 sack) of cement will create a slurry with a density of 15.8 ppg.

The slurry density is adjusted by varying either the mix water ratio or use of additives. Most slurry densities are within the range 11–18.5 ppg.

Additives to adjust density include:

Density reducing materials

- Bentonite (SG 2.65)—reduce a 15.8 ppg to a 12.6 ppg slurry with 12% bentonite
- Diatomaceous earth
- Gilsonite (SG 1.07)
- Pozzolan (SG 2.5)—50:50 mix with 2% bentonite will create a 13.3 ppg slurry

Density increasing materials

- Barites (SG 4.25)
- Ilmenite (SG 4.6)
- Hematite (SG 5.02)

Mix Water

The mix water ratios detailed above are dependent on:

- The need for a pumpable slurry
- A minimum amount of free water if allowed to stand or settle

Reducing the mix water ratio has the following effect:

- Causes an increase in slurry density, compressive strength, and viscosity
- Slurry becomes harder to pump
- Less volume of slurry is built per sack of cement used; i.e., yield decreases

During a typical cementing operation, a lead and tail slurry are often utilized. The difference between these is due to a reduction in the amount of mix water being used. An increase in water content for the tail slurry will permit longer pumping and setting time, but results in a lower compressive strength and additional free water. The free water can be reused by the addition of bentonite in the slurry to bind the free water.

Thickening Time (Pumpability)

Thickening time is the time available for a slurry to be mixed, pumped, and displaced into the annulus before it starts to thicken and set up. This time will depend on the additives used (retarders to increase time and accelerators to reduce time), and the conditions downhole (an increase in temperature, pressure, and fluid loss will all reduce thickening time).

Thickening time is determined during laboratory testing. The time to reach 100 Bearden Units (Bc) is recorded as the thickening time. Pumpability will normally cease at around 70 Bc.

Compressive Strength

A compressive strength of about 500 psi minimum, including a safety factor, is felt necessary for supporting a casing string and withstanding differential pressures prior to drilling ahead. For production casing or liner strings, a compressive strength around 2,000 psi is often required for perforating.

The *waiting on cement* (*WOC*) period enables the cement strength to develop fully. The time period is dependent on the bottomhole temperature, pressure, ratio of mix water, and time elapsed since mixing. Accelerators (e.g., $CaCl_2$) can reduce the WOC time to less than 3 hours.

Water Loss

The cement setting process is the result of a chemical reaction resulting in dehydration. Thus, it is important that any water loss be controlled until the cement is in place to ensure that it remains pumpable. The amount of water loss acceptable would depend on the type of job being conducted.

Squeeze job—this would require a controlled water loss (typically 50–200 ml) to enable cement slurry to be pumped into the perforations before a significant, impermeable filter cake was created.

Primary cementing—water loss is less critical and would typically be of the order 250–400 ml.

Liner job—fluid loss of around 50 ml.

Horizontal hole—fluid loss less than 50 ml.

Permeability

Once set, cement has a permeability of less than 0.1 millidarcy (tight sandstones are around 1–10 millidarcies). Disturbances during setting, e.g., gas percolation or pressure testing, can increase this by several orders of magnitude.

CEMENT ADDITIVES

Most cement slurries will contain some additives to improve the individual properties, depending on the job. Additives could be required to:

- vary the slurry density.
- change the compressive strength.
- accelerate or retard setting time.
- control filtration and fluid loss.
- reduce slurry viscosity.

They may be in dry/granular or liquid form or may be blended in with the cement.

Quantities of dry additives are usually expressed in terms of percentage by weight of cement (% BWOC). Liquid additives are usually expressed in terms of volume by weight of cement (gal/sx).

Type of Additive	Example	Typical Dowell Product
Accelerator	$CaCl_2$ NaCl	S1 D44
Retarder	Calcium lignosulphonate CMHEC Saturated salt solution	D13, D81 D8, D120
Density increase	Barites Hematite	D31 D76
Density decrease	Bentonite Diatomite Pozzolan	D20 D56 D61
Friction reducer	Polymers Calcium lignosulphonate	
Fluid loss	Organic polymers CMHEC	Flac D59, Flac D60 D8

Accelerators

Reduce the WOC time (time to reach 500 psi compressive strength).

Used on shallow wells with low temperatures

Common additives:

Calcium chloride 1.5–2.0%
Sodium chloride 2.0–2.5%
Seawater

These will act as retarders in higher concentrations.

Retarders

Used on deeper sections where the higher temperatures promote more rapid setting.

If the static BHT is above about 260°F, the effect of the retarder should be measured by pilot testing.

Calcium lignosulphonates 0.1–1.5%
Saturated salt solution

Density Reducing

Used to reduce slurry weight where there is concern about exceeding fracture gradient.

Also reduces the compressive strength and increases the thickening time.

Allows more mix water to be used (creates greater slurry volume)—hence termed "extenders."

Prehydrated bentonite 2–20%. Reduce compressive strength and sulphate resistance
Pozzolan 50:50 mix with Portland cement. Decrease in compressive strength, increase in sulphate resistance
Diatomaceous Earth 10–40%

Density Increasing

Used when cementing in overpressured zones

Barites $BaSO_4$. Used for densities up to 18 ppg
Hematite Fe_2O_3. Densities up to 22 ppg
Graded sand 40–60 mesh. Gives a 2 ppg density increase

Fluid Loss Additive

Used to prevent dehydration of slurry and premature setting. Also reduces free water.

Carboxymethyl hydroxyethyl cellulose CMHEC 0.3–1%

Dispersants (Friction Reducing)

Added to improve the flow properties. Reduces the viscosity enabling turbulent flow to be achieved at a lower circulating pressure—less chance of incurring losses.

Polymers 0.3–0.5 lb/sx of cement
Salt 1–16 lb/sx of cement
Calcium lignosulphonate 0.5–1.5 lb/sx

CEMENT TESTING

Cement recipes must be tested according to API 10 specifications.

Initially, a formulation would be designed to suit the cement job proposed; e.g., a surface hole (conductor) slurry would differ from a liner recipe in terms of its water loss requirements, setting time, etc.

A freshly mixed sample comprising cement, mix water, and chemicals from the rig, would then be tested in the lab BEFORE the actual job takes place to ensure that there are no contamination problems. Since the test work requires a minimum of 24 hours to complete, it is important that fresh samples are dispatched to the lab from the rig as soon as possible.

Compressive Strength

This used to be the unconfined pressure required to crush a 2" cube of cement. A series of cement cubes would be made by using molds and allowing to set. Periodically, one of the cubes would be removed and tested to destruction. A more recent test involves use of acoustic and ultrasonic waves. The Ultrasonic Cement Analyzer (UCA) continuously monitors the strength development of a setting cement sample under simulated downhole temperature and pressure conditions. A chart printout records the setting history.

Water Content

Ideally, a cement slurry should have a viscosity (consistency) that will enable it to displace mud efficiently while permitting a strong bond to form between cement and casing. This means that the slurry should set without any free water forming. Free water is water that is squeezed out of the setting cement and creates pockets or a surface layer on the cement.

Maximum water—will provide a set volume with 1.5% free water maximum. The free water is determined by allowing a sample of freshly mixed (20 mins) slurry to stand in a measured cylinder.

Normal water—will provide a slurry that has a consistency of 11 B_cs (Bearden units—units of consistency) after 20 mins of mixing.

Minimum water—will provide a slurry with a consistency of 30 B_cs after 20 mins mixing.

Note: cement testing uses Beardon units for measuring viscosity because these are based on torque or drag.

Thickening Time

This is measured using a high-pressure/high-temperature thickening time tester (consistometer). It comprises a rotating cylindrical slurry container with a stationary paddle, the whole lot being enclosed within a pressure chamber. It is able to simulate well conditions with BHSTs up to 500°F and in excess of 25,000 psi. The container with slurry rotates at a standard rate as the temperature and pressure are ramped up at a predetermined rate. The torque created at the paddle shaft due to the setting cement is measured on a strip recorder. Pumpability limit or thickening time is reached when the slurry consistency reaches 70–100 B_cs.

Slurry Density

This is typically measured using a pressurized balance. A sample of cement is decanted into the sample chamber and a screw lid attached. Further, slurry can be injected through the nonreturn valve in the cap using a hand pump. This subjects the slurry to sufficient pressure to eliminate errors due to entrapped air bubbles.

Water Loss

The fluid loss test measures the filtrate generated over 30 minutes through a filter press dressed with a 325-mesh screen. The test can be conducted at 100 or 1,000 psi and at temperatures up to 400°F, and with either freshly mixed slurry or one that has been on the thickening tester for a while. With no additives, all neat cement slurries have a fluid loss in excess of 1,000 ml. With long chained polymer additives at concentrations of 0.6 to 1% by weight of cement (BWOC), the fluid loss can be reduced to 50–150 ml.

Permeability

Can be measured using a permeameter but generally is not a prime driver in designing a cement slurry.

Rheology

Cement rheology is determined using a six-speed rheometer equipped with the appropriate rotor sleeve, bob, and torsion spring. After recording the dial readings corresponding to the six preset rotary speeds (600, 300, 200, 100, 6, and 3 rpm), the various rheological parameters can be calculated—PV, YP, n, and K values.

SPACERS

During displacement, the slurry will become contaminated in part with residual mud and filter cake from the drilling operation. The effect of contamination is to alter the various properties of the cement. The effects of contamination are minimized by pumping various spacers ahead of the main slurry.

Prior to pumping any slurry, a series of chemical wash/spacers is usually pumped comprising base oil (for OBM), detergent washes, scavenger mud (to reclaim valuable drilling fluid), and a viscosified pill. The purpose of these spacers is to:

- Physically separate mud from cement—must be no compatibility problems
- Remove mud/wall cake from the annulus—turbulent flow regime preferred
- Leave the casing and formation water wet—surfactants
- Provide less hydrostatic head; i.e., reduced pump pressures—oil or water

Spacer Characteristics

- Some fluid loss control and a shear-thinning characteristic (reduce pump pressures).
- Turbulent flow is the preferred regime to generate efficient displacement and erosion of wall cake.
- A minimum contact time of 10 minutes is deemed sufficient and will determine the volume pumped.
- Under laminar flow conditions, the density and frictional pressure loss of the spacer should be greater than that of the displaced fluid.

EQUIPMENT

Casing Shoe

Run on the bottom of the casing string. Rounded profile to assist when running into open hole. Known as a float shoe if run with a ball or poppet valve.

Float Collar

Usually placed 2 or 3 joints above the shoe and acts as a stop for the cement plugs. The float collar ensures that there will be cement sealing the last few joints of casing when pumping ceases; i.e., when the plug is "bumped." Some drilling programs permit an additional displacement of up to half the shoe track in an attempt to correct for pump efficiency error and observe a plug bump. This also minimizes the volume of cement to be drilled afterwards.

The float collar also contains a ball valve, which prevents the cement in the annulus from flowing back into the casing when the displacement has finished. A flow (or backflow) test is conducted after pumping to confirm that it is holding. When running casing, since the float will prevent backflow, it is usual to have to periodically fill the casing (every 5 joints). Failure to do this can lead to a collapsed casing string.

Centralizers

These are either of the hinged metal rib variety or the solid spiral body type and both serve to centralize the casing in the hole.

Advantages of centralized pipe:

- Improved displacement efficiency—minimal eccentricity (figs. 5.1 and 5.2)
- Reduced differential sticking risk
- Prevent keyseating problems
- Reduced drag in directional wells

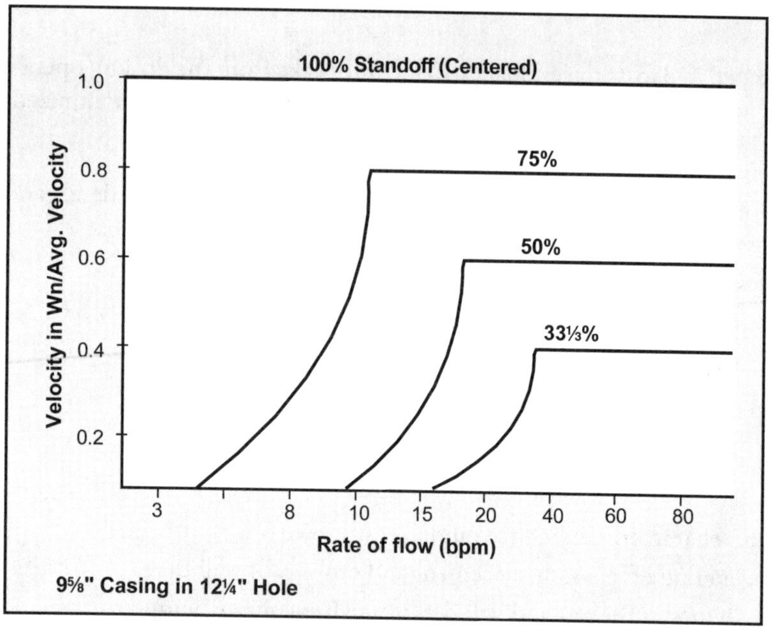

Figure 5.1 Influence of standoff on mud removal

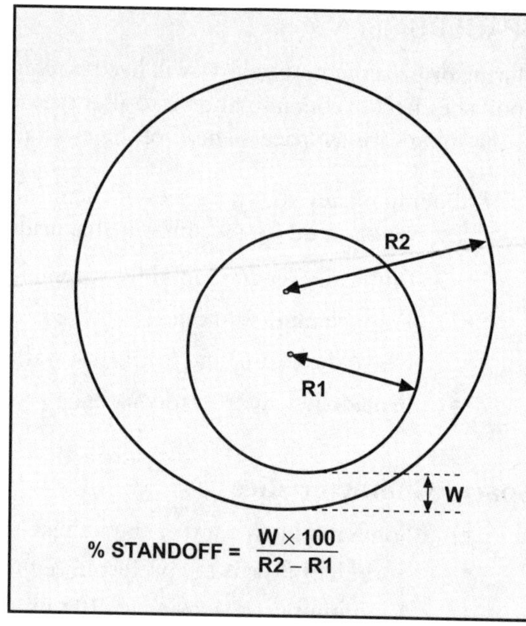

Figure 5.3 Standoff

The centralizers are clamped to the casing using a hinge and pin mechanism while a stop collar serves to locate them in position. Spacing and quantity of centralizers is dependent on hole angle, casing weight, and mud weight. The suppliers can provide an optimum spacing program utilizing API's recommended standoff criteria (fig. 5.3). Typically, the centralizers might be concentrated on the critical, higher angle sections of the shoe and just below the hanger while the remainder of the casing would have them spaced very sporadically.

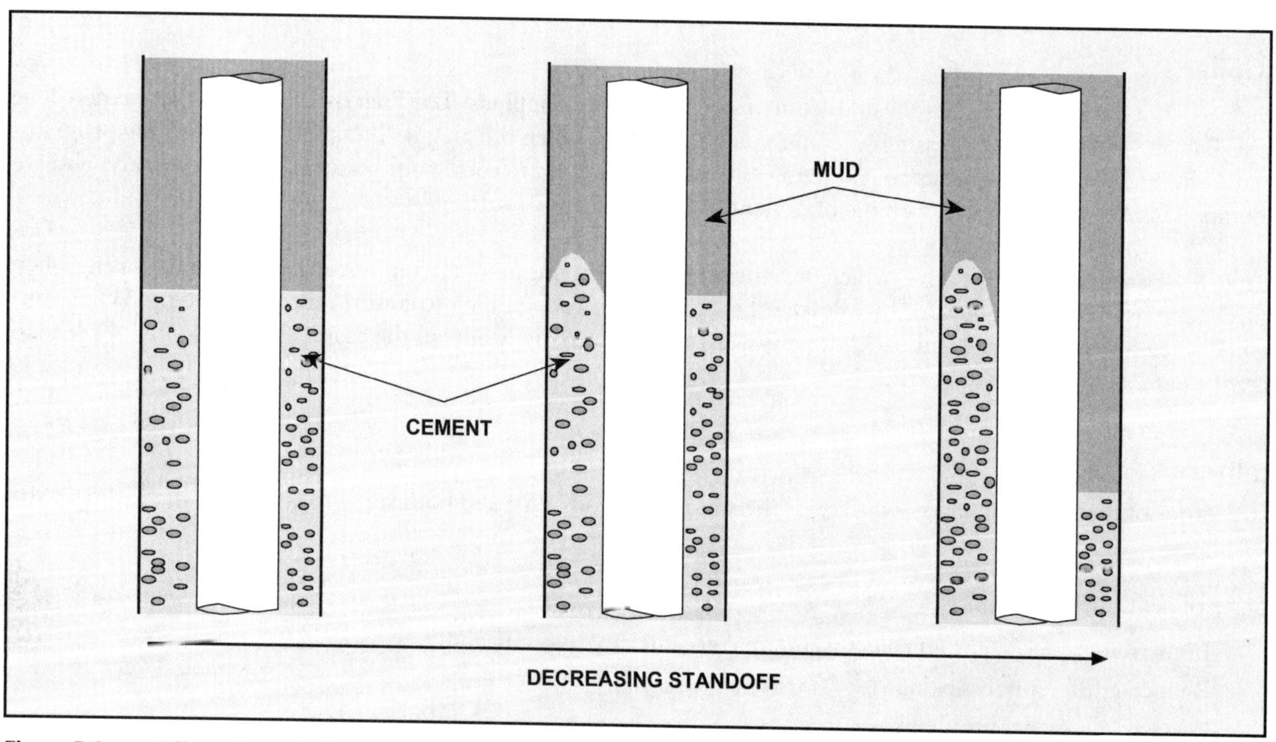

Figure 5.2 Effect of standoff on mud displacement

Scratchers

Wire brushes that can be clamped onto the casing and secured with stop collars. Used to physically dislodge wall cake, gelled mud, and other debris.

Cement Heads

The cement head connects the discharge line from the cement unit to the top of the casing (fig. 5.4).

For a full-bore application, the casing is run back to the rig floor and the plugs are loaded into the surface cement head. Launching involves removing the retainer and pumping the plug downhole.

For a subsea system, the casing is run and landed on drill pipe while the plugs are stored in a special plug basket under the casing hanger. The plugs are launched by dropping either a dart or nitrile ball which seats in the plug and allows it to be pumped down.

Cement Plugs

Cement plugs are used to separate the cement slurry from the spacer or mud and prevent contamination. On long casing strings, additional plugs are pumped ahead of and between the spacer train to minimize contamination caused by different flow regimes within the different spacers, and to maximize their effectiveness when they exit into the annulus.

Plugs are normally manufactured out of rubber. Various proprietary devices are used to "lock" plugs together to make them easier to drill out (often termed PDC drillable).

The bottom plug has a thin diaphragm in its center. After it lands out on the float collar, the diaphragm ruptures when a predetermined differential pressure is reached. It is normally dropped ahead of the spacer or cement.

The bottom plug has a solid center.

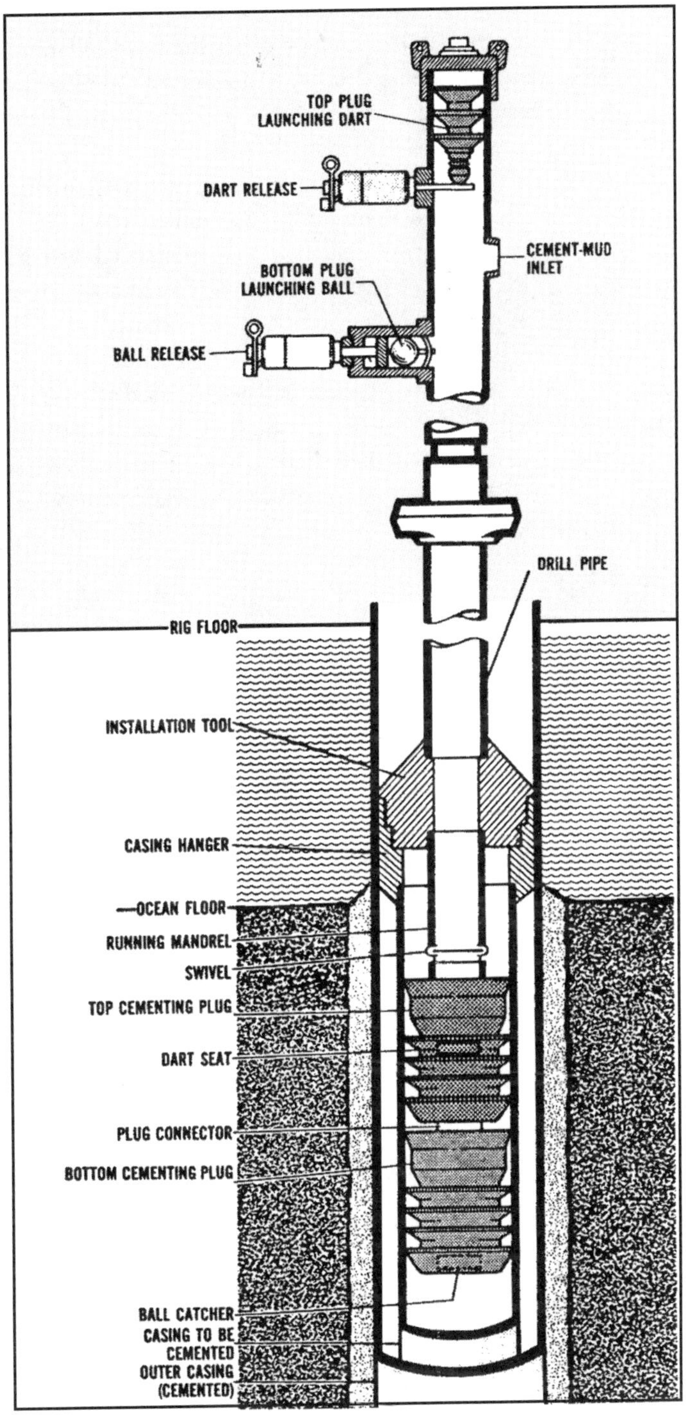

Figure 5.4 Subsea cement system

CEMENTING PRACTICES

Primary Cementing

- Ensure that a computer simulation of the cement job has been performed to establish minimum and maximum flow rates and ECDs.
- Condition the mud to reduce rheology (YP, gels) prior to final trip out to run casing.
- Confirm that the plugs are correctly placed in cement head—bottom (diaphragm) plug below top (solid) plug.
- Run the casing in until a few feet off bottom. Break circulation if required on the way in.

- Circulate at least one casing volume to ensure that there is nothing to plug the shoe and to remove any trip gas.
- Pump spacers, release bottom plug, and pump cement slurry (lead and tail).
- Release top plug, clear cement line, and begin to displace.
- Displacement rate should be altered depending upon what is in the annulus (mud, spacer, or cement). Most spacers and cement require turbulent flow (if possible) to maximize mud removal and reduce mud contamination.
- When the bottom plug reaches the float collar, the diaphragm should rupture, allowing continued pumping.
- The displacement volume to land the top plug should have been calculated previously.
- Displacement rate should be reduced as plug bump is approached to prevent excessive pressures and any shock as plug lands.
- If no bump is seen, it is common practice to further displace up to half of the shoe track (note that some operators have adopted a pump till bump philosophy).
- All mud returns should be closely monitored throughout for losses, which could be evidence of fracturing formation.
- If losses are observed, the displacement rate can be adjusted to reduce the ECD; i.e., annular pressure losses.
- The plug should be bumped with about 1,000 psi differential having first confirmed that the casing burst pressure less safety margin would not be exceeded.
- If required, the pressure can be increased at this point and a casing pressure test performed (note that it is necessary to confirm the pressure rating of all components before performing this test).
- The pressure should then be bled off to confirm that the float valve is functioning and supporting the differential back-pressure due to heavy cement in the annulus.

Stage Cementing

Used in applications where long sections of casing require cementing but concerns exist because of

- long pumping times.
- high pump pressures.
- excessive hydrostatic pressure due to column of cement—exceed fracture gradient.

First Stage

Repeat of primary cementing operation above.

Second Stage

This necessitates the inclusion of a DV collar in the original casing string at a predetermined depth. The first stage places cement in the annulus from the bottom up to the DV collar. The ports on the DV collar can then be opened by dropping a special dart (bomb) and shearing the retaining pins (1,000–1,500 psi). Circulation is then established through the DV collar. The above primary cementing procedure can then be repeated but without any pipe reciprocation. Further stages could be included if necessary.

Inner String Cementing

Conventional cementing approach with large diameter casing will result in

- large displacement volumes.
- extended displacement duration.
- significant volume of cement remaining in shoe track.

As an alternative, the casing can be cemented through tubing or drill pipe. A special float shoe is used which allows the drill pipe to stab in providing a hydraulic seal. The casing is run as normal and then the inner string is run and stabbed into the float shoe. The cement job proceeds as before using smaller drill pipe plugs. After displacement and confirmation that the float shoe is holding back the differential pressure, the pipe can be retrieved.

Care needs to be taken with this technique as the possibility of collapsing the casing is significantly increased.

Liner Cementing

A liner string usually comprises a shoe and float collar, as with larger casings, along with a liner hanger (hydraulic or mechanically set) to secure the upper end. The entire assembly is run on drill pipe and then the hanger typically set 300–500 ft within the previous casing. Once set, mud is circulated to ensure an obstruction-free cementing path around the liner. Prior to cementing, the running tool is backed off the liner hanger to guarantee removal of the drill pipe afterwards.

Liner cement recipes usually contain extra additives for fluid loss control, retarding, possible gas blocking, etc. Since the mix proportions are critical and there is no lead slurry, it is usually batch mixed prior to the job. This guarantees the quality and density throughout the job.

A typical liner cementation operation would proceed as below:

- Position liner at required depth
- Circulate bottoms up—ensure low rheology (minimal YP and gels); rotate liner
- Set liner hanger
- Release setting tool and slack off weight (10–20 K lb)
- Pump spacer
- Pressure test surface lines
- Pump premixed slurry
- Release plug
- Pump spacer
- Displace cement out of liner into annulus—rotate liner if possible
- Pump-down plug releases liner wiper plug
- Both plugs pumped down liner until they latch into landing collar
- Bump plugs with 1,000 psi
- Bleed off and check for backflow
- Pick up, position tail pipe at top of liner, and circulate excess cement out from above liner

Squeeze Cementing

Use of hydraulic pressure to force cement into an annulus or formation. Usual applications:

- Seal off gas or water producing zones to improve production
- Repair casing failures
- Seal off lost zones
- Remedial work on primary cement jobs; e.g., top-up jobs
- Prevent vertical reservoir fluid migration into production zone
- Prevent fluids escaping from abandoned zones

To pump cement into a formation, a permeability of 500 darcies would be required. Since this does not occur normally, use several techniques to compensate.

High-Pressure Squeeze

- Formation is broken down first and then cement is squeezed (dense, impermeable formations preferred).
- Use solids-free fraccing fluid. Mud filter cake buildup would prevent fraccing.
- Since the overburden generally provides the maximum principal stress (acting vertically), fractures initiated would be vertically oriented; i.e., pushing the rock apart horizontally against the direction of the minimum principal stress.
- Once fracturing has occurred, cement would be spotted against the fracture zone and then pumped away into the formation after closing in the well.
- The injection pressure should gradually rise as the cement fills up the fractures.

Low-Pressure Squeeze

- Here the pressure always remains below fracture pressure.
- Perforations should be flushed clean—free from mud and other plugging materials.
- An injectivity test using water should be conducted first to confirm feasibility of a squeeze.
- A buildup of pressure would force fluid from the cement into the pores leaving a filter cake to form on the surface gradually inhibiting the process.
- As the injection process ceases in one location, it can commence at a different site and will continue until an impenetrable seal has blocked off the loss zone.
- Fluid loss additives are important. The use of neat cement alone would result in dehydration of the slurry due to the high fluid loss of neat cement. This in turn would create bridging before all the permeable zone could be sealed.
- Preferred slurry properties: fluid loss 5–200 ml; water: solids ratio of 0.4 by weight.

Running Squeeze

- Cement is pumped slowly and continuously until final pressure is obtained. Used for repairing damaged casing.

Hesitation Squeeze

- Pumping is stopped periodically to allow the slurry to dehydrate and create a filter cake. Usually pumping in increments of 0.25–0.5 bbl every 10–15 minutes.

Bradenhead Squeeze

- Cement is pumped through drill pipe/stinger (no packer), spotted and squeezed after closing the BOPs.
- Since the cement cannot move up the annulus, it is forced into any loss zones.
- Low-pressure squeeze option.
- Difficult to place cement accurately.
- Cannot be used for selectively squeezing perforations.
- As the casing is pressured up, restricted by burst specification.

Packer Squeeze

- Packer enables cement squeeze to be more accurately targeted.
- Since the annulus is sealed off, can use higher pressures (not limited by casing burst).
- Setting depth important—too high and cement contaminated with mud and excess fluid pumped into formation before cement. Too low—risk of cementing packer in.
- Packer normally set 30–50 ft above zone of interest with or without a tailpipe.
    - ❖ Drillable packer (e.g., Halliburton's EZSV or Fasdril)
        - Single use only
        - Back-pressure valve prevents backflow after squeezing
    - ❖ Retrievable packer
        - Multiple use
        - If backflow occurs after releasing packer, reset and apply squeeze again.

Cement Plugs

These are used to fill a section of hole and prevent movement of fluids within (fig. 5.5). Typical applications are:

- Abandoning depleted zones
- Seal off lost circulation zones
- Provide a kickoff platform for sidetracks
- Isolating a zone for formation testing
- Abandoning an entire well—provision of barriers (government regulations specify that plugs are required to seal off production zones, aquifers, etc.)

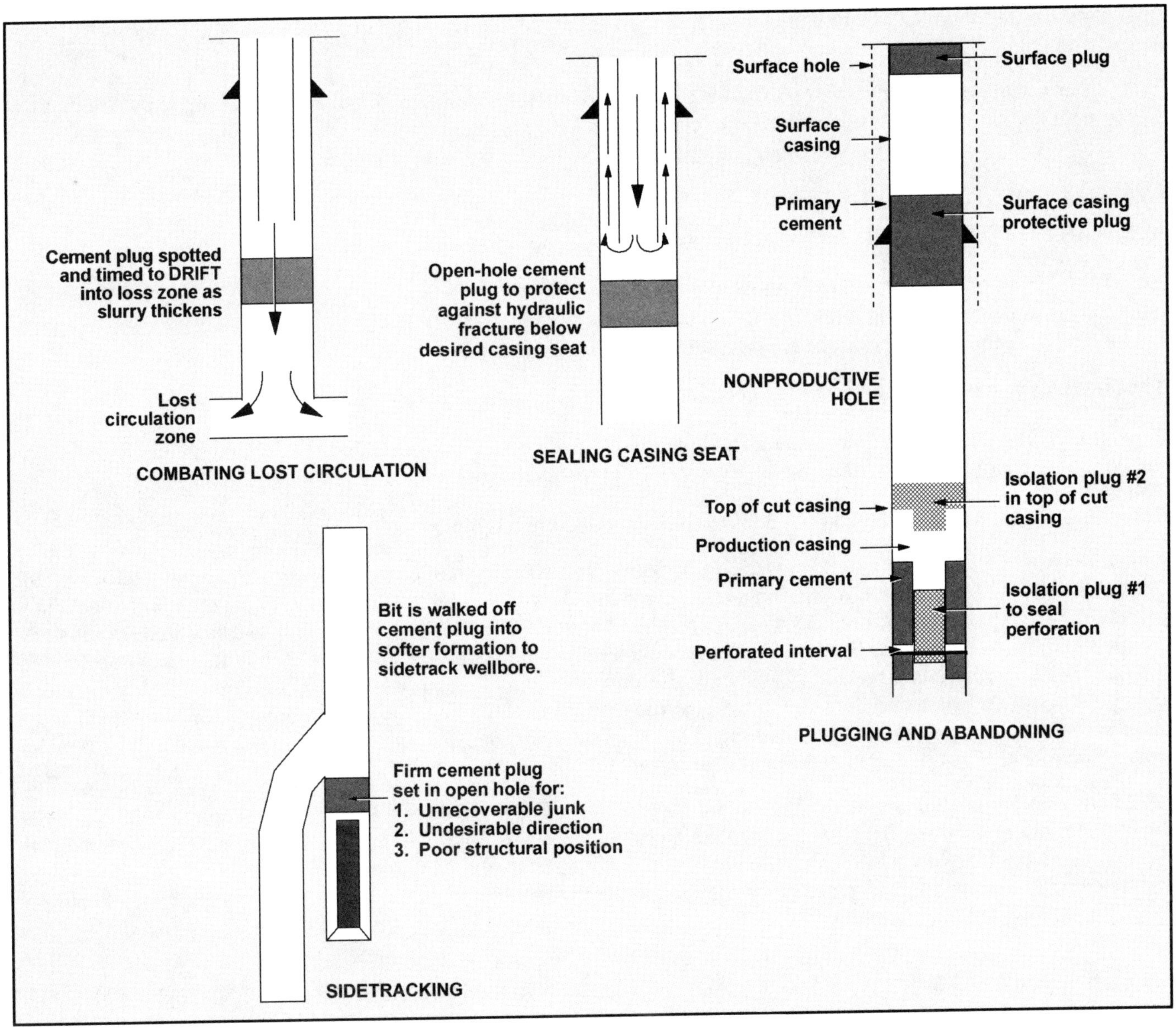

Figure 5.5 Cement plugs

The biggest problem setting plugs is mud contamination which can be minimized by:

- Use a gauge section of hole
- Use a plug volume sufficient to allow for some contamination—typically 500 ft height
- Condition the mud beforehand
- Use a preflush ahead of the cement
- Use a weighted slurry which contains less water

Plug Placement

Balanced plug—attempt to displace sufficient cement out of the drill pipe such that the column of cement in both pipe and annulus are of equal height. The drill pipe or stinger can then be withdrawn, leaving the plug in place.

Bridge plug—this can be set on depth and a 500 ft cement plug spotted above it. This method gives better depth control and reduced risk of contamination.

Dual plug—setting an initial balanced plug which can then be tagged to give a reference base upon which a second plug can be spotted (height of plug dependent on position of initial plug).

Note: When setting a series of cement plugs, it is advisable to pump a wiper dart or ball after each plug to ensure that the pipe/stinger does not become plugged itself with cement.

EVALUATION OF CEMENT JOB

A cement job has failed and will require remedial work if any of the following situations exist:

- The cement does not fill the annulus to the required height.
- The cement does not provide a seal at the shoe.
- The cement does not isolate undesirable formations.

The effectiveness of the job (and hence the need for additional work) can be measured by various means:

Temperature survey—running a thermometer inside the casing to detect the top of cement. The hydration process of setting cement is exothermic (gives out heat) and is detectable from within the casing.

Radiation log—radioactive tracers can be added to the cement before it is pumped (Carnolite, an example).

Cement bond log (CBL)—this is a sonic log capable of both detecting the top of cement and determining the quality of the cement sheath (fig. 5.6). It is run on wireline, emits sonic signals, and must be centralized to yield credible results. These pass out through the casing and are picked up by a receiver some 3 ft away. Both the transit time and amplitude of the signal are used to indicate the cement bond quality. Since the speed of sound is greater in casing than in the formation or mud, the first signals to return are those of the casing. If the amplitude of this signal (E1) is large, this indicates that the pipe is free (poor bond). When cement is firmly bonded to the casing and formation, the signal is attenuated (weakened) and is characteristic of the formation behind the casing. The signal can also indicate where the cement is bonded to the casing but not the formation. The effect of channelling can also be detected.

The CBL usually gives an amplitude curve and a variable density log (VDL) which indicates the strength of the signals by the intensity of dark and light streaks. The casing signals show up as parallel lines. A good bond is demonstrated by

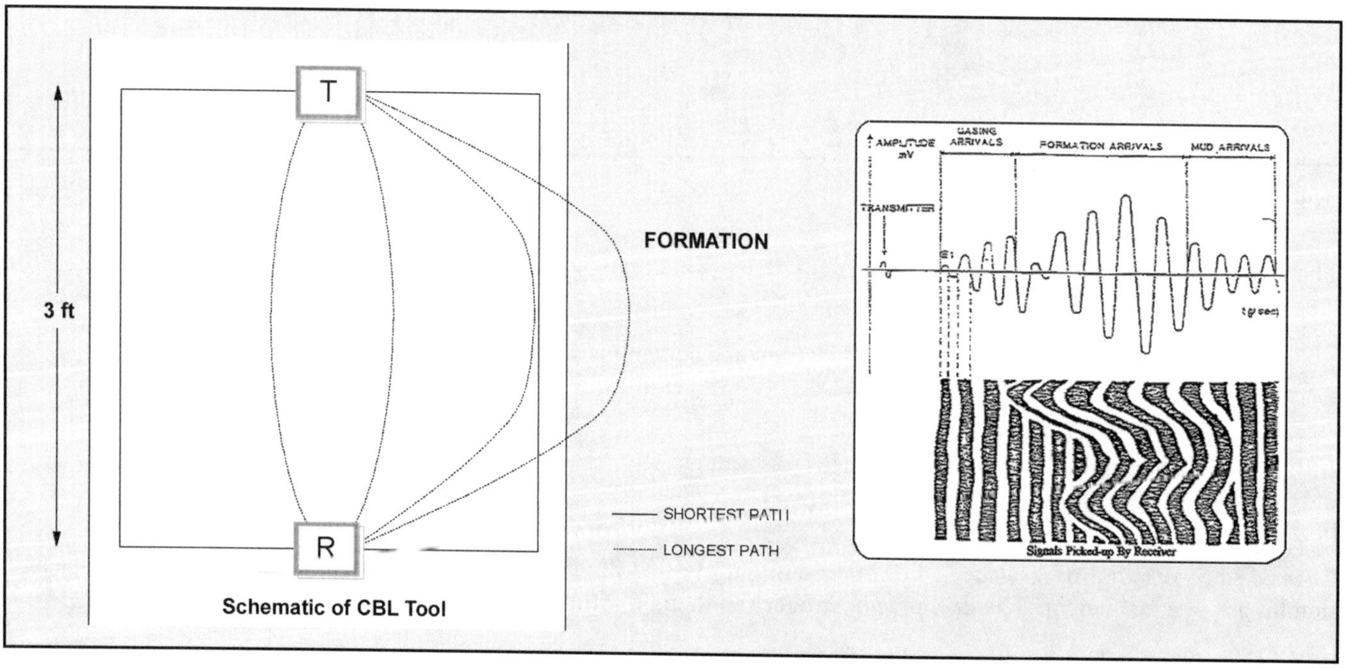

Figure 5.6 Cement bond log

wavy lines. There is no standard API scale to measure the effectiveness of the CBL and many factors can result in false interpretations:

- During the setting process, the velocity and amplitude of the signals vary significantly. It is recommended that the CBL is not run until 24–36 hours after the cement job to give realistic results.
- Cement composition affects signal transmission.
- The thickness of the cement sheath will cause changes in the attenuation of the signal.

The CBL will react to the presence of a microannulus (a small gap between the casing and cement). This usually heals with time and is not a critical factor. Some operators recommend running a CBL under pressure to eliminate this effect (the casing will balloon out under pressure and occupy any microannulus).

One of the limitations of the CBL is that it only gives a one-dimensional view of the cement bond at a given depth. An alternative tool that can be run is the Cement Evaluation Tool (CET) which uses ultrasonic transducers and the principles of casing thickness resonance to give a full radial picture of the cement bond around the full circumference of the casing. This is extremely useful in identifying if a channel is present and, on directional wells, the exact orientation of this channel.

CEMENTING CALCULATIONS

The principal calculations required for a cement job are:

- The amount of slurry required to fill the annulus outside the casing to the programmed height.
- The amount of mud needed to pump to displace the cement; i.e., bump the top plug.

In all cement calculations, it is necessary to know the yield per sack of the cement being used to be able to confirm that sufficient material is on site for the job (including contingency). The yield/sack depends on the amount of additives in the cement and the required final slurry density.

Schematics are invaluable in clarifying the volumes required including details regarding annular capacities (open hole and cased hole), different grade casings, section lengths, etc.

Example

A 7" liner is to be set as shown in the schematic in Figure 5.7.

Calculate the following:

- The amount of water per sack required to give a 16 ppg slurry
- The yield in cuft/sack
- The slurry volume required
- The tonnage of cement blend required
- The mud displacement to latch the wiper plug
- The mud displacement to bump the plug
- Required thickening time

Assumptions:

- 30% open hole excess volume
- Static bottomhole temperature 270°F
- Slurry formulation
- Class G + 35% BWOC silica flour
- D603 @ 0.4 gallons per sack
- D109 @ 0.09 gallons per sack
- Fresh water

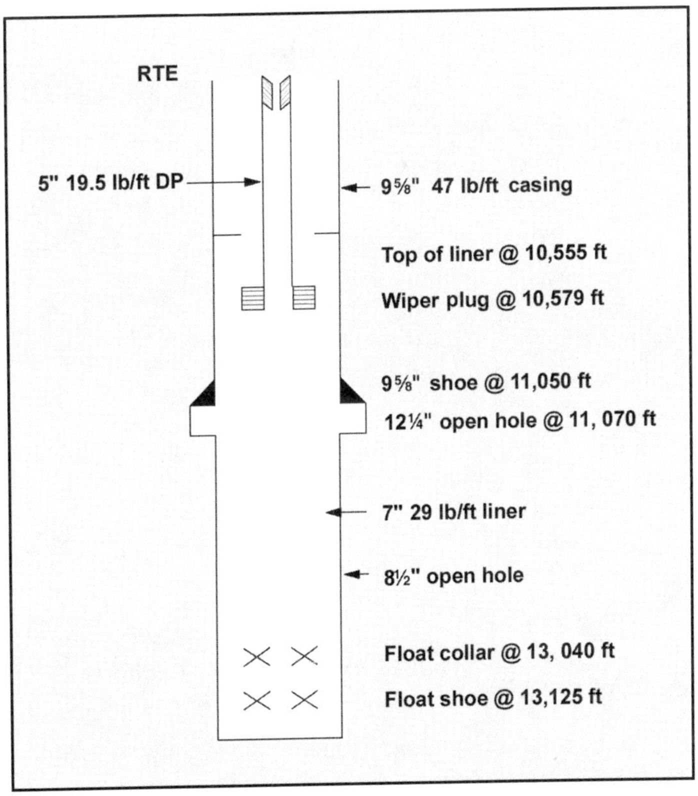

Figure 5.7 Schematic

Note:

BWOC = by weight of cement
D603 is a liquid fluid loss additive
D109 is a high-temperature liquid retarder
Fresh water is used as the mix water as seawater would accelerate the thickening time

Calculations:

- Calculate the amount of water per sack required to give a 16 ppg slurry.
- Using a variation of the equation density = mass/volume, it is possible to calculate the amount of water required.
- First, it is necessary to calculate the combined weight and volume of the slurry's components per sack of dry cement.
- This is best done in tabular form as shown below:

Material	Weight (lb)	Absolute Volume (gal/lb)	Volume (gal)
Cement	*94**	*0.0382**	**3.59**
Silica flour	*32.9**	*0.0456**	**1.50**
D603	<u>3.6</u>	*0.110**	*0.40**
D109	<u>0.9</u>	*0.096**	*0.09**
Water	<u>Y/0.12</u>	*0.12**	Y
TOTAL	**131.4 + Y/0.12**		**5.58 + Y**

From cementing tables (example Halliburton Red Book—*Technical Data, Physical Properties of Cementing Materials and Admixtures*), read off the absolute volume for all the slurry's components.

- One sack of cement weighs 94 lb
- 35% BWOC silica flour weighs 35% × 94 lb = 32.9 lb
- All of the figures <u>*in italics marked with a* *</u> are taken from the slurry formulation.
- All of the figures <u>underlined</u> are calculated by dividing the volume by the absolute volume to give weight.
- All of the figures in **bold type** are calculated by multiplying the weight by the absolute volume to give the volume.
- Y is the amount of water required.

So, for a 16 ppg slurry, the totals can be represented as:

$$16 = \frac{131.4 + Y/0.12}{5.58 + Y}.$$

Rearranging this gives

$$16 \times (5.58 + Y) = 131.4 + 8.33Y$$
$$Y = 5.49 \text{ gal/sack.}$$

The yield in cuft/sack

The yield is the volume of slurry obtained by mixing 1 sack of cement with the specified additives and mix water, expressed in cuft/sack of cement.

This is the total volume from the table above converted from gallons to cubic feet.

Hence Yield = (5.58 + 5.49) gal × 0.1337 cuft/gal
Yield = 1.48 cuft/sack

The slurry volume required

The slurry volume required is the sum of the following:

- Shoe track volume
- Liner/8½" open-hole volume
- Liner/12¼" open-hole volume
- Open-hole excess
- Liner/casing volume

Cementing tables are invaluable for these calculations as they have capacities and volumes precalculated.

Shoe track volume = (13,135 ft – 13,040 ft) × 0.0371 bbl/ft
= 3.52 bbl

Liner/8½" volume = (13,135 ft – 11,070 ft) × 0.0226 bbl/ft
= 46.67 bbl

Liner/12¼" volume = (11,070 ft – 11,050 ft) × 0.0982 bbl/ft
= 1.96 bbl

Open-hole excess = (46.67 bbl + 1.96 bbl) × 0.3
= 14.59 bbl

Liner/casing volume = (11,050 ft – 10,555 ft) × 0.0256 bbl/ft
= 12.67 bbl

TOTAL VOLUME = 3.52 + 46.67 + 1.96 + 14.59 + 12.67 bbl
= 79.41 bbl or 445.9 cu ft

The tonnage of cement blend required

The tonnage of cement blend is calculated by first working out the total number of sacks of cement required (total slurry volume divided by the yield), converting to a tonnage, and then adding 35% (to allow for the silica flour).

Sacks of cement required = 445.9 cu ft/1.48 cu ft/sack
= 301.3 sacks

Tonnage of cement required = 301.3 sacks × 94 lb/sack/2,205 lb/metric ton
= 12.84 tons of cement

Tonnage of blend required = 12.84 tons × 1.35
= 17.33 tons

Knowing the total number of sacks of cement required, it is now possible to calculate the total quantities of additives required, the number of tanks of mix fluid required (including an allowance for any dead space).

The mud displacement to latch the wiper plug

The mud displacement to latch the wiper plug is simply the volume of the drill pipe to the wiper plug.

Remember to use an average ID of the drill pipe (by calipering a number of joints and taking the average ID and make an allowance for the tool joint internal upset). Do not assume that the drill pipe capacity is as per the quoted tables. This is especially important when setting balanced cement plugs.

For this example a DP capacity of 0.0179 bbl/ft is assumed

Volume to latch wiper plug = 10,579 ft × 0.0179 bbl/ft
= 189.4 bbl.

For liner cement jobs, the ability to observe the drill pipe dart latching the liner wiper plug is a useful tool in allowing a recomputation of the total displacement volume if required.

The mud displacement to bump the plug

The mud displacement to bump the plug is the capacity of the liner from the wiper plug to the float collar.

$$\text{Volume to bump plug} = (13{,}040 \text{ ft} - 10{,}579 \text{ ft}) \times 0.0371 \text{ bbl/ft}$$
$$= 91.3 \text{ bbl.}$$

Required thickening time

The required thickening time is the total time to mix, pump, and displace the slurry. Some assumptions are required, but it is always useful to perform this check and compare it against the thickening time determined from the laboratory test. If there is insufficient or too much thickening time, then a new slurry formulation is required.

Assuming the following

- Slurry mix rate of 3 barrels per minute
- Displacement rate of 8 barrels per minute
- Contingency of 30 minutes (allowing for breakdowns, problems with bulk cement delivery)

$$\text{Required thickening time} = \text{Total slurry volume divided by } 3 + \text{Total displacement volume divided by } 8 + 30 \text{ minutes}$$
$$= (79.41/3) + [(189.4 + 91.3)/8] + 30$$
$$= 92 \text{ minutes.}$$

The slurry mix rate should be determined for the actual cement unit in use and the displacement rate should be modified, according to the displacement schedule calculated to ensure optimum displacement efficiency. A minimum displacement rate should be determined if losses are encountered and the displacement rate is reduced.

Useful Equations and Conversions

$$\text{Cubic feet} = \text{Barrels} \times 5.6146$$
$$\text{Gallons} = \text{Cubic feet} \times 7.4805$$
$$\text{Gallons} = \text{Barrels} \times 42$$
$$\text{Barrels per linear foot} = (D^2 - d^2) \times 0.0009714$$
$$\text{Cubic feet per linear foot} = (D^2 - d^2) \times 0.005454$$

where

D = diameter of hole or inside diameter of largest casing in inches

d = outside diameter of casing or liner being cemented in inches

1 bag of cement = 94 lbs

1 bag of cement = 1 cubic foot.

CHAPTER 6
Drill Bits

BIT SELECTION

Bit performance is measured by the total length and time drilled before the bit has to be pulled and replaced. Minimum cost per metre (or foot) is the primary objective. Careful review of offset well data must be undertaken when selecting a bit for any particular hole section.

Primary considerations when selecting a bit type are:

- Geology
  - ❖ Formation properties
    - Compressive strength

      Refers to the intrinsic strength of the rock which is based upon its composition, method of deposition, and compaction. It is important to consider the 'confined' or 'in situ' compressive strength of a given formation. Many bit manufacturers now provide a supplementary rock strength analysis service as an aid to bit selection.
    - Elasticity

      Affects the way in which a rock fails. A rock that fails in a plastic mode will deform rather than fracture.
    - Abrasiveness
    - Overburden pressure

      Affects the amount of compaction of sediments and therefore the rock hardness.
    - Stickiness
    - Pore pressure

      Affects mud weight requirements which, in turn, can affect penetration rates.
    - Porosity and permeability
  - ❖ Formation changes within a given hole section

    Changes in formation during one bit run can have a significant effect on bit performance. The formations to be drilled and the predicted depths of formation changes will be given in the drilling program and will form the basis of bit selection. It is important to remember the difference between exploration and appraisal/ development drilling in that:
    - For appraisal/development drilling, much will be known about the properties of the predicted formations and bit selection will be based upon offset bit performance along with electric log data (sonic, gamma ray, mud log data, core samples, etc.).
    - For exploration drilling, little may be known of the drillability of the formations that are likely to be encountered and so a more conservative bit program will be developed. In such situations, it is prudent to load out a wider variety of bit designs to cover all eventualities.
- Hole size and casing program
- Directional profile of well path and steerability of bit design
- Drive type (rotary/rotary steerable/mud motor/turbine)
- Drilling fluid properties
- Hydraulics
- Rig capabilities

ROLLER CONE BITS

Three-cone rock bits, or roller cone bits (fig. 6.1), were first introduced in the 1930s by Hughes Tool Company. Roller cone bits are comprised of steel cutters mounted on the bit body in such a manner that they are free to rotate. Most rock bits have three cones, although designs using two and four cones do exist.

Roller Cone Bit Features

Rock bits have three main elements

- Cutters (or cones)
- Bearings
- Bit body

The Cutters

The cutting elements on a rock bit are circumferential rows of teeth extending from each cone and interfitting between rows of teeth on the adjacent cones. These are either machined out of the cone steel forgings (milled-tooth bits), or are pre-manufactured from harder tungsten carbide and assembled in pockets machined into the cones (insert bits). Tungsten carbide insert bits were originally designed to drill extremely hard or abrasive formations, such as chert and quartzite, which were unsuitable for drilling with softer formation milled-tooth designs. However, because of their superior durability, there are now also insert bit designs suitable for drilling softer formations economically. The teeth can be a variety of shapes and sizes, depending on the intended application, and are responsible for actually crushing or gouging the formation as the bit rotates. The crushing comes from the high weight on the bit used driving the teeth into the rock as the cones and bit rotate (fig. 6.2).

Cone Offset

The gouging action of a tricone bit is the result of offsetting the cones on the bit so they do not rotate about their true centers. *Offset* is the horizontal distance between the centerline of the bit and a vertical plane through the centerline of the journal (fig. 6.3). The degree of offset is referred to as skew. If the cones are forced to rotate about an axis other than their true geometrical axis of rotation, they will slide or drag along the hole bottom, occasionally thereby producing a drag cutting mechanism in addition to the crushing system. In general, the greater the offset distance on the bit, the higher the degree of gouging/scraping cutting action it has. Softer formation rock bits have more offset than those bits designed to drill hard rock, where there may actually be no offset at all and the bit removes formation purely by a crushing action.

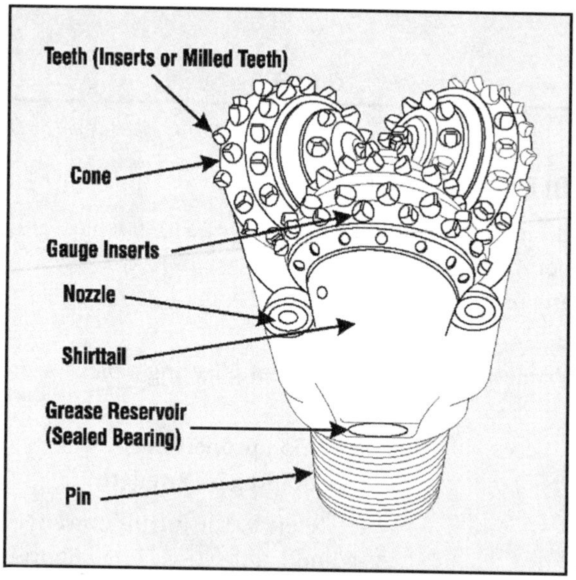

Courtesy of Reed-Hycalog

Figure 6.1 Roller cone bit nomenclature

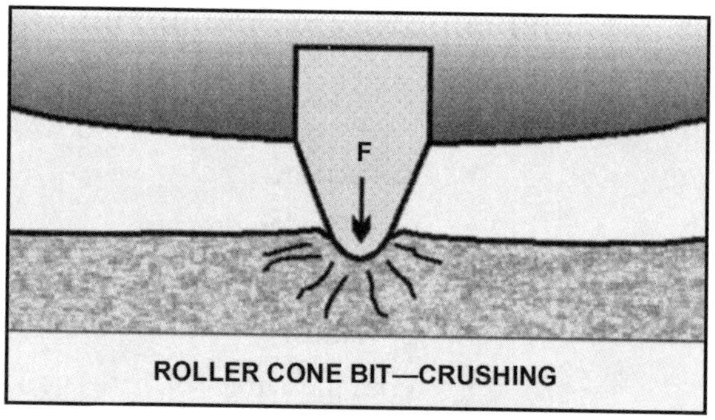

Courtesy of Reed-Hycalog

Figure 6.2 Roller cone bit cutting action

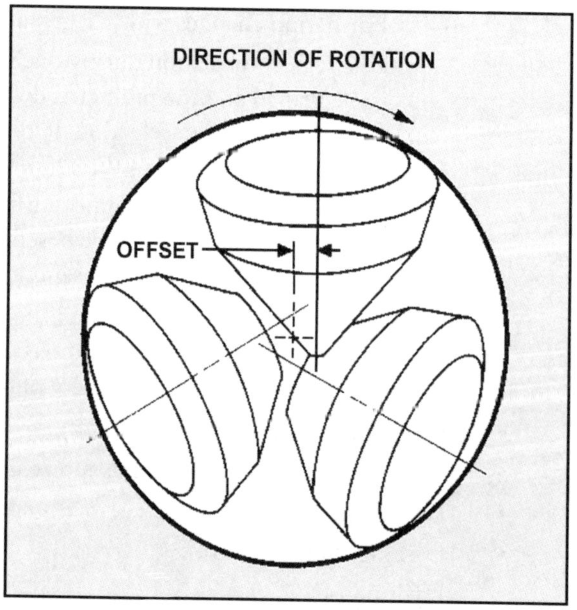

Figure 6.3 Cone offset

The Bearings

The bearings allow the cones to rotate about the bit body. Bearings for premium oilfield rock bits are sealed and lubricated to ensure longer life in the harsh downhole environment. Nonsealed roller ball bearings are produced and are primarily used for top hole sections where trip time is short and high rotary speeds are desirable. The most common premium rock bit bearing today, however, is the sealed journal bearing. This contains no rollers, just a solid journal pin mated to the inside surface of the cone or a bushing, which fits between the cone and the journal (fig. 6.4).

It is designed so that all the bearing elements are uniformly loaded and high weights on bit and rotary speeds can be used. A sealed reservoir of lubricant is held within the bit body to lubricate the bearing.

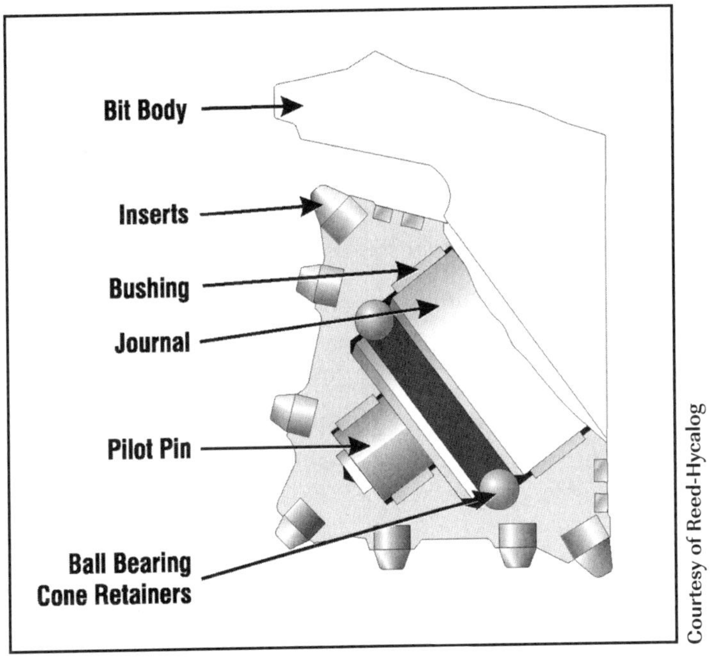

Figure 6.4 **Rock bit bearing cross section**

Tricone Bit Selection

In many cases, roller cone drill bits can be run in the same applications where fixed-cutter bits, primarily PDC bits, are run, particularly premium large-diameter tooth-type motor bits and high-speed premium insert bits (some of which incorporate metal seals). Roller cone bits, in general, drill slower than PDC bits and have a shorter life in terms of footage that can be drilled. However, on a per bit basis, they are priced lower than PDC bits. The choice of which type of bit to run often depends on the results of a cost per foot analysis.

The following basic guidelines should be used as an aid to roller cone bit selection—

- Shale has a better drilling response to rpm.
- Limestone has a better drilling response to weight on bit.
- Bits with roller bearings can be run at a higher rpm than bits with journal bearings.
- Bits with sealed bearings can give longer life than bits with open bearings.
- Milled-tooth bits with journal bearings can be run at higher weights than milled-tooth bits with roller bearings.
- Fixed-cutter bits can run at higher rpm than roller cone bits.
- Bits with high cone offset may wear more on gauge.
- Bits with high cone offset may cause more hole deviation.

Applications where roller cone bits tend to be used in preference to fixed-cutter bits include:

Exploration wells where there is insufficient information to determine whether or not the formations to be drilled are too hard to be drilled with PDC bits. Another factor which favors roller cone bits in exploratory wells is size of cuttings. Geologists sometimes prefer that PDC bits not be run because the cuttings generated by PDC bits in potentially producing formations tend to be much smaller than those made by roller cone bits.

Short intervals where the longer life of a higher cost PDC cannot be leveraged into a lower cost per foot.

High risk situations where there is a high probability of damaging the bit (such as drilling out cement equipment that contains metal parts).

Low cost drilling areas where the value of the time saved by a faster drilling PDC bit is not sufficient to offset the higher bit price.

Extremely hard formations where PDC bits have not yet been demonstrated to drill economically.

Highly faulted areas with hard stringers where it is extremely difficult to predict when an extremely hard streak (particularly one containing chert nodules) will be encountered.

The IADC Roller Cone Bit Classification System

This classification scheme provides a method for categorizing roller cone bits according to their design features and intended applications (see Table 6.1). The classification code for an individual bit contains four characters. The first three characters are numeric and the fourth is alphabetic.

First Character—Cutting Structure Series (1–8)

The series numbers describe general formation characteristics. Numbers 1–3 refer to milled-tooth bits, and 4–8 cover insert bits. Within the groups, the formation becomes harder as the number increases.

Second Character—Cutting Structure Types (1–4)

Each of the above series is divided into four types, or degrees, of hardness. Type 1 refers to bits designed for the softest formation in a particular series and Type 4 to bits for the hardest.

Third Character—Bearing/Gauge

Seven categories exist for bearing design and gauge protection.

Fourth Character—Features Available (Optional)

Sixteen alphabetic characters are used to indicate "Features Available." These include special cutting structures, hydraulic configurations, and body gauge protection.

-

FIXED-CUTTER BITS

Unlike roller cone bits, no uniform classification system exists which relates bit style to formation drillability. There is an IADC classification but it does not relate to formation drillability.

Fixed-Cutter Bit Types

Although some modern bit styles incorporate more than one diamond type (e.g., natural diamond and TSP), the bits are traditionally classified according to the nature of their diamond cutter. The three types of fixed diamond cutters are: natural diamond, thermally stable polycrystalline (TSP) diamond, and polycrystalline diamond compact (PDC).

PDC Bits

Cutting structure made of manufactured diamonds which are thermally stable up to 1,292°F (700°C). PDC bits cut the formation in a shearing action.

Unlike the relatively small diamonds used in (natural) diamond and even TSP bits, the PDC can be attached to the body as large, sharp cutting elements. The PDC cutting elements are bonded to a tungsten carbide substrate (providing further impact resistance) which is fixed into the body/blades of the bit. The body can be of the steel or matrix type.

PDC bits currently represent a large majority within fixed-cutter bits.

Natural Diamond Bits

Cutting structure made of natural diamonds, stable up to about 1,562°F (850°C) which require good cooling, and are sensitive to shock load

Cutting mechanism: plowing/grinding

Body: matrix only

TABLE 6.1
IADC Codes—Roller Cone Bits

	SERIES	FORMATION	TYPE	BEARINGS & GAUGE						
				1	2	3	4	5	6	7
				Standard Roller Bearings	Roller Bearing, Air Cooled	Roller Bearing, Gauge Protected	Sealed Roller Bearing	Sealed Roller Bearing, Gauge Protected	Sealed Friction Bearing	Sealed Friction Bearing, Gauge Protected
MILLED-TOOTH BITS	1	Soft formations with low compressive strength and high drillability including gumbo clays, redbeds, top hole clays, and unconsolidated formations with occasional sharp sand	1							
			2							
			3							
			4							
	2	Medium to medium-hard formations with high compressive strength	1							
			2							
			3							
			4							
	3	Hard, semiabrasive, and abrasive formations including firm sandy shales, softer shales, gypsum, salt, and chalk.	1							
			2							
			3							
			4							
TUNGSTEN CARBIDE INSERT BITS	4	Soft formations with low compressive strength and high drillability including hard shale, siltstone, limestone, sand, and shaley limestone	1							
			2							
			3							
			4							
	5	Soft-to-medium formations with low compressive strength	1							
			2							
			3							
			4							
	6	Medium-hard formations with high compressive strength including hard abrasive limestone, dolomite, silty shale, and quartzitic material or streaks	1							
			2							
			3							
			4							
	7	Hard, semiabrasive, and abrasive formations including hard abrasive sandstone, dolomite, silty shale, and quartzitic material or streaks	1							
			2							
			3							
			4							
	8	Extremely hard and abrasive formations including chert or granite	1							
			2							
			3							
			4							

TSP Bits

Cutting structure is made of manufactured diamonds, which exhibit higher resistance to temperature stable up to 1,832–2,192°F (1,000–1,200°C) than natural diamonds which might contain inclusions. Advantage over natural diamond: TSP diamonds can be orientated in the bit body and are self-sharpening like PDC cutters when they start to wear.

But TSP diamonds are more difficult to bond to support material than PDC; therefore, like natural diamonds, they are used for matrix bodied-type bits only.

Cutting mechanism: Mainly plowing/grinding like natural diamonds and, to a lesser extent, shearing.

Impregnated Diamond Bits

Impregnated diamond bits (commonly called impregnated or 'impreg' bits) contain sharp natural diamond grit mixed (in various concentrations) through a tungsten carbide matrix.

The diamonds used in these bits are generally much smaller than those used in conventional natural diamond bits. Larger-sized natural diamonds are placed on the gauge to maintain the hole size during the bit run. TSP diamonds are sometimes used in conjunction with the diamond grit for specific applications where higher rates of penetration are required.

Impregnated diamond bits drill in a similar fashion to natural diamond cutter bits but, when the diamonds become worn and torn out of the matrix, new ones are continually being exposed. This gives them the ability to drill the hardest, most abrasive formations at high rpm, which makes them particularly useful when turbines are used.

PDC Bit Technology

Bit Nomenclature (figs. 6.5 and 6.6)

Cone

The cone of the bit provides a degree of stability when the bit is drilling.

Nose

The nose is the first part of the bit to encounter any change in formation when drilling a vertical hole. Because of this, it is desirable to have a relatively large number of cutters set on the nose.

Taper

The taper length is usually governed by the cutter density requirement and the application. A bit designed for harder formations would therefore tend to have a more extended taper than one designed for softer formations. However, an alternative way to achieve a high cutter density without increasing the taper is to increase the number of blades. PDC bits used in directional drilling applications will generally have a shorter taper.

Outer Diameter Radius (ODR)

The ODR refers to that region of the bit profile where the radius at the end of the flank leads into the gauge of the bit. This region of a bit is extremely important, especially in motor or turbine applications where rotating speeds are high, as the cutters must withstand the effects of high velocity due to their radial position on the face of the bit.

Although the angular velocity of cutters at the bit gauge is identical to that of cutters within the cone, the tangential velocity is greater since it is a function of radial location.

Gauge

Length

Generally standard, gauge length is a compromise between stability and directional responsiveness.

- Long gauge provides increased bit stability.
- Short gauge design is used for increased directional responsiveness, ultra short for sidetracking capability.

Protection

Maintaining the full gauge diameter is crucial to avoid undersized hole. If the bit is used for directional applications, especially if a motor or a turbine is to be used, reinforced protection is necessary.

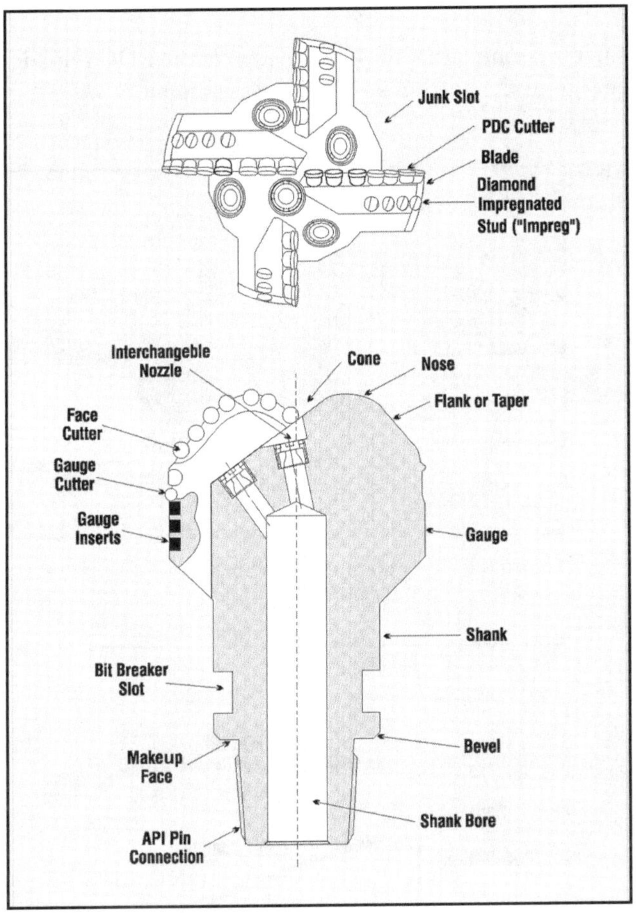

Figure 6.5 Steel-Bodied Bits

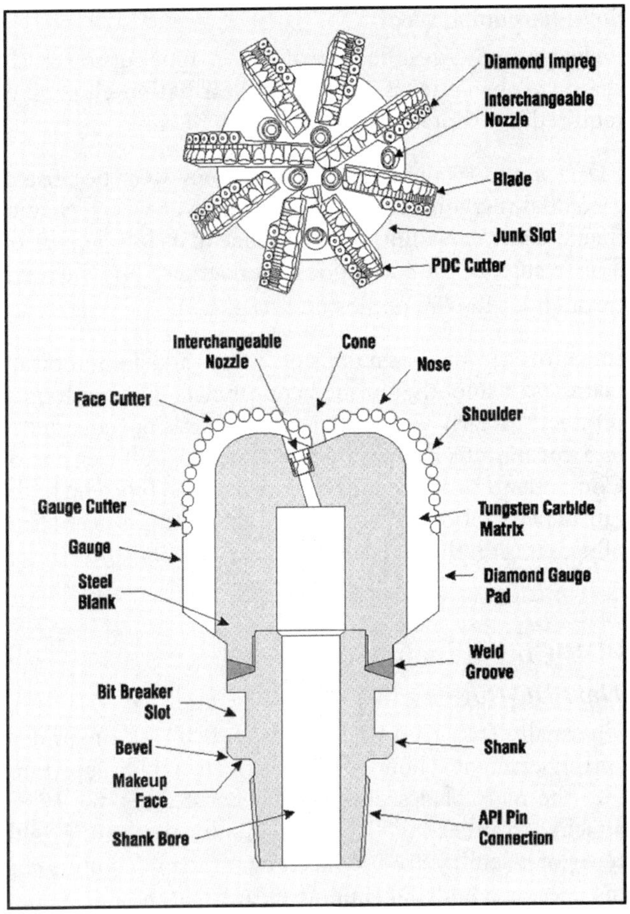

Courtesy of Reed-Hycalog

Figure 6.6 Matrix-Bodied Bits

Natural diamonds are used for gauge protection on matrix-bodied bits, but can also be used in tungsten carbide inserts in the gauge of steel-bodied bits.

Steel-bodied bits use tungsten carbide inserts.

In both cases, diamond impregnated elements may also be positioned to the rear of the gauge cutters and face cutters to back them up and to help reduce bit related torque by limiting the depth of cut of the primary cutters. Gauge cutters should be of a preflattened shape.

PDC Bit Cutting Action

PDC bits drill by cutting the formation in shear, much like the cutting action of a lathe (fig. 6.7). The vertical compressive loads cause the rock to fail in shear along a failure plane approximately 40° to horizontal.

Courtesy of Reed-Hycalog

Figure 6.7 PDC bit—shearing

A bit's cutting action plays a key role in the amount of energy required to drill through a given formation. This characteristic is generally presented in terms of "specific energy" which is defined as the amount of energy required to cut a unit volume of formation. A bit which fails the rock in shear directly, rather than using high compressive loads to cause the rock to fail in shear along its natural failure plane, has a lower specific energy. As a general rule, shear strength is approximately one-half the compressive strength. However, this relationship can vary depending on specific rock type.

Self-Sharpening Wear

To keep the energy efficiency of a shear-type cutting mechanism high, it is essential that the cutting edge of the PDC remains sharp. As the cutter is used and a wear flat develops, the specific energy of the cutting system increases as more weight is required to maintain a constant depth of cut.

PDC cutters retain a sharp edge as they wear because the tungsten carbide directly behind the diamond layer wears away more rapidly than the polycrystalline diamond, due to its lower abrasion resistance. This results in the formation of a diamond lip which remains sharp throughout the life of the cutter (fig. 6.8).

In contrast to this, diamonds on a natural diamond bit dull with use, taking on a smooth, polished appearance. The teeth on a roller cone product similarly wear and, in the process, become dull. This results in a cutting mechanism that becomes less efficient as the bit drills. Consequently, roller cone bits and diamond bits tend to drill at a lower rate of penetration as they wear, while PDC bits maintain a higher rate of penetration throughout the total interval drilled (fig. 6.9).

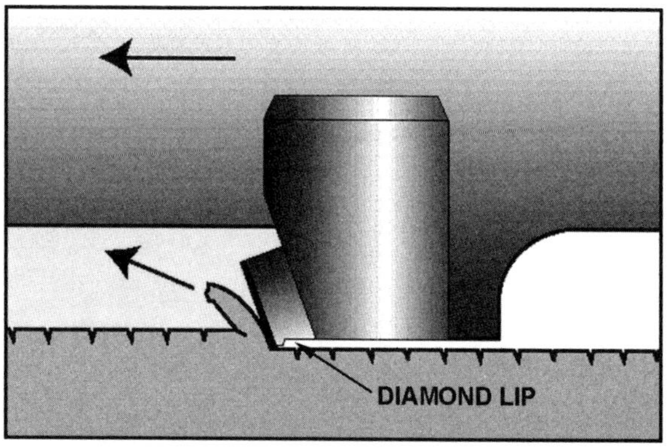

Courtesy of Reed-Hycalog

Figure 6.8 Self-sharpening wear

PDC Cutter Technology

The PDC Cutter

Thermally stable up to 1,292°F (700°C), the manufactured PDC cutting element is bonded to a tungsten carbide substrate that is fixed into the body/blades of the bit. In most cases, the PDC cutters are attached to either angled post or cylindrical substrates, although other types of assembly are produced by bit manufacturers (fig. 6.10). For instance, a cylindrical support means that greater cutter density can be achieved compared with standard post cutters, as conical cutters can be placed closer together in the bit body.

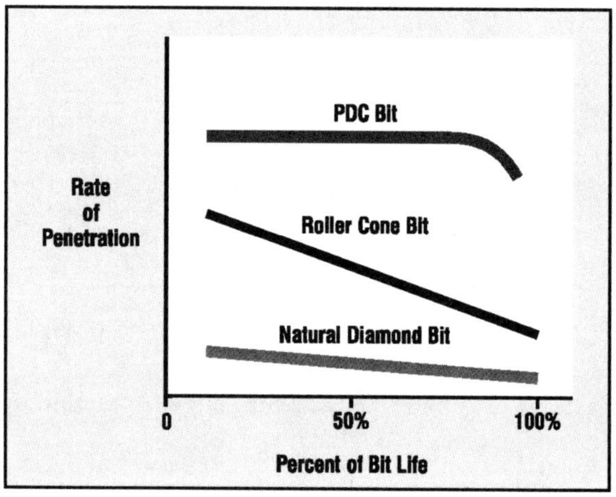

Figure 6.9 Effect of cutter wear on penetration

Cutter Density

Generally speaking, the harder or more abrasive the formation, the higher the cutter count. However, a higher cutter count also makes a bit more costly (particularly since PDC components constitute a high percentage of the total bit cost) and, in general, cause the bit to drill at a slower rate of penetration.

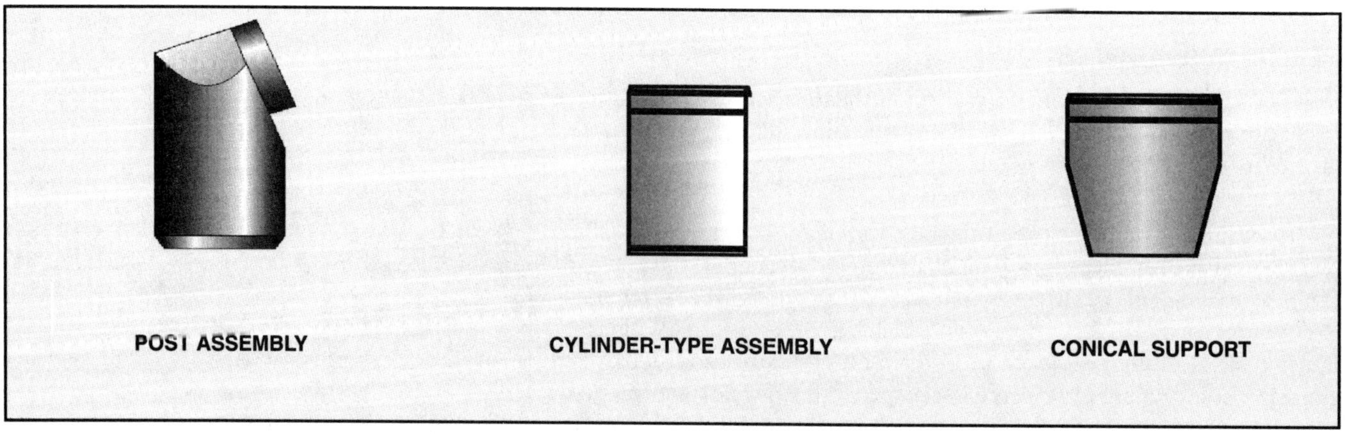

Courtesy of Reed-Hycalog

Figure 6.10 PDC cutter assemblies

Cutter Size

A variety of PDC cutter sizes are available ranging from 8 mm to 50 mm in diameter. Larger PDC cutters are more aggressive, generate more torque, and are more susceptible to impact damage than bits with smaller cutters, and so are better suited to soft formations.

- *8 mm* cutters have been used on products for harder formations. However, smaller cutters often mean lower ROP and higher WOB. Also useful in directional applications as the reduced point loading resulting from the distribution of the WOB over a larger number of cutters produces less bit face torque.
- *13 mm* cutters are the industry standard size. They are the most suitable for medium-to-hard formation as well as abrasive rock.
- *16 mm* cutters are often associated with medium-soft to medium-hard formation.
- *19 mm* cutters are generally associated with fast drilling in soft-to-medium formation. Large diameter cutters are proven to perform well in low compressive strength, highly elastic formations, which tend to deform rather than fracture.
- *24 mm* and above are associated with soft formations. Space is limited on the bit face and, by using large cutters, cutter redundancy is limited. When one cutter fails, the bit may have to be pulled out of the hole (POOH). Additionally, as large cutters wear, they provide a large surface of contact which increases heat causing damage to the diamond layer.

Recent developments in bit technology include combining two different cutter sizes (e.g., 13 mm and 16 mm) as well as using different designs of cutter in the same bit.

Cutter Distribution

Cutters are positioned across the bit face so as to ensure the most efficient use of the PDC elements, allowing maximum bit life. For harder or more abrasive formations, cutter redundancy is optimized.

Cutter Orientation

The orientation of a PDC cutter in the bit body, and hence, the angle at which the cutter engages the formation, has a significant effect on the performance of a PDC bit (fig. 6.11). The orientation of a PDC cutter is defined by its back-rake and side-rake angles, both of which can be positive, negative, or zero.

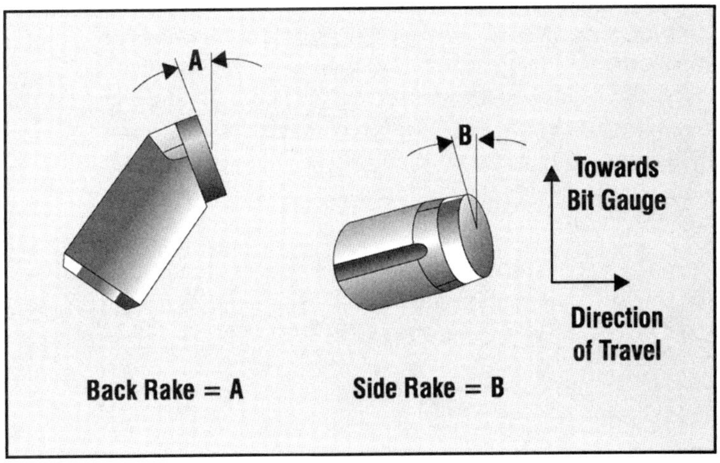

Figure 6.11 Cutter orientation

Back Rake

The angle from vertical of the PDC cutting element as it is presented to the formation. This controls the aggressiveness and life of the cutter.

The back-rake angle is said to be more aggressive when the cutter is positioned such that a given weight on bit results in a greater depth of cut. The smaller the back-rake angle, the more aggressive the cutter. Therefore, a cutter with a back-rake angle of 5° will be more aggressive than a cutter with a back-rake angle of 30°. In general, a more aggressive back rake will make the bit more suitable for drilling softer formations at high rates of penetration. If the back rake is too aggressive, then drilling harder formations might result in chattering of the cutter and the possible initiation of bit whirl.

Cutters with a larger back-rake angle are less aggressive and therefore better suited to drilling harder formations. They also generate less torque for a given weight on bit allowing for improved steerability in directional applications. In most cases, a PDC bit will be designed such that the cutters are arranged with varying degrees of back rake, radiating from the most aggressive in the cone of the bit to the least aggressive at the ODR.

Side Rake

A measure of cutter skew relative to a line at 90° to the direction of travel of the bit. Side rake can be used to mechanically direct cuttings either towards the cone of the bit (negative side rake) or outwards towards the junk slots (positive) to aid cleaning of the bit face. However, as increasing side rake results in the reduction of the effective operating width of the cutter, its use has somewhat limited application.

Cutter Design —General

Since PDC drill bits first became available, bit manufacturers have focused on ways to make their products both more impact and abrasion resistant. The following features are among those readily available on PDC bit designs:

Chamfered Edges

Chamfered or beveled edges around the circumference of the diamond table improve impact resistance and, when taken to extremes, are said by some manufacturers to help reduce torque by effectively increasing the cutter back rake.

Multiple Diamond Layers

GeoDiamond's Twin Edge™ cutters feature a second PDC disc positioned within the carbide support behind the primary cutting element.

Diamond Impregnated Supports

The Security DBS range of FI PDC bits have primary PDC cutting elements, the supports of which are composed of a series of diamond impregnated disks.

Cutter Geometry

The PDC diamond table attached to the cutter post is susceptible to a variety of modes of failure. These include impact damage in the form of chipped and broken teeth, and spalled diamond layers as a result of poor heat transfer through the cutter. Such cutter damage obviously has an impact upon bit performance. In recent times, bit manufacturers have been addressing this by focusing on the internal geometry of the PDC cutting elements.

On a standard PDC cutter, the bond between the diamond layer and the carbide substrate forms a simple, planar interface (fig. 6.12).

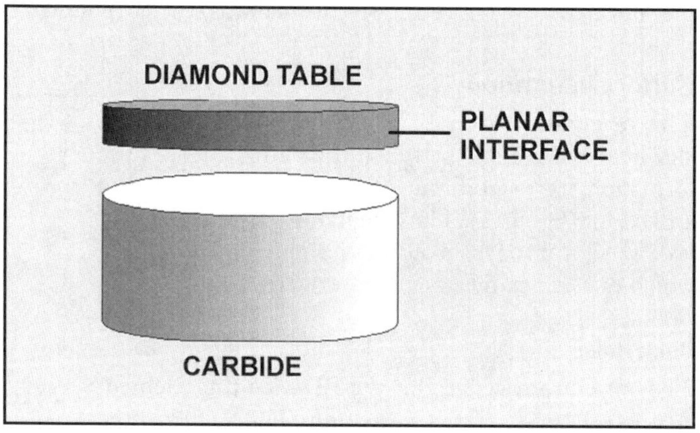

Figure 6.12 Standard PDC cutter

By designing a cutter with a nonplanar interface (fig. 6.13) between these two elements, superior impact and abrasion resistance result. This is due to the improved mechanical locking and reduced stress between diamond table and carbide and substantially increased diamond volume. In recent years, such innovations have allowed PDC bits to be run in formations which were previously thought suitable only for insert roller cone designs or natural diamond bits.

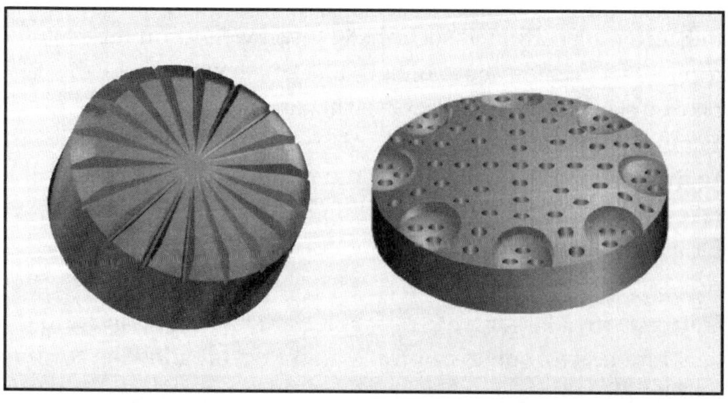

Courtesy of Reed-Hycalog

Figure 6.13 Examples of nonplanar cutter substrates

Most PDC bit manufacturers are now able to supply PDC cutters at this time. Examples include GeoDiamond, Security DBS, and Reed-Hycalog (fig. 6.14),

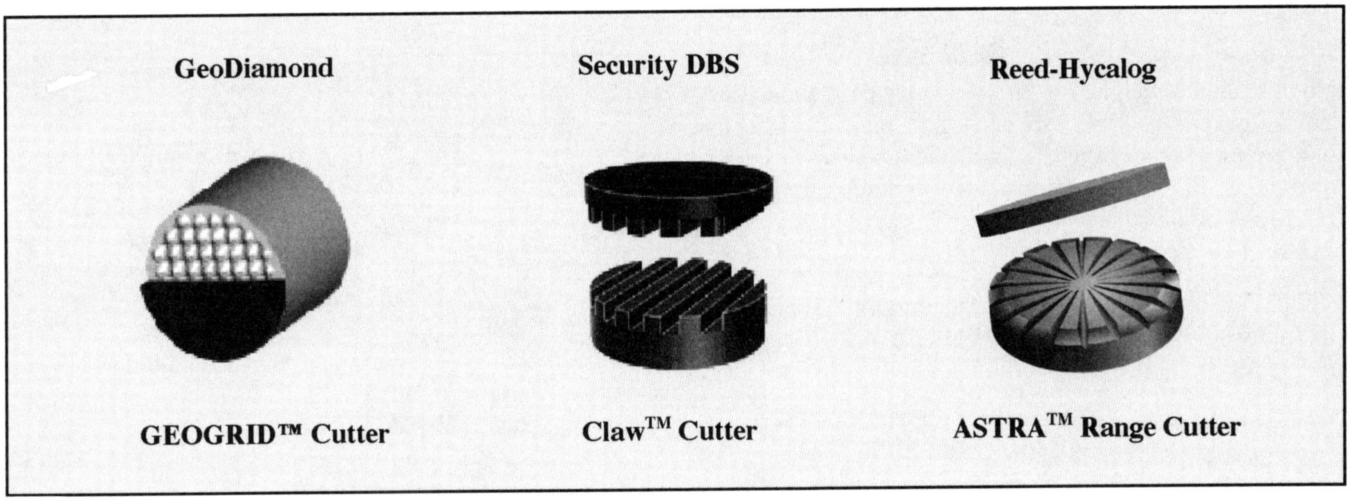

Figure 6.14 PDC bit manufacturers

Fixed-Cutter Bit Applications and Design Characteristics

PDC bits are highly suitable for soft-to-medium-hard, generally nonabrasive, formations of homogeneous composition. Improvements in cutter technology and bit design have, however, been extending the range of formations that can be drilled with PDC bits in recent years. Conglomerate, chert, and volcanic rocks are usually not considered PDC drillable.

In contrast, TSP and natural diamond bits perform well in medium-to-hard formations such as limestone, dolomite, anhydrite, mildly abrasive sands, interbedded hard sandstone, and brittle silty shales.

TSP and natural diamond bits are effective in harder (medium-to-hard), more abrasive formations than PDC bits but are not as effective in softer formations.

Due to their cutting mechanism, shearing as opposed to the crushing/gouging action produced by conventional roller cone bits, PDC bits require considerably less WOB.

Fixed-cutter bits are known to perform better in oil-based mud than in water-based mud whereas roller cone bits are less effective.

When the drilling parameters are optimized for a given formation, considerable ROP improvement can be expected when compared with conventional roller cone bits. However, these bits are a lot more expensive than conventional roller cone designs. For these reasons, a thorough economic evaluation must be performed.

Fixed-cutter bits are also a good option for the following applications.

High Rotary Speeds

Often associated with motor but particularly with turbine due to the inability of tricone bit bearing seals to tolerate very high rotary speeds (fig. 6. 15). Fixed-cutter bits also carry less risk of leaving junk in the hole. Fixed-cutter bits designed specifically for turbine applications are built with an extended profile, generally parabolic with a long taper and a nose close to the central axis of the bit. This allows for increased cutter redundancy in high wear areas.

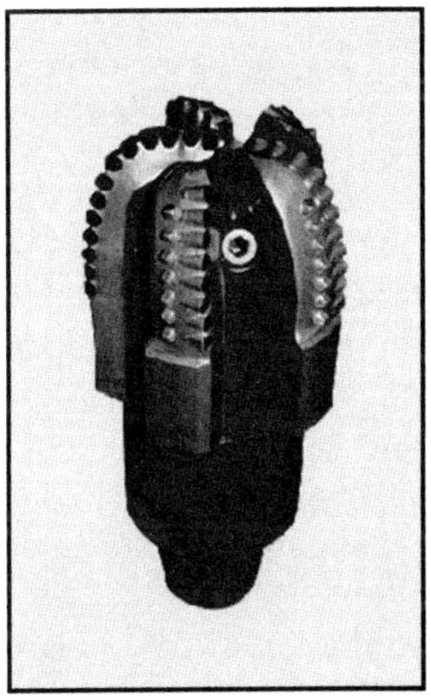

Courtesy of Reed-Hycalog

Figure 6.15 PDC turbine bit

Slimhole Drilling

Fixed-cutter bits drill more efficiently than roller cone bits at low weight on the bit. For this reason, PDC and natural diamond bits are often favored for coiled tubing drilling where transfer of weight to the bit face is limited.

Directional and Horizontal Drilling

When reduced weight on bit is required for directional drilling purposes, PDC bits can again be more effective than roller cone designs (fig. 6.16). However, in certain formations, PDC bits may produce too much torque when steering is involved. In this case, tricone bits may be the preferred option.

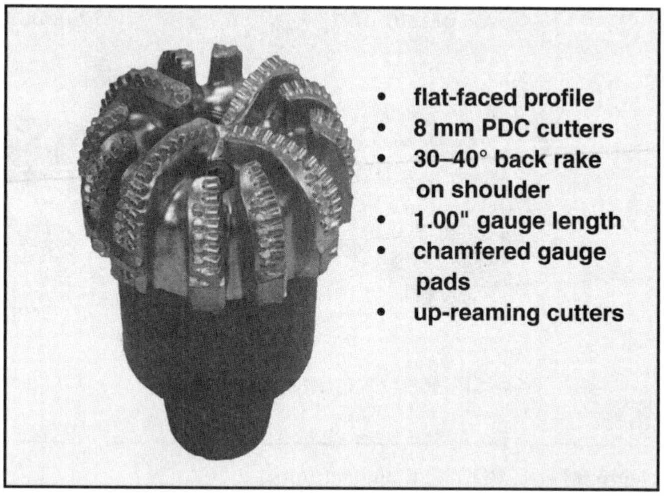

Figure 6.16 Example of PDC bit for directional applications

When selecting a PDC drill bit for a directional application, the following design features should be considered:

- *Cutter Size*
  Smaller diameter PDC cutters produce less reactive torque than, say, 19 mm cutters and so aid steerability. In general, cutters of 13 mm diameter and smaller are the preferred option. The use of torque-reducing features are of particular importance when bits are to be run in conjunction with a mud motor which will stall if the bit produces too much.
- *Cutter Orientation*
  High degrees of back rake on PDC cutters make bit designs less aggressive, therefore helping maintain tool face control.
- *Bit Profile*
  A flat face profile, incorporating a relatively shallow cone with a sharp breakover from the nose to the shoulder of the bit, reduces point loading on individual cutters by allowing better distribution of WOB. This also reduces torque and makes the bit more steerable.
- *Gauge Length*
  Gauge length is very important when selecting bits for directional applications. If a lot of steering is likely to be required, then a short gauge length, 2.50 inches or less, will provide better responsiveness. However, if drilling long, horizontal sections, a bit with a slightly longer, heavily protected gauge, may be preferred.
- *Additional Design Features*
  Up-reaming (or back-reaming) cutters: positioned at the lower end of the gauge pads, they provide extra, lateral cutting action.

  Chamfered lead edge on gauge pads.

Bicentric and Eccentric Bit Designs

Bicentric or eccentric bits are designed such that their pass-through diameter is smaller than the diameter of the hole that they actually drill (fig. 6.17). This is achieved by designing the bit so there is an asymmetry in the structure, such as an enlargement in the body to one side of the axis. In use, this enlarged side will rotate with the bit and cut a full-gauge hole (or slightly over gauge depending on the design and degree of eccentricity). However, with no rotation, the asymmetry allows the bit to pass through a narrower diameter hole than the one just drilled. Some designs feature a smaller pilot bit section followed by a larger diameter, reaming section.

Bicentric and eccentric bits have a number of applications and can be of benefit when drilling sloughing shales or creeping salt formations.

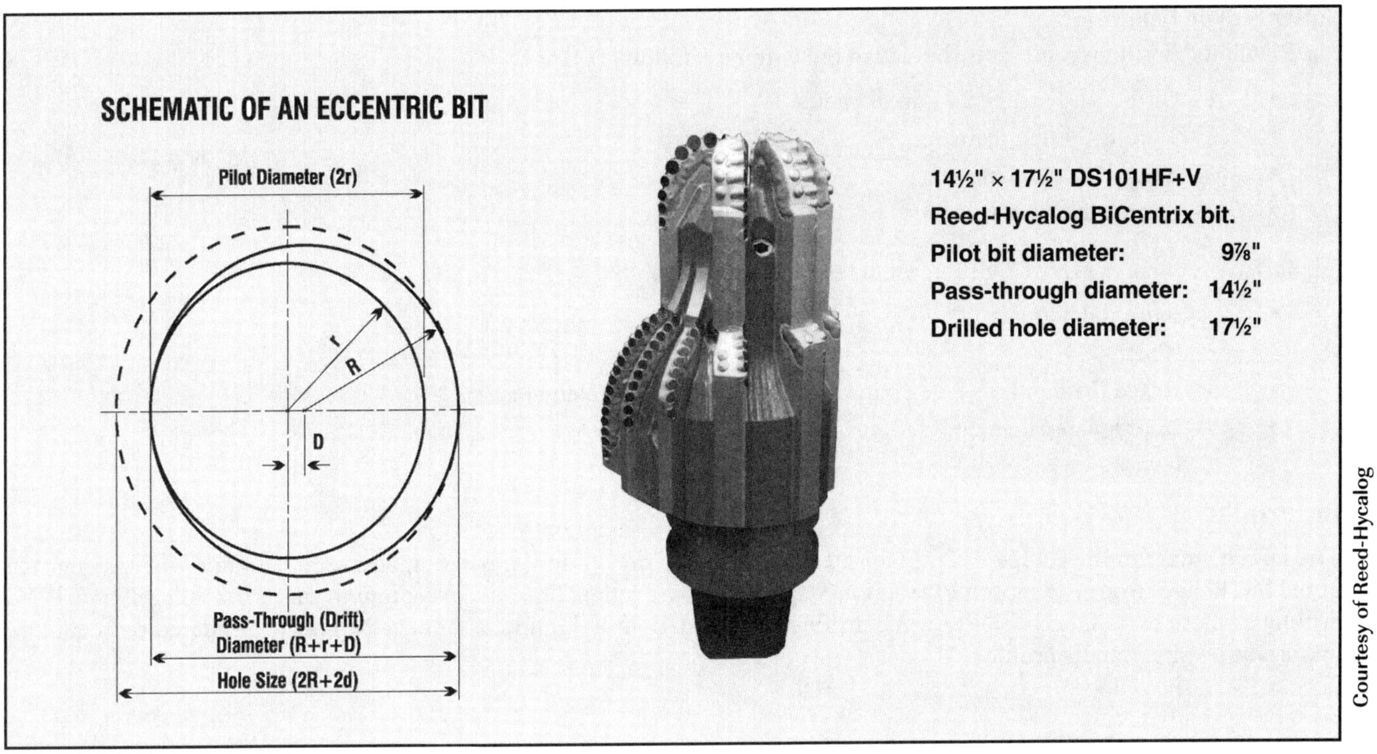

Figure 6.17 Bicentric and eccentric bits

Fixed-Cutter Bit Classification

Unlike roller cone bits, no uniform classification system exists that relates bit style to application. Fixed-cutter IADC codes are intended only to provide a means for characterizing the general physical appearance of fixed-cutter drill bits. Fixed-cutter bits with similar or even the same IADC code, may have significantly different performance capabilities.

The IADC fixed-cutter bit classification system is represented by a four figure coding system. The four characters describe body material, cutter density, cutter size or type, and bit profile, respectively.

Body Material

M—matrix
S—steel

Cutter Density

For PDC bits, this character relates to total cutter count, including standard gauge cutters, and ranges from 1 (light set) to 4 (heavy set).

For surface-set diamond bits (natural diamond, TSP, or impregnated diamond designs), the numbers 6 to 8 are used to designate cutter density. In this case, however, the character represents the size of the diamonds used in the bit design rather than cutter count.

- 6 – diamond sizes larger than 3 stones per carat
- 7 – 3 stones per carat to 7 stones per carat
- 8 – smaller than 7 stones per carat.

In essence, the character is a rough indication of how hard or abrasive the intended application would be. That is, a surface-set diamond bit with an 8 representing cutter density would have smaller diamonds and would be intended for harder and/or more abrasive formations than would a diamond bit coded as a 6.

Cutter Size or Type

For PDC bits, this digit represents the size of the cutters as follows:

- 1 – cutters larger than 24 mm diameter
- 2 – 14 mm–24 mm diameter
- 3 – 8 mm–13 mm diameter
- 4 – 8 mm cutters and smaller

For surface-set bits, the third digit represents the type of diamond:

- 1 – natural diamonds
- 2 – TSP
- 3 – mixed diamond type; i.e., natural diamonds and TSP elements
- 4 – impregnated diamond designs

Bit Profile

The last character in the coding system indicates the profile of the bit design based upon overall length of the cutting face of the bit. Ranges from 1 (flat profile) to 4 (long tapered turbine style). The only exception to this is for 'fishtail' type PDC drill bits, whose cleaning capabilities while drilling soft formations at high rates of penetration is considered to be a more important feature than its profile.

BIT HANDLING AND MAKEUP PROCEDURES

Care should be exercised when handling tricone bits fitted with tungsten carbide inserts and fixed-cutter bits of all types. Under no circumstances should the cutting structure of a fixed-cutter bit be allowed to contact any steel surfaces on the rig.

Prior to initial makeup, all bits should be gauged with an appropriately sized gauge ring; i.e., calibrated 'Go' and 'No Go' gauge rings. Because roller cone bits and diamond bits are manufactured to different tolerances, it is important that bits be graded using the appropriate gauge ring. A 'Go' gauge, designed for grading roller cone bits, will show an in-gauge fixed-cutter bit to be undergauge. A 'No Go' gauge should always be used to grade a fixed-cutter bit.

The bit serial number should be recorded, together with the bit type and diameter. The bit should be closely examined for damage, blocked nozzles, etc. If needed, TFA (total flow area) can then be modified using the appropriate spanner to change the nozzles. In all cases, check that nozzles are properly fitted. Natural diamond bits, impregnated bits, and TSP bits have a fixed TFA which cannot be modified at the rig site.

Tricone bits should be made up to the drill string using a correctly sized bit breaker. Fixed-cutter bits should be 'walked' by hand onto the bit sub until the tool joints shoulder. The correct makeup torque should then be applied.

BIT RUNNING PROCEDURES

Roller Cone Bits

Running In

- Make the bit up to proper torque.
- Lower the bit slowly through ledges and doglegs.
- Lower the bit slowly at liner tops.
- Roller cone bits are not designed for reaming. If reaming is required, then it should be undertaken with light weight and low rpm.
- Protect nozzles from plugging.

Establish a Bottomhole Pattern

- Rotate the bit and circulate when approaching bottom. This will prevent plugged nozzles and clear out hole fill.
- Lightly tag the bottom with low rpm.
- Gradually increase the rpm.
- Gradually increase weight on bit (WOB).

Before Rerunning Green Bits

- Make sure the bit is in gauge.
- Check any bit for complete cutting structure.
- Check any sealed-bearing bit for effective seals.
- Soak any sealed-bearing bit in water or diesel to loosen formation packed in the reservoir cap equalization ports.
- Regrease 14¾" diameter and larger open-bearing bits.

Fixed-cutter bits

Preparation

- Before running a diamond bit into the hole, have a junk basket run on the previous bit.
- After the previous bit is pulled, inspect it for junk damage and other wear, then gauge it.
- If the previous bit appears OK, the bit may be prepared to be run into the hole.
- Check O-ring and install nozzles, if appropriate, using the correct nozzle spanner as supplied by the bit manufacturer.
- Check for cutter damage.
- Check that the bit is within tolerance on diameter and that there is no foreign material inside it.

Running the Bit (Rotary Assembly)

- Handle the fixed-cutter bit with care. DO NOT set the bit down without placing wood or a rubber pad beneath the diamond cutters.
- A correct bit breaker should be used and the bit should be made up to the correct torque as determined by the pin connection size.
- Care should be taken in running the bit through the rotary table and through any known tight spots. Hitting ledges or running through tight spots carelessly may damage the cutters or gauge.
- Reaming is not recommended; however, if necessary, pick up the string and run the maximum fluid possible. Rotate at about 60 rpm. Advance the bit through tight spots with no more than 4,000 lb weight on bit (WOB) at any time.
- As hole bottom is approached, the last three joints should be washed down slowly at full flow and with 40 to 60 rpm to avoid plugging the bit with fill.
- Once the bottom is located, the bit should be lifted just off bottom (0 to 1 foot if possible) and full volume circulated while slowly rotating for about 5 to 10 minutes.
- After circulating, ease back to bottom and establish the bottomhole pattern.
- When ready to start drilling, increase the rotary speed to about 100 rpm and start cutting a new bottomhole pattern with approx. 1,000 to 4,000 lb WOB.
- Cut at least one foot in this manner before determining optimum bit weight and rpm for drilling.
- Determine optimum ROP by conducting a drill-off test.

Running the Bit (PDM and Turbine)

- Start the pumps and increase to the desired flow rate when approaching bottom.
- After a short cleaning period, lower the bit to bottom and increase WOB slowly.
- After establishing a bottomhole pattern, additional weight may be slowly added.
- As weight is increased, pump pressure will increase, so the differential pressure and WOB must be kept within the recommended downhole motor specifications.
- Drill pipe should be slowly rotated to prevent differential sticking.
- All other operating practices are as per standard practices.

Drill-Off Tests

Drill-off tests are performed in order to ascertain the optimum combination of weight on bit and rotary speed to maximize penetration rate. They should be done

- at the start of a bit run.
- on encountering a new formation.
- if a reduction in ROP occurs.

Drill-Off Test Procedure

1. Maintain a constant rpm. Select a WOB (near maximum allowable).
2. Record the time to drill off a weight increment, i.e., 5,000 lb.
3. Reapply the starting weight and record the length of pipe drilled during step 2.
4. From steps 2 and 3 the penetration rate may be found.
5. Repeat steps 2 and 3 at least four times. The last test should be at the same value as the first. This repeat test will determine if the formation has changed or not.
6. Plot seconds to drill off versus bit weight.
7. Plot penetration rate versus bit weight.
8. Select the bit weight which produced the fastest ROP. Maintain this WOB constant and repeat the above but varying the rpm.
9. Plot ROP versus rpm and select the rpm which resulted in the fastest ROP. This is the optimum rotary speed.
10. These values for WOB and rpm obtained will result in optimum progress for the particular formation and bit type.

Drilling Out Float and Shoe Equipment

If using a PDC bit, ensure that the float equipment is PDC drillable. Drilling out should be performed with high WOB and low rpm.

BIT RELATED DRILLING DYNAMICS

Downhole vibrations can be extremely detrimental to bit performance and may also result in damage of downhole tools such as MWD/LWD sensors and mud motors. Downhole vibrations are associated largely with the interaction between drill bits and the formation being drilled and are, in the main, the result of the nature of the somewhat aggressive cutting structure of PDC designs. It should be remembered, however, that such vibrational problems are not restricted to PDC bit runs and that roller cone bit assemblies can be subject to many of the same problems.

There are three main forms of downhole vibration which, whether occurring independently of one another or together, can impede overall drilling performance:

- Axial vibrations
- Lateral vibrations
- Torsional vibrations

Axial Vibrations

Often referred to as *bit bounce*, axial vibrations take the form of periodic vertical motion of the bit in the direction of the bit's central axis. As a bit vibrates up and down at the bottom of the hole, the weight applied to each individual cutter changes. The resulting depth of cut for the cutters then varies, ranging from a minimum when the bit is in the up position to a maximum when the bit is again on bottom. Variations in the depth of cut translate into variations in torque. These torque fluctuations can be a cause of torsional vibration at the bit, leading to the potential for slip-stick behavior. Bit bounce is generally of more concern when running roller cone bits.

Lateral Vibrations

Lateral vibration, sometimes referred to as *bit whirl*, is the periodic sideways movement of the bit in the "x-y" plane. Bit whirl is a specific phenomenon that occurs when the downhole dynamic forces cause the bit's instantaneous center of rotation to move from its geometric center. When a PDC bit whirls, it cuts a characteristic multilobed bottomhole pattern versus the concentric circles seen in the bottomhole pattern of a smooth running bit.

When an individual PDC cutter 'grabs' the formation, the bit's instantaneous center of rotation is shifted to the point of cutter/formation contact. This in turn creates a backward whirling motion that imparts impact loading on the PDC cutters on the side opposite the center of rotation. Whirl induced damage is generally caused by these high impact loads coming from behind the cutters (fig. 6.18).

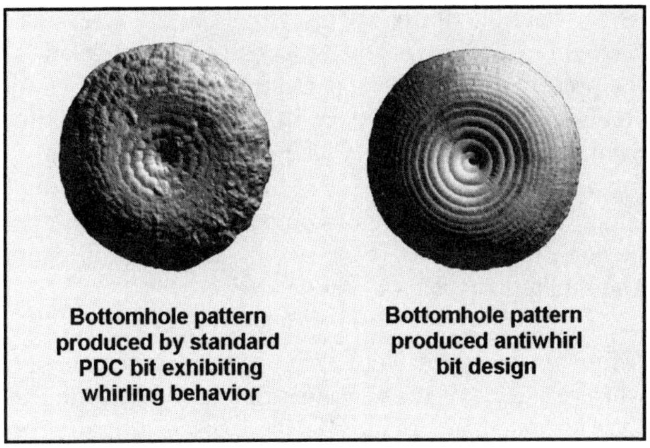

Courtesy of Reed-Hycalog

Figure 6.18 Whirl induced damage

Whirling occurs when an imbalance force is introduced, such as when

- drilling an inclined hole.
- formation hardness changes and the borehole is not perpendicular to the formation bedding planes.
- the drill string is vibrating due to inadequate stabilization.
- the sum of the forces on the individual cutters has a lateral component.

The first three of these conditions are the result of factors beyond the control of the bit designer. However, in an effort to overcome the latter, a great deal of emphasis is placed on the force and mass balancing of PDC bit designs.

Using the latest in CAD techniques, the bit designer can control the direction and relative magnitude of the forces on individual PDC cutters. These individual cutter forces can be summed and resolved into their resultant components producing a resultant axial force (weight-on-bit force), torsional force (torque), and radial force for a known set of initial conditions. The radial, or 'out-of-balance', force is virtually directly proportional to the axial force and is therefore usually expressed as a percentage of the applied WOB in a particular direction.

Cutter force balancing is often confused with the dynamic mass balancing of the bit design as a whole. As a bit is rotated, a force is generated which is influenced by the mass distribution of the bit. This force is proportional to the mass of the bit, the distance between the center of mass and the axis of bit rotation, and the square of the speed of rotation. As most PDC bit designs are generally symmetrical in shape, there will not usually be any great distance between the center of mass and the designed axis of rotation. The effect of imbalance as a result of bit mass distribution is therefore relatively insignificant when compared with the force generated by the cutting structure.

One of the ways in which bit manufacturers attempt to overcome the problem of out-of-balance forces is to arrange for the sum of the load forces generated by the cutters to have a certain value directed through a large, low friction gauge pad designed to slide along the borehole wall (fig. 6.19). If the value of this resultant cutter force is always higher than the unavoidable lateral forces, then the total force will always pass through the gauge pad and the bit will be self-stabilizing while drilling. The direction of this force is limited to a small arc on the bit's circumference and continually pushes this area of the bit against the formation, The continuous contact against the formation reduces downhole vibrations and prevents the bit 'walking' around the hole; i.e., bit whirl.

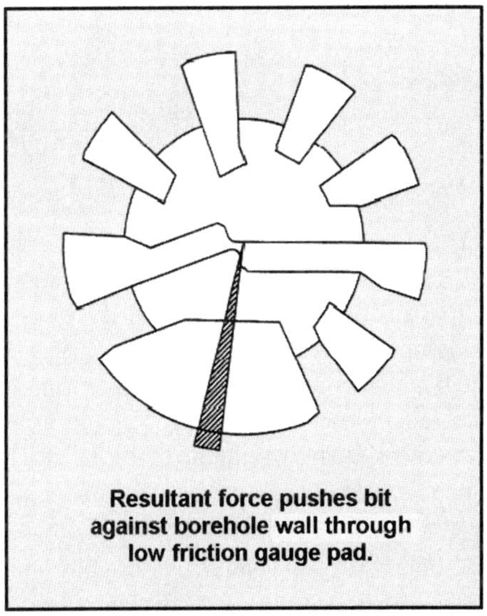

Courtesy of Reed-Hycalog

Figure 6.19 Schematic of antiwhirl bit design

Other manufacturers attempt to limit the effect of lateral vibrations by arranging PDC cutters such that cutters on one blade directly track those on the preceding blade (as opposed to being conventionally arranged in a spiral configuration, radiating from the cone of the bit). This approach is taken by Security DBS in its Trac-Set range of PDC bits.The resulting grooves of formation which remain uncut on each rotation of the bit are said to restrict lateral movement. Diamond impregnated backup studs, positioned directly behind the PDC cutters on each blade and familiar on many bit designs, are also said to help reduce lateral vibrations.

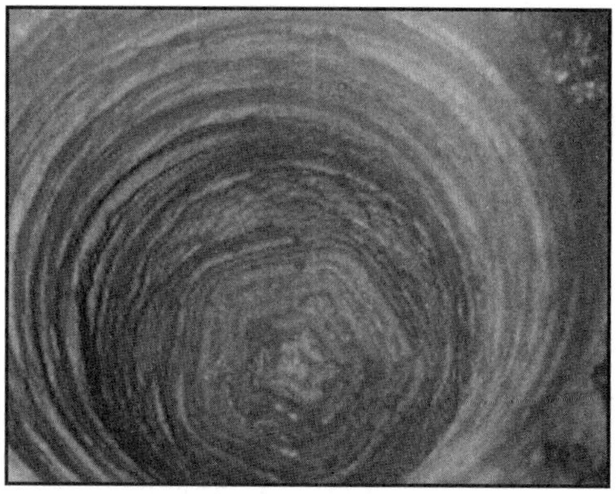
Courtesy of Reed-Hycalog

Figure 6.20 Bottomhole profile of conventional PDC bit in experiments conducted by Reed-Hycalog

Although the approach of each bit manufacturer towards the problem of bit whirl may vary, all aim to ultimately improve the stabilization of the bit design. One bit manufacturer, Reed-Hycalog, introduced a range of PDC bits, the steerability of which is said to approach that of roller cone bit designs (fig. 6.20). This is achieved by designing the bit with a 360° full contact gauge ring that prevents the outermost cutters from biting into the formation (figs. 6.21 and 6.22). The resulting restriction in lateral movement helps to stabilize the bit, produces a smoother torque response, and reduces bit whirl.

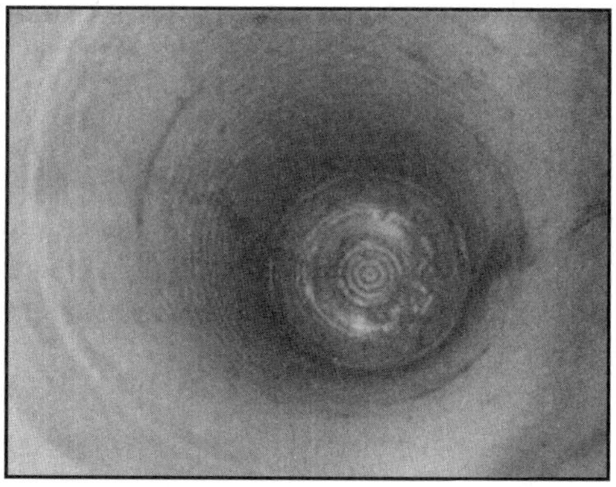
Courtesy of Reed-Hycalog

Figure 6.21 Bottomhole profile resulting from use of PDC bit with 360° full contact gauge ring

As well as improving directional responsiveness, use of the bit helps to produce a smoother borehole.

Torsional Vibrations and Slip Stick

Put simply, torsional stress is caused when one end of an object is twisted while the other end is held fixed or twisted in the opposite direction. Fluctuations in downhole torque can lead to the development of torsional vibrations.

There are two basic classes of torsional vibration to which the drill string is subjected:

- Transient vibrations which correspond to changes in downhole conditions; e.g., interbedded lithology types.
- Stationary vibrations which are self-induced through actions upon the drill string such as frictional forces between the pipe and the borehole wall, changes in weight on bit, or stabilizers hanging up.

Courtesy of Reed-Hycalog

Figure 6.22 12¼" 'Steering Wheel™' bit design by Reed-Hycalog. Note the full contact gauge ring.

Drill string torsional vibrations occur frequently. When they become severe, they can escalate into slip-stick oscillations whereby the bit may briefly stop turning until sufficient torque is developed at the bit to overcome static friction. When the stalled bit breaks free, it does so at rotational speeds from two to ten times the surface rotational speed. At the point of breaking free, a torsional wave then travels from the drill bit up the drill string to the surface. The rig reflects this wave back down to the bit, which again stalls. This cycle will be repeated unless the drilling parameters are adjusted to interrupt it. During severe torsional vibrations, it is possible for the bit to spin backwards. In hard, compacted rock formations this can damage the bit in a very short time.

Drill string torsional vibrations can be identified by cyclic oscillations of the surface torque, the drive motor current, and sometimes the rotary speed. Cyclic variations in standpipe pressure might also be observed.

Slip stick causes torque and rotational speed oscillations along the entire length of the drill string. These oscillation periods, and cyclic torque behavior measured at the surface, are good indicators of torsional vibrations in the drill string. The characteristics of the oscillations depend on the length and weight of the drill pipe, the mechanical properties of the drilling system, the surface rotational speed, and the nature and location of the downhole friction.

By using the maximum, the minimum, and the average values of the surface torque, one may deduce the extent of the slip-stick motion of the bit.

Reducing weight on bit and/or increasing rotary speed can help remedy the problem. In severe cases, modification of the lubricating properties of the mud system may be required.

Drill string torsional vibrations, and in particular slip-stick oscillations, are detrimental to the life of the drill string and the bit. Cyclic torque oscillations can lead to premature fatigue failure of drill pipe. There are tremendous advantages to be realized by reducing or eliminating slip-stick oscillations. The 'stick' phase of slip stick increases the probability of having a stuck pipe or a twist off, while the 'slip' phase is damaging to the bit due to the high rotational speeds that can be attained by the bit and could lead to backing off of connections.

Prediction and Monitoring of Downhole Vibrations

Software packages are now available which, when used appropriately, can help to predict the likelihood of the occurrence of downhole vibrations. Part of the directional driller's BHA design software, Sperry-Sun's 'Whirl™' module can predict, for a given BHA and hole geometry, the combinations of weight on bit and rotary speed that are likely to excite the BHA at resonant frequencies, and thus induce bit whirl. This provides the opportunity at the BHA design stage to either change the design to be more tolerant or to issue the driller with a set of parameters that he should avoid.

A number of service companies have developed downhole tools which enable vibration levels to be measured in real time. An example of such a tool is Sperry-Sun's MWD Drill String Dynamics Sensor. The tool consists of a triaxial accelerometer package mounted in a modified Dual Gamma Ray tool, thus eliminating the necessity for an additional MWD sub in the string. The accelerometers are oriented with the Z axis along the drill string, the X axis aligned laterally, and the Y axis at 90° to the other two axes but tangential to the drill string. This configuration allows the tool to monitor axial, lateral, and torsional vibrations as they occur. Real-time displays of average, peak, and instantaneous acceleration data can be used to interpret and analyze downhole vibrations, thus indicating the onset of slip-stick behavior, bit whirl, and bit bounce, and affording the opportunity to take corrective action.

DRILLING PROBLEM IDENTIFICATION

The three major sources of information while drilling are pressures (differential and circulating), torque, and penetration rate:

Differential Pressure

Reduced differential pressure indicates one or more of the following:

- Reduced flow rate
- Washout in the pipe
- Extreme erosion of the bit (not usual)
- Reduced WOB

An increase in differential pressure indicates one or more of the following:

- Increased flow rate
- Cutters have worn so that the bit face is in contact with the bottom of the hole.
- Excessive WOB
- Large depth of cut: formation softer than expected

Circulating Pressure

(Standpipe pressure with bit off bottom)

Increased circulating pressure could be due to one or more of the following:

- Heavier mud weight or inadequate mud properties
- Plugged or partially plugged bit nozzles
- Increased flow rate
- Annular restriction

Decreased circulating pressure could be due to one or more of the following:

- Lighter mud weight
- Washout
- Reduced flow rate
- Air in the mud
- Pump malfunction

Torque

Increasing torque:

- Hole angle changing
- Washout
- Formation change
- Poor mud properties
- WOB increased
- Poor hole cleaning
- Bearing failure on tricone bit

Decreasing torque:

- Formation change
- Rotary speed change
- WOB decreased
- Improvement in mud properties
- Hole angle straightening out

Irregular or varying torque:

- Reaming with stabilizer
- Dry drilling
- Bit balled up
- Drilling sand formation
- Junk in hole
- Washout
- Excessive WOB
- Rotary speed change

Penetration Rate (ROP)

An increase in ROP may indicate:

- Formation change and/or
- Drilling close to balance (overbalance reducing)

A decrease in ROP may result from one or more of the following:

- Worn bit
- WOB, rpm, or hydraulics not optimized
- Formation change
- Crooked hole
- Washout
- Overbalance increasing

Varying ROP indicates one or more of the following:

- Formation layers
- Bit wearing out
- Bit balled up
- Washout
- Inconsistent transference of WOB

DULL BIT GRADING

An important aspect of performance improvement is the reporting of all information pertaining to the bit run. Both the drilling engineer and the bit manufacturer alike will use this information to

- improve bit selection on future wells.
- improve bit design.

Along with the final dull grade of the bit itself, the following factors should all be accurately recorded:

- Footage drilled
- Penetration rate
- Average, maximum, and minimum drilling parameters (WOB, rpm, flow rate)
- Drilling fluid properties (type, density, viscosity, etc.)
- Drive type (rotary, rotary steerable system, motor, etc.)
- Percentage steering for run
- Inclination, azimuth, and build/turn rate
- Formation type
- Formation tops

IADC Dull Grading System

The IADC dull bit grading chart allows for eight factors to be recorded. The chart applies to both roller cone and fixed-cutter bit types, although different wear codes exist for each.

CUTTING STRUCTURE				B	G	REMARKS	
INNER ROWS	OUTER ROWS	DULL CHAR.	LOCATION	BRNG/ SEALS	GAUGE 1/16"	OTHER CHAR.	REASON PULLED

Cutter/Cutting Structure Wear

For both roller cone and fixed-cutter bits, wear is measured on a scale from 0 (zero wear) to 8 (total loss of cutting structure).

When grading a PDC bit, it is important to remember that the cutters should be graded on the condition of the *visible diamond table*, regardless of cutter shape or exposure. For example, if at the start of the bit run a PDC cutter has 50% of its diamond table exposed above blade height and then, after the run, all of the exposed 'usable' diamond has been worn, the correct wear grade for that cutter should be '4'—equating to 50% worn (fig. 6.23). A common mistake would be to grade such a cutter as an '8'.

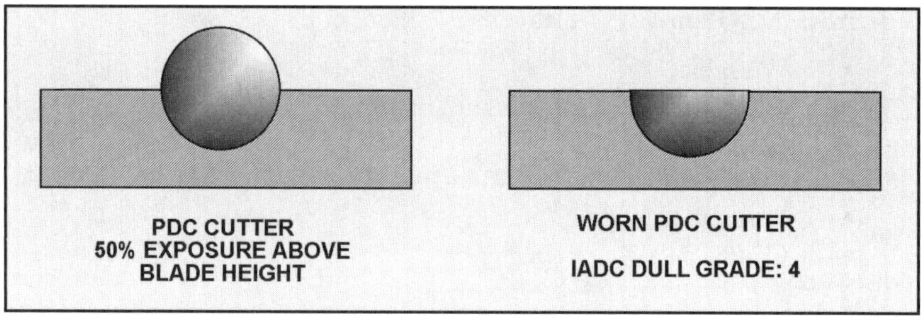

Figure 6.23 Cutter or cutting structure wear

For natural diamond, TSP, and impregnated diamond bits, wear is determined by comparing the initial visible cutter height (or, in the case of impregnated designs, initial blade height) with the amount remaining after the bit run. It is therefore important to remember to inspect and measure the cutting structure before the bit is run.

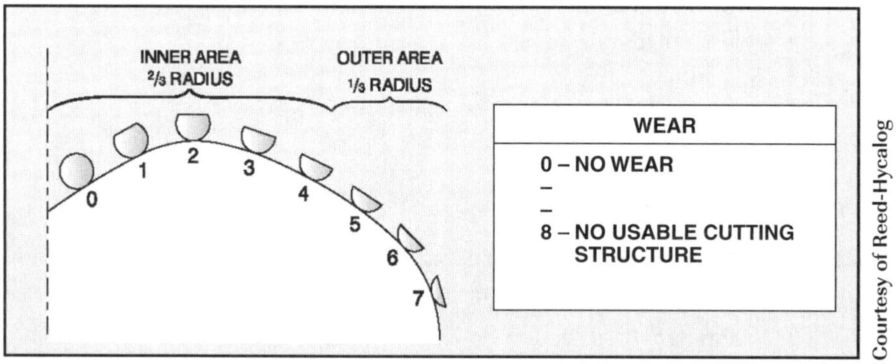

Figure 6.24 IADC cutter wear chart

The amount of cutter wear is entered in two boxes on the IADC chart, with the figures representing the average amount of wear of the cutters in the inner and outer rows (fig. 6.24).

Roller Cone Bit Location Codes

When entering the location of dull characteristics for roller cone bits, the following codes, along with the number of the cone (i.e., 1–3), should be used:

N = Nose row
M = Middle row
G = Gauge row
A = All rows

Fixed-Cutter Bit Location Codes (fig. 6.25)

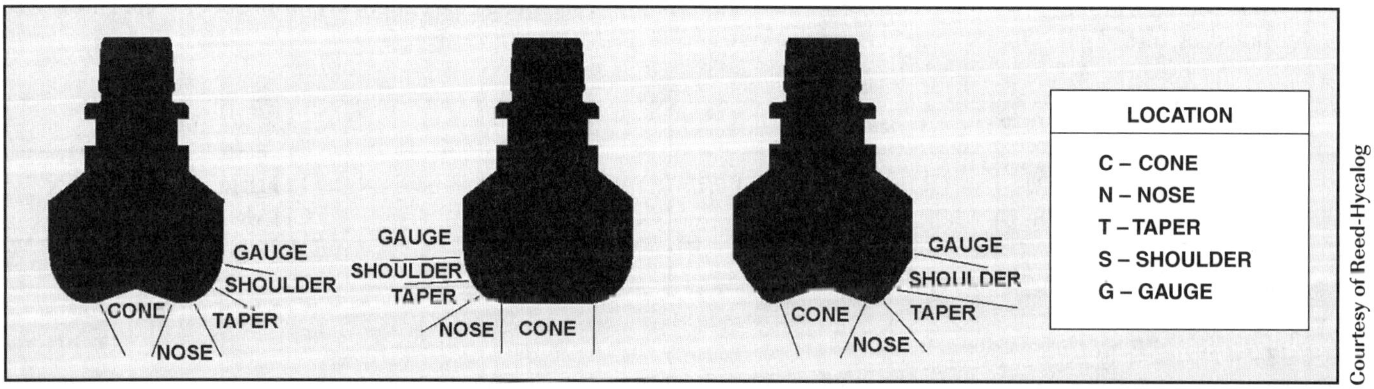

Figure 6.25 Fixed-cutter bit location codes

IADC Dull Grading Codes (fig. 6.26)

CUTTING STRUCTURE				B	G	REMARKS	
INNER ROWS	OUTER ROWS	DULL CHAR.	LOCATION	BRNG/ SEALS	GAUGE 1/16"	OTHER CHAR.	REASON PULLED

CUTTER WEAR

0 - NO WEAR
•
•
8 - NO USABLE CUTTING STRUCTURE

DULL CHARACTERISTICS

*BC - BROKEN CONE
BF - BOND FAILURE
BT - BROKEN TEETH / CUTTERS
BU - BALLED UP
*CC - CRACKED CONE
*CD - CONE DRAGGED
*CI - CONE INTERFERENCE
CR - CORED
CT - CHIPPED CUTTERS
ER - EROSION
*FC - FLAT CRESTED WEAR
HC - HEAT CHECKING
JD - JUNK DAMAGE
*LC - LOST CONE
LN - LOST NOZZLE
LT - LOST CUTTERS
NR - NOT RERUNABLE
OC - OFF-CENTER WEAR
PB - PINCHED BIT
PN - PLUGGED NOZZLE / FLOW PASSAGE
RG - ROUNDED GAUGE
RO - RING OUT
RR - RERUNABLE
*SD - SHIRTTAIL DAMAGE
*SS - SELF SHARPENING WEAR
*TR - TRACKING
WO - WASHED OUT - BIT
WT - WORN TEETH / CUTTERS
NO - NO MAJOR / OTHER DULL CHARACTERISTICS

* Roller Cone Bit Codes

WEAR LOCATION

C - CONE
N - NOSE (ROW)
T - TAPER
S - SHOULDER
G - GAUGE
A - ALL AREAS / ROWS
M - MIDDLE ROW
H - HEEL ROW

BEARING WEAR

NONSEALED BEARINGS
0 - NO LIFE USED
•
•
8 - ALL LIFE USED

SEALED BEARINGS
E - SEALS EFFECTIVE
F - SEALS FAILED

X - FIXED CUTTER BITS

GAUGE WEAR

I - IN GAUGE
1/16" - 1/16" UNDERGAUGE
2/16" - 1/8" UNDERGAUGE
•
•

REASON PULLED

BHA - CHANGE BOTTOM-HOLE ASSEMBLY
DMF - DOWNHOLE MOTOR FAILURE
DSF - DRILL STRING FAILURE
DST - DRILL STEM TEST
DTF - DOWNHOLE TOOL FAILURE
LOG - RUN LOGS
RIG - RIG REPAIR
CM - CONDITION MUD
CP - CORE POINT
DP - DRILL PLUG
FM - FORMATION CHANGE
HP - HOLE PROBLEMS
HR - HOURS
PP - PUMP PRESSURE
PR - PENETRATION RATE
TD - TOTAL DEPTH / CSG. DEPTH
TQ - TORQUE
TW - TWIST-OFF
WC - WEATHER CONDITIONS
WO - WASHOUT - DRILL STRING

Figure 6.26 IADC dull grading codes

The Gauge

Because roller cone bits and diamond bits are manufactured to different tolerances, it is important that bits be graded using the appropriate gauge ring. A 'Go' gauge, designed for grading roller cone bits, will show an in-gauge fixed-cutter bit to be undergauge. A 'No Go' gauge should always be used to grade a fixed-cutter bit.

Used Bit Conditions/Causes/Remedies Tables

TABLE 6.2
Roller Cone Bits

Condition of Dull Bit	Possible Causes	Possible Remedies
Excessive bearing wear	Excessive rotary speed Excessive rotating time Excessive WOB Excessive sand in circulating system Unstabilized drill collars Improper bit type	Slower rotary speed Reduced rotating hours Lighter WOB Removal of sand from circulating fluid Stabilize drill collars Use harder formation bit type with stronger bearing structure
Excessive broken teeth	Improper bit type Improper break-in procedure used for new bit Excessive WOB for type used	Use harder formation bit type having more teeth Proper break-in procedure used for new bit Lighter WOB
Unbalanced tooth wear	Improper bit type Improper break-in procedure used for new bit	Use of different bit type based upon the rows of teeth which are excessively worn on the dull bit Proper break-in procedure used for new bit
Excessive tooth wear	Excessive rotary speed Improper bit type Use of nonhardfaced type	Slower rotary speed Use of harder formation bit type having a greater number of teeth Use of bit type having hardfaced teeth
Fluid cut teeth & cone	Excessive circulation rate of fluid Excessive sand in circulating system	Reduction in circulation fluid rate Removal of sand from circulation fluid
Excessively undergauge	Improper bit type Excessive rotating time	Use of bit having greater gauge protection Reduce rotating hours
Skidded due to balling	Excessive WOB Improper bit type Insufficient fluid circulation rate	Lighter WOB Use of softer formation bit type having teeth more widely spaced Increase fluid circulation rate

TABLE 6.2
Fixed-Cutter Bits

Condition of Dull Bit	Possible Causes	Possible Remedies
Polished diamonds	Rotating in a hard formation without making hole	Add WOB or select a bit with smaller diamonds or cutters
Shattered diamonds	Improper stabilization Abnormal vibrations Improper handling	Correct the stabilization Reduce the vibration by changing the RPM Proper handling
Sheared diamonds	Improper break-in Broken nose stones	Clean bottom of hole Correct break-in procedure
Burned (flattened) diamonds	Inadequate cooling	Increase the circulation rate to improve cooling
Burned bit	Overheating as a result of plugging and/or balling up	Take care of proper hydraulics
Heat checking of PDC supports or bit body (matrix)	Overheating	
Worn throat	Fragments of hard formation rolling in the throat	Proper hydraulics
Loss of gauge	Long intervals reamed with insufficient cooling due to clogged junk slots	When reaming, minimize RPM and bit weight
Junk damage	Junk in hole just ahead of the first diamond bit used	Clean bottom on previous bit run with a junk basket in the string

BIT RUN ECONOMICS

Although drill bits contribute only a fraction of the overall equipment cost, they can be the most critical element in the calculation of drilling economics. The cost of a PDC or diamond bit may be many times that of a milled-tooth roller cone bit and it becomes evident that use of the more costly bit design must be economically justified through superior performance.

The accepted method of assessing bit performance in terms of economics is to calculate the cost per foot drilled. As a PDC bit is considerably more expensive than roller cone products, it is apparent the PDC bit must make up this additional cost by drilling faster and/or further. The following formula is used to calculate the cost per foot drilled:

$$C = \frac{R\,(T + D) + B}{F}$$

where

C = drilling cost per foot ($/ft)
R = rig operating cost ($/hr)
T = trip time (hr)
D = time spent drilling (hr)
B = bit cost ($)
F = footage drilled (ft).

The drilling cost per foot formula is valid for any bit type. The formula can be used after a bit run with actual performance data to calculate an actual cost per foot, or it can be used before a bit run with assumed values to compute a projected cost per foot. Projected cost per foot for a proposed bit is usually compared to the actual costs per foot achieved on offset wells.

When proposing a PDC bit in an area where roller cone bits are usually run, it can be useful to perform a break-even analysis to identify the performance level (in terms of footage and hours) a PDC bit would have to achieve to have a total drilling cost equal to roller cone bits.

Following is an example of break-even analysis for a PDC bit:

Performance data from an offset well

Total rotating time	=	100 hrs
Total trip time	=	45 hrs
Rig operating cost	=	$500 hr
Total bit cost	=	$16,000
Total footage drilled	=	3,750 ft

Using the drilling cost equation, the cost per foot achieved in this section of the offset well is calculated to be $23.60/ft. If a PDC bit can result in the same value, it will break even. If better performance can reasonably be expected, the use of a PDC can be justified economically. Break-even analysis requires the performance of the PDC bit be estimated. This performance can be calculated in two different ways.

1. Footage Assumed

If the footage is assumed to be equivalent, in this case 3,750, we must calculate the rate of penetration of the bit necessary to achieve a cost-per-foot of $23.60/ft or less.

The following formula is used:

$$ROP_{BE} = \frac{R}{C - (R \times T + B)/F}$$

where

ROP_{BE} = break-even penetration
C = offset cost-per-foot ($/ft)
T = trip time for PDC bit ($)
B = cost of proposed bit ($)
F = assumed footage (ft)
R = rig operating cost ($/hr).

Using the offset well performance data previously presented, and the following assumptions for the proposed bit, a break-even rate of penetration can be computed:

T = 10 hr
B = $28,000
F = 3,750 ft

In this example, the proposed bit would be required to drill the interval with a penetration rate of at least 34 ft/hr to justify its use.

2. Penetration Rate Assumed

If a penetration rate is assumed, we can calculate the minimum footage the bit must drill to save the operator money. The derived formula in this case is:

$$F_{BE} = \frac{R \times T + B}{C - R/ROP}$$

where

F_{BE} = break-even footage (ft)
ROP = assumed penetration rate (ft/hr).

Assuming a penetration rate of 45 ft/hr, the proposed bit must drill 2,287 ft to break even against the offset well.

CHAPTER 7
Hydraulics and Hole Cleaning

INTRODUCTION

Hydraulics planning is part of the overall drilling optimization process.

It involves a calculated balance of the various components of the circulating system to maximize ROP and keep the bit and hole clean while remaining within any constraints of the wellbore, surface, and downhole equipment.

CONSIDERATIONS FOR HYDRAULICS PLANNING

Maximizing ROP

Cuttings removal from the bottom of the hole is related to the fluid energy dissipated at the bit (bit hydraulic power). It has been shown that bit hydraulic horsepower is optimized when the pressure differential (pressure drop) across the bit is equal to two-thirds of the total system pressure (pump pressure). Maximizing hydraulic horsepower can be used to increase penetration rate in medium-to-hard formations.

Hole Cleaning

In soft formations or deviated holes, hole cleaning is often the dominant factor. There is little point in maximizing the ROP by selecting nozzles that optimize bit hydraulic horsepower or impact force, if the resulting flow rate is insufficient to lift the cuttings out of the hole. In these instances, it is preferable to determine a suitable flow rate first and then optimize the hydraulics.

Annulus Friction Pressure

In slim-hole or deep wells, the annulus friction pressure needs to be considered. Too high an annulus friction pressure increases the *equivalent circulating density* (*ECD*) and can lead to lost circulation, differential sticking, or hole instability.

Erosion

Soft, unconsolidated formations are prone to erosion if the annulus velocity and, therefore, flow rate are too high or the annulus clearance is small leading to the possibility of turbulent flow. In these instances, a reduction in flow rate will be required to minimize erosion.

Lost Circulation

If heavy lost circulation is anticipated and large quantities of LCM could be pumped, it may be necessary to install larger bit nozzles to minimize the risk of bit plugging.

FACTORS THAT AFFECT HYDRAULICS

The rig equipment, drill string and downhole tools, wellbore geometry, mud type, and properties are all factors that can affect hydraulics.

Rig Equipment

The single biggest factor of the rig equipment is the pump pressure limitation and volume output of the mud pumps in use. Increasing pump liner sizes increases the volume output but decreases the maximum allowable pump pressure. Most high-pressure pipe work from the mud pumps to the kelly or top drive is rated at a pressure higher than the pump rating.

Drill String and Downhole Tools

The main effect of the drill string is the frictional pressure drop or parasitic pressure losses that occur within the drill pipe and drill collars.

For a given flow rate and depth, a drill string with a smaller ID will have higher parasitic pressure losses. Hence, one of the benefits of using 5½" DP instead of 5" DP is that, for the same flow rate, there is more pressure available at the bit to optimize nozzle sizes, or for the same parasitic pressure loss in the drill string there is more available flow rate.

The addition of downhole tools can have an effect on the available or allowable pressure drop across the bit.

Negative pulse MWD tools require about 800 psi pressure drop below the MWD tool for adequate transmission of real-time data.

Most adjustable stabilizers require 450 to 800 psi off-bottom pressure drop below the tool for operation or actuation.

Steerable motors with PDC bits have a total pressure drop of 1,200–1,500 psi. This pressure drop is made as follows:

- Off-bottom differential pressure 200 psi
- Drilling pressure drop 600–800 psi for high-torque low-speed motors
- Drilling pressure drop of 1,000–1200 psi for high-speed motors
- Stalling buffer 400–500 psi

Steerable motors with tricone bits have a total pressure drop of 400–600 psi due to the lower reactive torque of a tricone bit. This pressure drop is made up as follows:

- Off-bottom differential pressure 200 psi
- Drilling pressure drop 200–400 psi for high-torque low-speed motors
- No stalling buffer, due to the lower reactive torque

Wellbore Geometry

The deeper the hole, the higher the parasitic pressure loss within the drill string, the less available pressure there is at the bit to optimize nozzle sizes.

The bigger the hole diameter, the lower the annulus velocity for a given flow rate, the harder it is to effectively remove the cuttings from the hole.

Conversely, the smaller the hole diameter, the higher the annulus friction pressure, the bigger the effect on ECD.

Mud Type and Properties

Mud density and rheology directly affect the pressure losses with the circulating system. Mud rheology and the different available models will be discussed later.

GENERAL RULES OF THUMB

Flow rate

- Flow rate should be maintained at 30 to 60 gpm per inch of bit diameter
- Flow rate must not be reduced to achieve more horsepower
- Too low a flow rate will cause bit balling and reduce effective hole cleaning

Hydraulic Horsepower

Maintain 2.5 to 5 hydraulic horsepower per square inch of bit diameter.

Maximum hydraulic horsepower should be considered when there is sufficient available pump horsepower.

Bit Pressure Drop

Design hydraulics for 50%–65% pressure drop across the bit.

If parasitic pressure losses are greater than 50% of the available pump pressure, then optimize for jet velocity.

Jet Velocity

Jet velocity affects chip hold-down and penetration rate.

Keep jet velocity >250 ft/sec.

HYDRAULIC CALCULATIONS

In order to optimize bit hydraulics and, therefore, bit performance, the entire hydraulic system must be considered.

The following factors will be discussed:

- Circulating pressure
- Flow rate
- Optimum annular velocity
- Pressure losses in the system
- Pressure drop across the bit
- Hydraulic horsepower at the bit
- Jet velocity and jet impact force

Selecting Pump Pressure and Flow Rate

A flow rate and resultant circulating pressure sufficient to allow good hole cleaning and adequate power at the bit, but which do not exceed the maximum allowable surface pressure, must be selected. The optimum flow rate will depend on total system pressure losses and whether bit hydraulics is to be optimized on the hydraulic horsepower or the jet impact force model. Once the optimum flow rate has been selected, the following factors must be considered:

- Are the pumps able to pump the desired flow rate?
- Is the desired flow rate within the operating range of any downhole tools in the string?
- Does the optimum flow rate exceed the desired minimum annular velocity?
- Does the optimum flow rate exceed the desired maximum annular velocity?

Annular Velocity

When designing the hydraulic program, the annular velocity must be considered. It is important to avoid solids retention in the annulus as the subsequent increase in mud density and hydrostatic head could cause mud losses to the formation. An optimum value for annular velocity is selected between upper and lower limits.

Velocity of fluid in the annulus is at its lowest in places where the annular cross-sectional area is at its highest. Since the area around the drill pipe typically has the largest cross-sectional area in the hole, the annular velocity is at its lowest around the drill pipe. Conversely, the annular velocity will be at its highest around the drill collars. The annular velocity around the drill pipe should be calculated to determine whether or not it is high enough to effectively clean the hole. However, in certain soft formations prone to erosion, the annular velocity around the drill collars should also be calculated. If it is found to be too high, the circulation rate should be reduced or smaller drill collars used. For a given drill pipe and hole size, the annular velocity can only be changed by varying the flow rate.

The fluid velocity in the annulus around the drill string is usually given by:

$$V_a = \frac{24.51 \times Q}{ID^2_{HOLE} - OD^2_{DP}}$$

where

V_a = annular velocity, ft/min
Q = flow rate, gpm
ID_{HOLE} = hole diameter or ID of casing, in.
OD_{DP} = outside diameter of either the pipe or collars, in.

Minimum Annular Velocity

The minimum annular velocity is determined by the cuttings carrying capacity of the drilling fluid. If the well is not efficiently cleaned, there will be a buildup of cuttings in the annulus resulting in an increase in mud density. The consequent increase in hydrostatic pressure could cause fluid losses to the formation. In deviated holes, cuttings beds may develop resulting in an increased risk of stuck pipe. The annular velocity should therefore, in relation to cuttings generation, be sufficient to maintain a fluid density below the fracture pressure of the formation.

Another factor governing the lower limit of annular velocity is the rate of settling of the drilled cuttings in the annulus. If this slip velocity exceeds the annular velocity, then cuttings beds will develop and the density of the drilling fluid will increase. The cuttings carrying capacity of a mud system is highly influenced by its viscosity and gel strength.

Maximum Annular Velocity

It is also important to avoid eroding loose formations and soft shales, which may result in large washouts, hole problems, stuck pipe, and poor cement jobs. Therefore, the maximum annular velocity is limited by hole constraints and, in particularly sensitive formations, is often limited to 100 ft/min. However, the rate of erosion of soft formations is influenced more by the flow regime of the fluid in the annulus, rather than the actual velocity itself. Turbulent flow is far more erosive than laminar flow, and will destabilize sensitive formations. Therefore, it is usually desirable to have laminar flow around the drill pipe.

Once the operating limits for annular velocity have been determined, it is then possible to select pump pressure and flow rate based on surface equipment limitations and the desirability of maintaining laminar flow in the annulus or limiting ECD (equivalent circulating density).

Estimating Reynolds Number and Determining Flow Regime

The type of flow behavior of the fluid in the annulus is described by calculating its Reynolds number (Re). There are several methods of calculating the Reynolds number depending on which rheological model is used to describe the fluid behavior. Further information on rheological models is shown in Appendix 1.

All of the equations in this section assume that the Power Law model is being applied.

Before calculating the Reynolds number for a given section, the Power Law constants, n and K, and the effective, or apparent, viscosity, m_e, of the fluid must be derived. These will be calculated separately for pipe and annular flow.

Definitions

Consistency factor (K, eq cp)

Describes the viscosity of a fluid. It is identical in concept to the PV. Describes dynamic flow only.

Flow index (n, dimensionless)

Describes the numeric relationship between a fluid's shear stress and shear rate on a log or log plot. Describes a fluid's degree of shear-thinning behavior.

Effective viscosity (μ_e, cp)

Describes the viscosity of the fluid flowing through a particular geometry. Fluid flowing through the annulus will have a different effective viscosity to that flowing inside the drill pipe and, hence, different values for n and K.

Note: There are a number of different sets of equations used to define hydraulic parameters dependent on the rheological model selected. The different sets of equations are valid for fluid behavior in laminar and turbulent flow and differ only in their approach. Note that the examples shown below use the API method of calculation for Power Law fluids. The equations use the 600 and 300 rpm viscometer readings, which can be back-calculated from the plastic viscosity and yield point values as follows:

$\theta 300$ = Plastic viscosity + yield point
$\theta 600$ = Yield point + $\theta 300$
$\theta 3$ = 10 second gel (using a hand-crank viscometer)
$\theta 3$ = $\theta 3$ (using a FANN 6-speed viscometer)

Laminar Flow

Mud particles move in straight lines parallel to the pipe or borehole walls (fig. 7.1). Adjacent mud layers have distinct shear planes between them and move past each other with no intermixing. Particles nearest the walls are effectively stationary but, towards the center, move progressively faster as the effects of friction are reduced. Hence, fluid moving under laminar flow conditions has a conical velocity profile.

Turbulent Flow

In contrast to laminar flow conditions, the velocity profile of turbulent flow is almost flat. Turbulent flow occurs at higher flow rates when the shear planes between the mud layers are no longer discrete and the mud flows in a chaotic fashion.

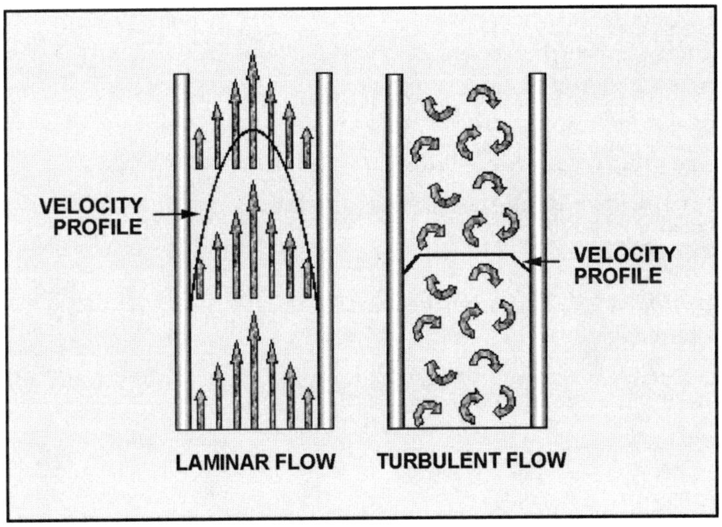

Figure 7.1 Velocity profiles

Pipe Flow Calculations

K and n

High shear rate n and K values can be back-calculated from the 600 and 300 rpm readings and are used for calculations inside the drill pipe.

$$n_p = 3.32 \times \log\left(\frac{\theta 600}{\theta 300}\right)$$

$$K_p = \frac{511 \times \theta 300}{511^{n_p}}$$

where

n_p = flow index in the drill pipe
K_p = consistency factor in the drill pipe, eq cp.

Fluid Velocity

$$V_p = \frac{0.408 \times Q}{ID^2_{DP}}$$

where

V_p = velocity of fluid inside the drill pipe, ft/sec
Q = flow rate, gpm
ID_{DP} = inside diameter of drill pipe or drill collars, in.

Effective Viscosity

$$\mu_{ep} = 100 \times K_p \times \left(\frac{96 \times V_p}{ID^2_{DP}}\right)^{n_p - 1} \times \left(\frac{(3n_p + 1)}{4n_p}\right)^{n_p}$$

where

μ_{ep} = effective viscosity inside the drill pipe, cp.

Reynolds Number

$$Re_p = \frac{928 \times ID_{DP} \times V_p \times W}{\mu_{ep}}$$

where

Re_p = Reynolds number for fluid in the drill pipe or drill collars
W = mud density, ppg.

Annular Flow Calculations

K and n

Low shear n and K values can be back-calculated from the 100 and 3 rpm readings and are used for annular calculations.

$$n_a = 0.657 \times \log\left(\frac{\theta 100}{\theta 3}\right)$$

$$K_a = \frac{511 \times \theta 3}{5.11^{n_a}}$$

where

n_a = flow index in the annulus
K_a = consistency factor in the annulus, eq cp.

Fluid Velocity

$$V_a = \frac{0.408 \times Q}{ID^2_{HOLE} - OD^2_{DP}}$$

where

V_a = velocity of fluid in the annulus, ft/sec
Q = flow rate, gpm
ID_{HOLE} = hole diameter or inside diameter of casing, in.
OD_{DP} = outside diameter of drill pipe or drill collars, in.

Effective Viscosity

$$\mu_{ep} = 100 \times K_p \times \left(\frac{144 \times V_p}{ID_{HOLE} - OD_{DP}}\right)^{n_p - 1} \times \left(\frac{(2n_p + 1)}{3n_p}\right)^{n_p}$$

where

μ_{ep} = effective viscosity inside the drill pipe, cp.

Reynolds Number

$$Re_p = \frac{928 \times (ID_{HOLE} - OD_{DP}) \times V_p \times W}{\mu_{ep}}$$

where

Re_p = Reynolds number for fluid in the drill pipe or drill collars
W = mud density, ppg.

Flow Regime and Critical Reynolds Number (Annulus)

To determine the flow regime, the value obtained for the Reynolds number is compared with the following figures, based on experiments conducted by observing the flow of a Newtonian fluid (in this case water) through a circular pipe:

Re	Flow Regime
< 2000	Laminar flow
2000–4000	Transitional flow
>4000	Turbulent flow

However, as drilling fluids do not conform to the properties of a true Newtonian fluid, the equations below have been developed to determine the critical Reynolds number at which the flow regime changes.

Laminar flow: $Re_c = 3{,}470 - 1{,}370n$
Transitional flow: $3{,}470 - 1{,}370n < Re_c < 4{,}270 - 1370n$
Turbulent flow: $Re_c = 4{,}270 - 1{,}370n$

where

Re_c = critical Reynolds number in an annular section
n = Power Law constant for the annulus.

Critical Flow Rate

Having derived the value of the critical Reynolds number, it is then possible to calculate the critical velocity, and hence the critical flow rate, beyond which laminar flow would begin to break down.

$$Vc = \left[\frac{(3{,}470 - 1{,}370\,n)\,(100)\,K\left(\frac{2n+1}{3n}\right)^{n}}{928\,W\,(ID_{HOLE} - OD_{DP})\left(\frac{144}{ID_{HOLE} - OD_{DP}}\right)^{1-n}}\right]^{\frac{1}{2-n}}$$

where

V_c = critical annular velocity
n = Power Law constant for annulus

W = mud density, ppg
ID_{HOLE} = hole diameter or ID of casing, in.
OD_{DP} = outside diameter of either the pipe or collars, in.

The critical flow rate can then be calculated:

$$Q_c = 2.45\, V_c\, (ID^2_{HOLE} - OD^2_{DP})$$

where

Q_c = critical flow rate, gpm
V_c = critical annular velocity, ft/sec
ID_{HOLE} = hole diameter or ID of casing, in.
OD_{DP} = outside diameter of either the pipe or collars, in.

System Pressure Losses

The standpipe pressure reading is a sum of all of the pressure losses within the system. Most such pressure losses represent wasted kinetic energy used in overcoming friction as the fluid circulates through restrictions in the system. They are sometimes referred to as being parasitic pressure losses. The pressure drop at the bit is, however, considered to be a useful loss of energy.

$$P_{total} = P_{surface} + P_{drill\ string} + P_{annulus} + P_{bit}$$

$$P_{drill\ string} = P_{drill\ pipe} + P_{collars} + P_{downhole\ tools}$$

Pressure losses are functions of circulation rate, mud density, viscosity, pipe and hole diameter, and system length. The maximum amount of friction loss that can be overcome is governed by surface equipment limitations. The general objective of optimizing hydraulics is to minimize system losses and maximize cleaning power at the bit. The table below illustrates the effect of changing certain variables within the hydraulic system.

Variables	System Losses	Horsepower at Bit
Lower flow rate	Decrease	Decrease
Larger flow area (e.g., pipe ID)	Decrease	Increase
Increased system length	Increase	Decrease
Lower mud weight	Decrease	Increase
Lower viscosity	Decrease	
Larger bit nozzle area	Decrease	Decrease

The general procedure for calculating system pressure losses is:

1. Determine the fluid velocity (or Reynolds number) at the point of interest.
2. Calculate the critical velocity (or Reynolds number) to determine flow regime.
3. Choose the appropriate pressure loss equation (dependent on choice of rheological model and flow regime).

Pressure Loss Equations

As has been previously noted, the choice of rheological model and the flow regime of the drilling fluid determine which set of pressure drop equations should be used in order to calculate system pressure losses.

For a Power Law fluid, the pressure lost in an interval is given as:

$$P_{Loss} = \frac{f \times V_2 \times W \times L}{25.81 \times D_{eff}}$$

where

P_{Loss} = pressure loss in the interval, psi
f = friction factor, dimensionless
V = either fluid velocity in the annulus or in the drill pipe, ft/sec
W = mud density, ppg
L = interval length, ft
D_{eff} = effective diameter, in.
D_{eff} drill pipe = ID_{DP}
D_{eff} annulus = $ID_{HOLE} - OD_{DP}$.

The friction factor used in this equation depends on the flow regime.

For turbulent flow:

$$f = \frac{a}{Re_b}$$

where

$$a = \frac{\log n + 3.393}{50}$$

$$b = \frac{1.75 - \log n}{7}.$$

For laminar flow:

$$f = \frac{16}{Re}.$$

Optimizing System Hydraulics

No matter which optimization process is used (the jet impact force or the hydraulic horsepower at the bit model), it is important to remember that the maximum jet impact force and the hydraulic horsepower conditions are only valid for a fixed depth. However, deviations of as much 20% from the ideal 65% or 49% of total surface pressure do not change the hydraulic horsepower or jet impact force at the bit significantly. This allows near maximum jet impact force or bit hydraulic horsepower to be maintained over a wide depth interval without having to change nozzle diameter. It is generally, however, best to optimize hydraulics for the end of the bit run to provide optimum bottomhole cleaning where it is most required.

Basic System

The most basic hydraulic system would consist only of drill pipe, collars, and a bit. The simple optimization process would include the following steps:

- Choose equipment to keep system losses at a minimum for the anticipated range of flow rates and depth.
- Decide which optimization method to use for bottomhole cleaning. For example, selecting the jet impact force model for the shallow part of the hole and the hydraulic horsepower method for greater depths.
- Determine optimum flow rate for the system at the end of the bit run.
- Adjust flow rate to meet minimum requirements or maximum constraints.

System with Downhole Tools

Most hydraulic systems will include an array of downhole tools such as mud motors, turbines, MWD or LWD sensors, and the like. These render the process of hydraulic optimization rather more complicated than in the basic case as they cannot be modeled by a set of simple equations.

There are two ways to optimize a system with downhole tools. The most straightforward method is to subtract the expected pressure losses for the downhole tools from the pump pressure. This information should be provided by the product supplier. The rest of the system is then optimized using the lower value for pump pressure. One of the drawbacks of this method is that it is difficult to predict what the pressure drop across the tools will be as the flow rate has not yet been determined.

The second method is to optimize the system as if there were no tools in the drill string. The pressure required for the tools is then subtracted from the available bit pressure. Results from this method give a higher flow rate with a corresponding larger nozzle area than method 1. As the flow rate is already known, this method gives an accurate pressure drop for the downhole tools. Both methods, however, provide maximum hole cleaning.

Optimizing Bit Hydraulics

The purpose of optimizing bit hydraulics is to provide maximum hole cleaning and optimum penetration rates with the minimum amount of horsepower.

Bit hydraulics will generally be optimized based on either jet impact force or hydraulic horsepower at the bit. The suitability of each method for a particular bit run will largely depend on previous experience in the area, borehole depth, and downhole conditions. If offset bit performance data is somewhat limited, and no conclusions can be drawn, then hydraulics should initially be optimized on hydraulic horsepower. As a general rule though, in the shallow, larger diameter hole sections where penetration rates are high and high volumes of cuttings are generated, higher flow rates are beneficial for effective hole cleaning. Optimizing hydraulics on jet impact force will provide 19.5% higher flow rates than the maximum hydraulic horsepower method. In the deeper, smaller diameter sections where penetration rates are lower and static and dynamic chip hold-down forces become the major hydraulic concerns, higher jet velocities and bottomhole pressure become critical. Under these conditions, maximizing hydraulic horsepower provides 14.3% higher jet velocities and 34.7% higher pressures than the jet impact force method.

Bit Pressure Drop

The pressure drop across a bit is defined as the difference between the pressure of the mud exiting the nozzles and the pressure of the fluid within the drill string immediately prior to entering the bit. If the bit pressure drop is extremely high for a given flow rate and mud weight, the fluid exiting the nozzles has a correspondingly high velocity. A lower pressure drop, on the other hand, under the same conditions of flow and mud weight, will result in fluid exiting the nozzles with lower velocity. Pressure drop is dependent on flow rate, mud weight, and the bit total flow area.

$$P_{Bit} = \frac{{}^{2} \times W}{12{,}031 \times C^{2}_{n} \times TFA^{2}}$$

where

P_{Bit} = pressure drop at the bit, psi
Q = flow rate, gpm
W = mud density, ppg
C_n = nozzle discharge coefficient, usually taken as being 0.95
TFA = total flow area, in.^{2}

Optimizing Hydraulic Horsepower at the Bit

This method assumes that cuttings removal depends on the amount of fluid energy dissipated at the bit. Optimum penetration rates will therefore be achieved when hydraulic *horsepower per square inch* (*HSI*) at the bit is maximized.

HSI provides a measure of the hydraulic power consumed at the bit and is a function of flow rate and bit pressure drop, as well as hole diameter, and will, therefore, increase as the flow rate is increased. However, as flow rate becomes higher, the *TFA* will eventually need to be increased to maintain a suitable pressure drop, in which case the *HSI* will once again fall.

Hydraulic horsepower per unit bit area is given by:

$$HSI = \frac{Q \times P_{Bit}}{1{,}714 \times A}$$

where

HSI = horsepower per square inch
P_{Bit} = bit pressure drop, psi
Q = flow rate, gpm
A = hole area ($\pi/4 \times$ hole diameter^2), in.^{2}

***HSI* is considered to be at a maximum when the pressure drop across the bit is 65% of the total surface pressure.**

Optimizing Jet Impact Force

This method of optimization makes the assumption that hole cleaning is maximized when the drilling fluid impacts the formation at maximum force. This method finds most applications in larger diameter, upperhole section formations are softer and cuttings removal benefits from the 'jetting' action.

The *jet impact force* (*JIF*) is the force which is exerted by the fluid exiting the nozzles when the bit is on bottom. It is a function of jet velocity, mud density, and flow rate. Jet velocity is the governing parameter in this method of optimizing bit hydraulics. The higher the jet velocity the better the hole cleaning effect. The accepted minimum value for optimized bottomhole cleaning is approximately 350 ft/sec.

$$JIF = \frac{Q \times W \times V_n}{1{,}932}$$

where

Q = flow rate, gpm
W = mud density, ppg
V_n = jet velocity, ft/sec

$$V_n = \frac{Q \times 0.32}{TFA}.$$

Jet impact force is maximized when the pressure drop across the bit is 49% of the total surface pressure.

ANNULAR HYDRAULICS AND HOLE CLEANING

The ability of a drilling fluid to lift cuttings is affected by many factors including fluid density and rheology, annulus size and eccentricity, annular velocity and flow regime, pipe rotation and cuttings density, and particle size and shape.

The relationship between the various parameters is complicated and no one theory or set of equations can satisfactorily combine all of the observed phenomena. Nevertheless, the monitoring of cuttings generation and transport rates is imperative for a successful drilling operation.

General Factors Affecting Hole Cleaning

- *Inclination*
  Vertical and near-vertical wells—in holes with an inclination less than 30°, cuttings are effectively suspended by fluid shear, and cuttings beds do not form. Hole cleaning is generally not problematic, providing that mud rheology is adequate.

Deviated wells (inclination greater than 30°)—cuttings tend to settle on the low side of the hole forming cuttings beds. These may either migrate uphole or slide downhole resulting in the annulus packing off.

- *Rheology*
  Laminar flow conditions—increasing mud viscosity improves hole cleaning. Particularly effective if low shear rheology and YP/PV ratio are high.

  Turbulent flow conditions—reducing viscosity will help remove cuttings.

- *Yield Stress*
  A measure of the low shear properties of a mud system. Yield stress governs the size of cuttings which can be dynamically suspended.

- *Mud Density*
  Mud density affects the buoyancy of drilled cuttings. A heavier mud system enables cuttings to 'float' more easily.

- *Flow Rate*
  In highly deviated holes, flow rate combined with mechanical agitation are important factors for effective hole cleaning. In vertical wells, increasing annular velocity and/or increased rheological properties will improve hole cleaning.

Cuttings Slip Velocity

The cuttings slip velocity is the velocity at which a drilled cutting will fall through the mud column under the influence of gravity. In the simplest case, in order to effectively remove drilled cuttings, the velocity of the fluid in the annulus must exceed the cuttings slip velocity. The situation is made more complicated by flow conditions and friction along with the many other factors mentioned above. For instance, under laminar flow conditions, particles in the center of the mud column will be moving at a velocity greater than the average annular velocity, and will be recovered at surface more quickly than expected. However, frictional forces between the borehole/casing wall and the fluid means that drilled cuttings in that region will be moving upwards at a rate less than the average annular velocity. In contrast, during turbulent flow conditions, providing that the fluid velocity exceeds the particle slip velocity, then solids will be removed continuously in all parts of the annulus. Turbulent flow, therefore, provides better hole cleaning than laminar flow, but is less desirable because of the increased chances of erosion.

As a general guide, it is recommended that slip velocity should be less than half of the annular velocity (averaged over the cross section).

As already noted, the relationship between the many factors affecting the rate of cuttings slip is complicated and researchers have developed a number of different methods of estimating its value, none of which are considered to be definitive. The most comprehensive methods are based on particle Reynolds numbers, drag coefficients, particle density, shape and size, and mud density and rheology. The following example, based on a correlation devised by Walker and Mayes, 1975, is a simplified method.

<u>Step 1</u>: Find the shear stress developed by the particle.

$$\tau_p = 7.9 \times \sqrt{T \times (20.8 - W)}$$

where

τ_p = particle shear stress, lb/100 ft^2
T = particle thickness, in
W = mud density, ppg.

The following table provides an approximation for the thickness and diameters of disk-shaped particles.

Expected Penetration Rate (ft/hr)	Particle Thickness (inches)	Particle diameter (inches)
≥ 60	0.3	0.6
30–60	0.2	0.3
15-30	0.1	0.4
≤ 15	0.1	0.3

Step 2: Determine the boundary shear rate.

The boundary shear rate is like a critical shear rate. Particle shear rates above this value are treated with calculations for the turbulent condition. Shear rates below use the laminar calculations. The turbulent or laminar condition of the particle is not related to the turbulent or laminar flow condition of the fluid in the annulus.

$$\gamma_b = \frac{186}{d_p \times W}$$

where

γ_b = boundary shear rate, sec^{-1}
d_p = particle diameter, in.
W = mud density, ppg.

Step 3: Find the shear rate developed by the particle using the laminar Power Law constants (n_a and K_a) for the mud.

$$\gamma_p = \frac{\tau_p^{\frac{1}{n_a}}}{K_a}$$

where

γ_p = particle shear rate, sec^{-1}
τ_p = particle shear stress from step 1, lb/100 ft^2
n_a = flow index in the annulus
K_a = consistency factor in the annulus, eq cp.

Step 4: Determine the slip velocity for the laminar or turbulent condition.

Laminar Condition

If $\gamma_p < \gamma_b$, the slip velocity is determined by:

$$V_s = 1.22 \times \tau_p \times \sqrt{\frac{\gamma_p \times d_p}{W}}$$

where

V_s = slip velocity, ft/min
γ_p = particle shear rate, sec^{-1}
τ_p = particle shear stress from step 1, lb/100 ft^2
d_p = particle diameter, in.
W = mud density, ppg.

Turbulent Condition

If $\gamma_p > \gamma_b$, the slip velocity is determined by:

$$V_s = \frac{16.62 \times \tau_p}{\sqrt{W}}$$

where

V_s = slip velocity, ft/min
τ_p = particle shear stress from step 1, lb/100 ft^2
W = mud density, ppg.

Cuttings Transport Velocity

The cuttings transport velocity for each different hole geometry is obtained by subtracting the slip velocity of the cuttings from the annular velocity in that particular section.

$$V_t = V_a \times V_s$$

where

V_t = cuttings transport, ft/min
V_a = annular velocity, ft/min
V_s = slip velocity, ft/min.

Cuttings Transport Efficiency

Perhaps more important than the actual cuttings transport velocity is the cuttings transport efficiency. This is simply the ratio of cuttings transport to annular velocity. Please note that the equation shown here is valid for hole angles less than 30°. Evaluation of hole cleaning for wells of higher inclination is much more complicated as the drilled cuttings may form a cuttings bed on the low side of the hole.

$$E_t = \frac{V_t}{V_a} \times 100$$

where

E_t = transport efficiency, %
V_t = cuttings transport, ft/min
V_a = annular velocity, ft/min.

Cuttings Concentration

Note: valid for hole angles less than 30°.

The concentration of cuttings in the annulus depends on the transport efficiency, as well as the volumetric flow rate and the rate at which cuttings are generated at the bit (*ROP* and hole size dependent). Experience has shown that cuttings concentration in excess of four or five volume % can lead to pack-off, tight hole, or stuck pipe incidents. When drilling soft formations, the cuttings concentration may easily exceed 5% by volume if penetration rates are not controlled. Some operators recommend a maximum cuttings concentration of 4% by volume.

Cuttings concentration is given by:

$$C_a = \frac{ROP \times ID_{HOLE}}{14.71 \times E_t \times Q} \times 100$$

where

C_a = cuttings concentration, vol %
ROP = rate of penetration, ft/hr

ID_{HOLE} = hole diameter, in.
E_t = transport efficiency, %
Q = flow rate, gpm.

When cuttings concentration exceeds 4 or 5% by volume, the effect on hydrostatic pressure and equivalent circulating density can be substantial. The change in hydrostatic pressure depends on the density of the cuttings as well as their concentration in that particular hole section.

The effective static mud density due to the cuttings concentration in that section of hole is given by:

$$W_e = \left(SG_c \times 8.34 \times \frac{C_a}{100}\right) + \frac{W(1 - C_a)}{100}$$

where

W_e = effective mud weight, ppg
SG_c = specific gravity of the cuttings
C_a = cuttings concentration, vol %
W = mud density, ppg.

The effect is most pronounced when drilling top hole sections. The following conditions cause an increase in cuttings concentration:

- large diameter holes drilled at high ROP
- pumps unable to provide sufficient annular velocities
- rapid mud system building rate may yield insufficient viscosity

It is clear that an increase in cuttings concentration in the annulus results in a corresponding increase in effective mud density.

Equivalent Circulating Density (ECD)

When the drilling fluid is circulating through the wellbore, the circulating pressure must be sufficient to overcome not only the friction losses through the drill string and bit, but also the hydrostatic pressure of the fluid in the annulus and the friction losses through the annulus. The pressure required to overcome the total friction losses in the annulus, when added to the hydrostatic pressure of the fluid, gives the *equivalent circulating density* (*ECD*) as follows:

$$ECD = \frac{\sum P_a + W}{0.052 \times TVD}$$

where

ECD = equivalent circulating density, ppg
$\sum P_a$ = sum of friction pressure losses in the annulus (corresponds to the pump pressure minus the pressure losses through the surface equipment, drill string, and bit), psi
TVD = true vertical depth of hole, ft
W = mud density, ppg.

The majority of drilling situations may not be limited by frictional *ECD*. Exceptions are in the case of drilling slim-hole wells. *ECD* is particularly aggravated by deep, slim holes using heavy mud weights close to the formation fracture pressure. The flow rates selected may be lowered to prevent loss of circulation.

Equivalent Circulating Density (ECD) with Cuttings

A complete description of the extra pressure to which the open hole is exposed combines the effect of friction (*ECD*) and the effect of cuttings loading (effective mud density).

$$ECD + \text{Cuttings} = \frac{\sum P_a}{0.052 + TVD} + \left(SG_c \times 8.34 \times \frac{C_a}{100}\right) + \frac{W(1 - C_a)}{100}$$

where

ECD = equivalent circulating density, ppg
$\sum P_a$ = sum of friction pressure losses in the annulus, psi
TVD = true vertical depth of hole, ft
SG_c = specific gravity of the cuttings
C_a = cuttings concentration, vol %
W = mud density, ppg.

HOLE CLEANING GUIDELINES

Guidelines for Vertical Holes

- Rheology is very important in transporting cuttings in vertical or deviated holes. Large diameter holes, in particular, cannot be cleaned by velocity alone. Providing that the mud rheology is correct, hole cleaning is not generally a problem in such wells. Annular velocity is generally greater than the cuttings slip velocity and so cuttings are effectively removed from the hole. To ensure that a low slip velocity is achieved, these wells are usually drilled with viscous, high-yield point mud systems.

  In vertical wells (fig. 7.2), annular velocity is generally much higher than the cuttings slip velocity and drilled cuttings are effectively removed from the annulus.

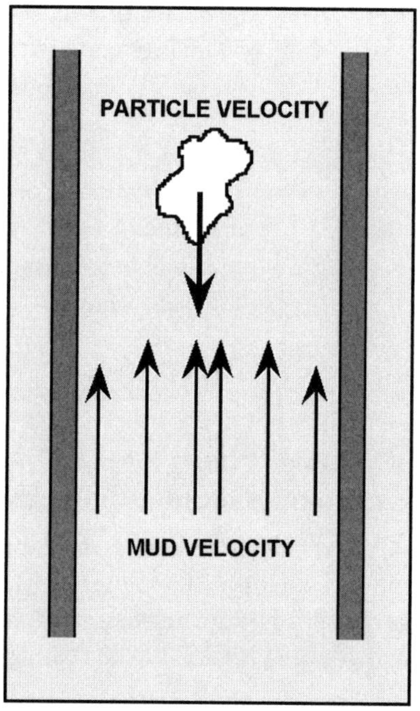

Figure 7.2 Vertical well

- Circulate at least 1.3 × bottoms up for vertical wells. Monitor returns at the shakers ensuring the rate of return has decreased to acceptable levels before tripping.
- Limit the use of high viscosity pills. Instead, adjust the properties of the active system to provide optimum hole cleaning. High weight pills should not be used in vertical holes.
- Reciprocate rather than rotate the pipe prior to tripping. This helps remove cuttings from stagnant zones near the wellbore wall.
- Pull through tight spots only if the pipe is free going down. Agree a maximum allowable overpull in advance and work up progressively toward it, ensuring that the pipe is free to move downward on each occasion. Stop and circulate if overpulls become excessive.
- Only back ream if absolutely necessary. Back reaming may result in hole pack-off and stuck pipe. It can also mask the onset of potentially serious hole problems, which may have been detected at a much earlier stage had back reaming not taken place.

Guidelines for Deviated and Extended-Reach Wells

The measures noted above are also valid for the effective cleaning of deviated wells. However, the geometry of deviated and extended-reach holes makes adequate hole cleaning rather more difficult than for vertical or near-vertical wells. Many of the problems encountered are associated with the nature of the cuttings beds, which form on the low side of the hole.

In deviated holes (fig. 7.3), cuttings will 'slip' to the lower side of the wellbore. In this situation, the velocity of the drilling fluid has to be higher in order to keep the cuttings moving up toward the surface.

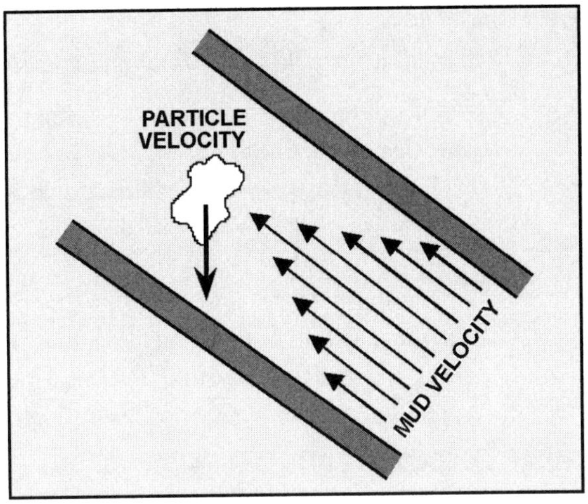

Figure 7.3 Directional well

Low-Angle Sections (10°–40°)

Cuttings Bed Characteristics

Cuttings beds form on the low side of the hole and are subject to 'particle recycling', which has a detrimental effect on hole cleaning (fig. 7.4).

1. Because of increasing inclination, the cutting is forced toward the low side of the annulus, where it travels downward due to a lack of lifting force in the flow (low velocity near the wall).
2. At some point, due to a higher shear stress, the cutting is lifted and reenters the high-velocity region at the middle of the annulus.
3. Then, it is swept upward and continues to travel until
4. its tendency to drop overcomes the lifting force in the flow and it is forced toward the low side of the annulus again.

This process can be repeated many times resulting in the cutting's shape being altered through grinding. Measures used to minimize this problem include viscous sweeps.

Compared with vertical sections, low-angle sections require higher annular velocities for efficient hole cleaning. For this type of well section, laminar flow dominates cuttings transport. Therefore, viscosity, yield point, and initial gel strengths have a significant effect on annular cuttings concentration and hole cleaning efficiency.

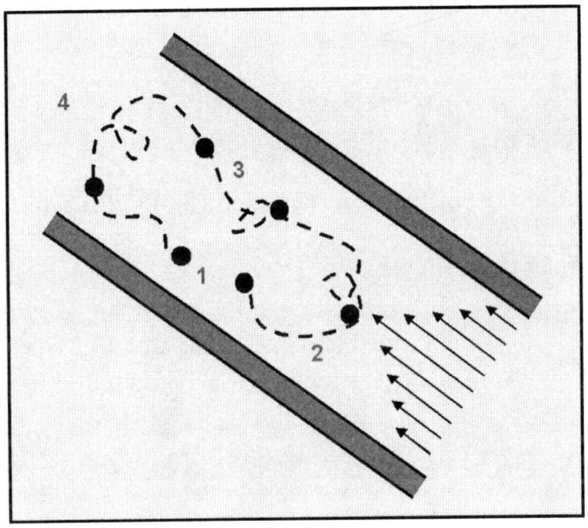

Figure 7.4 Particle recycling

Field Guidelines

- General use of laminar flow.
- Maintain a high yield point and gel strength to reduce the settling of cuttings when pumps are off.
- Always ensure that the hole is clean before turning the pumps off.
- Maximize YP/PV ratio.
- Use viscous sweeps to reduce the effects of particle recycling.

Critical-Angle Sections (40°–60°)

Cuttings Bed Characteristics

In well sections with inclinations greater than 40°, cuttings do not recycle as readily as in lower-angle sections. This is because gravity tends to hold them down on the low side of the hole.

Well sections with inclinations between 40° and 60° are considered critical, not only because a cuttings bed develops, but also because it is unstable and prone to sliding downward (avalanching).

The consequence of avalanching is an instantaneous buildup of cuttings around the drill pipe and/or BHA which, if not treated properly, can result in stuck pipe. Also,

- Turbulent flow exhibits a desirable, eroding effect on cuttings beds.
- Pipe movement (rotation/reciprocation) mechanically disturbs cuttings beds.

Field Guidelines

In the well sections over 40°, attention must focus on minimizing cuttings beds.

- For the range of intermediate inclinations (40°–60°), turbulent flow is recommended. Since cuttings transport in turbulent flow is not affected by rheological properties, lower mud parameters (i.e., YP, PV) may be used. However, static mud parameters, such as gel strength, are usually desirable even if turbulent flow is preferable. If turbulent flow cannot be used because of other adverse factors, like wellbore instability, annular velocity should be kept as high as possible.
- Rotating and/or reciprocating drill pipe has a mechanical, destructive influence on the cuttings bed (this influence is the main factor that provides a higher cleaning rate for higher rpm). As pipe rotation is typically governed by directional drilling needs, periodic wiper trips should be considered.
- Combination sweeps (low viscosity/high density) are effective at eroding the cuttings bed, and carrying the cuttings to the surface.

High-Angle Sections (>60°)

Cuttings Bed Characteristics

At high angles of inclination, the formation of a cuttings bed is almost instantaneous, and its thickness is governed primarily by annular velocity. The cuttings bed that forms at angles greater than 60° is stable. This means it will not avalanche. The waving, vortex-like character of turbulent flow has a destructive influence on the bed being formed. There is a tendency for cuttings to be withdrawn (lifted) from the bed and displaced upward in the annulus, where such a process may occur again. This kind of interaction, together with the flat velocity profile typical of turbulent flow, leads to better hole cleaning.

Pipe movement (rotation/reciprocation) mechanically disturbs cuttings beds.

Field Guidelines

- Turbulent flow is preferable in high-angle wells. Generally, the same recommendations as those described for intermediate-angle wells are applicable in this region for turbulent flow. However, the requirements to ensure a mud gel strength are less important. If turbulent flow cannot be achieved, the YP/PV ratio should be maintained as high as possible.
- Again, rotating and/or reciprocating drill pipe has a mechanical, destructive influence on the cuttings bed. It is this influence that provides a higher cleaning rate for higher rpm, particularly at high inclinations where a considerable cuttings bed is formed. Pipe rotation >120 rpm is highly recommended. Periodic wiper trips should be used if drill pipe rotation is restricted. The use of rotary steerable directional tools may prove helpful.
- Combination sweeps (low viscosity/high density) are effective at eroding the cuttings bed, and carrying the cuttings to the surface. The effectiveness of a cuttings-bed-maintenance program can be determined through several indicators.

Poor Hole Cleaning Indicators

A number of rig-site indicators can be used as a guide to how effectively the hole is being cleaned. These include:

- Shape and size of cuttings at the shakers—small, well-rounded cuttings may indicate extended periods of regrinding downhole—an indication of the presence of cuttings beds.
- Rate of return of cuttings versus expected volume.
- Increased torque and drag.
- High pickup weight.
- Poor weight transfer. A higher than normal surface weight is required to get a pressure-drop response from the mud motor.
- Difficulty orienting the motor, due to excessive friction between the cuttings and the drill string.
- Excessively ground cuttings, due to extended particle recycling and drill pipe interaction with the cuttings bed.

Effects of Mud Type on Hole Cleaning Efficiency

- In highly deviated wells, and for lower values of yield point and plastic viscosity, cleaning performance for both mud types is roughly the same. However, at higher values of yield point and plastic viscosity, water-base muds provide better cleaning. The general observation is that an increase in mud-yield point and plastic viscosity results in increased cuttings concentration for both muds.
- As a result from this higher cuttings concentration, torque requirements for both muds increase with increasing yield point and plastic viscosity at higher hole inclinations.
- Hole-cleaning performance of oil-base muds at critical angles (40° to 60°) is reduced by severe cuttings bed avalanching (due to reduced friction).

Hole Cleaning 'Aids'

Mud rheology, density, annular velocity, and pipe movement are seen as being key to the success of efficiently cleaning the wellbore. The importance of sound operational practices and vigilant monitoring of the hole condition cannot be underestimated. There are, however, tools available which, when used appropriately, can help to predict, and possibly alleviate, the onset of hole cleaning problems.

- *Hole Cleaning Software Packages*
  Can be used to predict the likelihood and potential location of the buildup of any cuttings beds. Should be used with caution, making sure that results are combined with past experience and offset data.
- *Cuttings Bed Impellers/'Enhanced Performance' Drill Pipe/Agitators*
  Lift the cuttings away from the low side of the hole into the area of higher annular velocity but, unless mud rheology and other conditions are sufficient, they are likely to drop out of suspension and form beds higher up the hole.
- *'Pressure While Drilling' Tools*
  In recent years, tools have been developed which enable downhole pressure measurements to be taken in real time. The 'Pressure While Drilling' (PWD) tools give an indication of any changes in equivalent circulating density. PWD tools have a number of potential applications including:
  - LOT
  - Lost circulation detection
  - Flow/kick detection and monitoring
  - Swab/surge information
  - Mud property monitoring
  - Measurements of differential pressure for overall drilling performance optimization
  - Underbalanced drilling
  - Hole cleaning and stability monitoring

By accurately measuring ECD while drilling, these tools allow engineers to assess the condition of the hole before serious hole cleaning problems occur.

APPENDIX 1
TFA Chart

$$TFA = \frac{n \times \pi \times D^2}{4}$$

where

n = number of nozzles
D = nozzle diameter in 32nds of an inch.

Total Flow Area (TFA) of Standard Nozzles (in.²)										
Nozzle Size (in.)	Number of Nozzles									
	1	2	3	4	5	6	7	8	9	10
7/32	0.038	0.075	0.113	0.150	0.188	0.225	0.263	0.301	0.338	0.376
8/32	0.049	0.098	0.147	0.196	0.245	0.295	0.334	0.393	0.442	0.491
9/32	0.062	0.124	0.186	0.249	0.311	0.373	0.435	0.497	0.559	0.621
10/32	0.077	0.153	0.230	0.307	0.383	0.460	0.537	0.614	0.690	0.767
11/32	0.093	0.186	0.278	0.371	0.464	0.557	0.650	0.742	0.835	0.928
12/32	0.110	0.221	0.331	0.442	0.552	0.663	0.773	0.884	0.994	1.104
13/32	0.130	0.259	0.389	0.518	0.648	0.778	0.907	1.037	1.167	1.296
14/32	0.150	0.301	0.451	0.601	0.752	0.902	1.052	1.203	1.353	1.503
15/32	0.173	0.345	0.518	0.690	0.863	1.035	1.208	1.381	1.553	1.726
16/32	0.196	0.393	0.589	0.785	0.982	1.178	1.374	1.571	1.767	1.963
17/32	0.222	0.443	0.665	0.887	1.108	1.330	1.552	1.773	1.995	2.217
18/32	0.249	0.497	0.746	0.994	1.243	1.491	1.740	1.988	2.237	2.485
19/32	0.277	0.554	0.831	1.108	1.384	1.661	1.938	2.215	2.492	2.769
20/32	0.307	0.614	0.920	1.227	1.534	1.841	2.148	2.454	2.761	3.068
22/32	0.371	0.742	1.114	1.485	1.856	2.227	2.599	2.970	3.341	3.712
24/32	0.442	0.884	1.325	1.767	2.209	2.651	3.093	3.534	3.976	4.418
26/32	0.519	1.037	1.556	2.074	2.593	3.111	3.630	4.148	4.667	5.185
28/32	0.601	1.203	1.804	2.405	3.007	3.608	4.209	4.811	5.412	6.013

APPENDIX 2
Rheological Models

INTRODUCTION

A rheological model is a description of the relationship between the *shear stress* (τ) experienced by the fluid and the *shear rate* (γ). Because the rheology of the drilling fluid directly affects the circulating system pressure losses, the more accurately the rheological models used to describe the drilling fluid represent the fluid, the more precise the hydraulic analysis can be.

Rheological Terms

Term	Symbol	Unit	Definition
Shear stress	τ	lb/100 ft²	The force per unit area required to move a fluid at a given shear rate. Measured on a viscometer by the dial deflection at a given shear speed.
Shear rate	γ	sec^{-1}	The change in fluid velocity divided by the channel width through which the fluid is flowing in laminar flow
Shear speed		rpm	The rotational speed of a standard oilfield viscometer
Viscosity	μ	Centipoise	The shear stress divided by the shear rate. For any fluid, can either be measured at a single point or over a range of shear values
Rheogram			Plot of shear stress versus shear rate (see examples later)
Apparent viscosity	μ_a	Centipoise	The viscosity used to describe a fluid flowing through a particular geometry
Yield point	YP τ_y	lb/100 ft²	The force required to initiate flow. The calculated value of the shear stress when the rheogram is extrapolated to the y axis at $\gamma = 0$ sec^{-1}. *Note:* YP is a time independent measurement and is usually associated with Bingham fluids.
Yield stress	τ_o	lb/100 ft²	The force required to initiate flow. The calculated value of the shear stress when the rheogram is extrapolated to the y axis at $\gamma = 0$ sec^{-1}. *Note:* YP is a time independent measurement and is usually associated in the Herschel-Buckley model as τ_o and in the Bingham model as YP. It can also be viewed as a zero times gel strength.
Gel strength		Centipoise	Time dependent measurement of a fluid's shear stress under static conditions. Commonly measured after 10 sec and 10 min intervals.
Plastic viscosity		Centipoise	Contribution to fluid viscosity under dynamic flow conditions. Related to the size, shape, and number of particles in a moving fluid. Calculated from the 600 and 300 rpm speeds using a rheometer.
Flow index	n	None	The numerical relationship between a fluid's shear stress and shear rate on a log/log plot. Describes a fluid's degree of shear thinning behavior. Used to describe pseudo-plastic fluid behavior.
Consistency index	K	eq cps lbs/100 ft² sec^n	The viscosity of a flowing fluid identical in concept to the PV. Viscous effects due to a fluid's yield stress are not part of the consistency index as the latter relates to dynamic flow only. Used to describe pseudo-plastic fluid behavior.

Newtonian Fluids

For a Newtonian fluid the ratio of shear stress to shear rate is a constant. This constant is the viscosity of the fluid (μ).

$$\mu = \frac{\tau}{\gamma}.$$

Typical Newtonian fluids are gases at ambient temperature and pressure and most simple fluids (like water).

Non-Newtonian Fluids

All other fluids are termed non-Newtonian. For these fluids the relationship between shear stress and shear rate can be defined as follows

$$\tau = A + B\gamma^n$$

where A, B, and n are constants depending on the model in use.

The following graph shows the variations in the most common models currently in use.

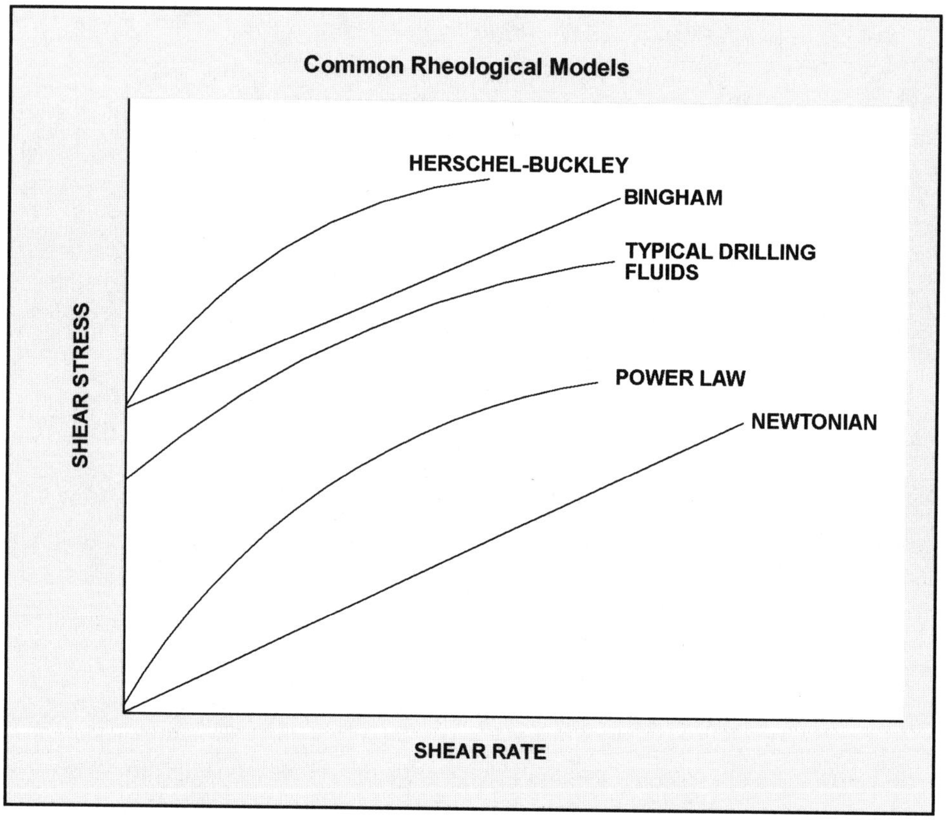

Bingham Plastic

For a Bingham plastic fluid (close approximation to most water-based muds)

$$\tau = YP + (PV \times \gamma)$$

where

$$PV = \theta 600 - \theta 300 \qquad \text{and} \qquad YP = \theta 300 - PV.$$

Power Law

For a Power Law fluid (approximation to oil and pseudo oil muds)

$$\tau = K\gamma^{n}$$

where

$$n = 3.32 \times \log\left(\frac{\theta 600}{\theta 300}\right)$$

$$K = \frac{511 \times \theta 300}{511^{n}}.$$

Herschel-Buckley

For a Herschel-Buckley (or modified Power Law or Yield Power Law) fluid (closer approximation to oil and pseudo-oil muds)

$$\tau = YP + K\gamma^{n}.$$

CHAPTER 8
Drill String Design

DRILL STRING COMPONENTS

The principal components of the drill string are as follows.

Kelly or Top Drive System (TDS)

It is not exactly part of the drill string but transmits and absorbs torque to or from the drill string while carrying all the tensile load of the drill string.

Drill Pipe (DP)

Transmits power by rotating motion from the rig floor to the bit and allows mud circulation.

Drill pipe is subjected to complex stresses and loads as is the rest of the drill string. Drill pipe should never be run in compression or used for bit weight except in high angle and horizontal holes where stability of the string and absence of buckling must be confirmed by using modelling software.

Heavy Weight Drill Pipe (HWDP)

HWDP makes the transition between drill pipe and drill collars, thus avoiding an abrupt change in cross-sectional area. It is also used with drill collars to provide weight on the bit, especially in 6" or 8½" holes where the buckling effect of the HWDP due to compression is minimal.

HWDP reduces the stiffness of the BHA. HWDP is also easier or faster to handle than DC and, more important, reduces the possibility of differential sticking.

Drill Collars (DC)

Provide weight on the bit, keeping the drill pipe section in tension during drilling. The neutral point should be located at the top of the drill collars section: 75 to 85% (maximum) of the drill collars section should be available to be put under compression (available weight on bit).

Other Downhole Tools

Include: stabilizers, crossover, jars, MWD, underreamer, etc.

They all have different functions, but have two major common points: their placement is crucial when designing the drill string and they introduce 'irregularity' in the drill string; i.e., different ID/OD and different mechanical characteristics (torsion/flexion, etc.), which must be taken into account when designing the drill string.

Drill Bit

See Chapter 6, Drill Bits.

DRILL STRING CONSIDERATIONS

Drill Pipe

The main factors involved in the design of a drill pipe string are

- collapse and burst resistance.
- tensile strength (tension).

- torque (torsion).
- resistance against crushing by action of the slips.
- presence of aggressive fluids, e.g., H_2S and CO_2, or resistance to corrosion.

The forces acting on the tubulars of the drill string include:

- Tension, the combined weight of drill collars and drill pipe plus any overpull. An overpull safety margin should be available to pull on a stuck string.
- Torsion, high torque values can be obtained in tight hole conditions. The recommended tool joint makeup torque should be used and not be exceeded.
- Fatigue in corrosive environment.
- Fatigue associated with mechanical notches.
- Cyclic stress fatigue while rotating through crooked holes. Dogleg severity of more than 3°/30 m (3°/100 ft) should be avoided if possible.
- Abrasive friction.
- Vibration at critical rotary speeds.

Different grades of steel are available to meet different hole requirements. The most common are G105 and S135. G105 is most commonly used in shallow or H_2S environments. S135 is considered a standard for offshore operations. U150 is a relatively new grade that is being used for deepwater operations.

Hardfacing (also called hard banding) of tool joints is performed to limit the degree of circumferential wear produced on the tool joint. Hardfacing is proven to be efficient, but it also can provide considerable casing wear, leading to a reduction in casing performance properties.

Care must be exercised in the use of hard-banding materials (generally from tungsten carbide). A smooth hard-banding weld, flush with the outside diameter of the tool joint, should be preferred. The recommended hard banding is ARNCO 200XT or ARMACOR M.

If new hard-banded pipe or pipe that has been recently rehard-banded is being used, every effort should be made to run this pipe in the open hole section. This will result in a degree of roughness being taken off the new surface finish and will minimize any adverse impact on casing wear.

Enhanced performance drill pipe is a 'stabilized' drill pipe that can be used in deviated/horizontal wells to

- help prevent differential sticking.
- reduce torque and drag.
- reduce wall contact and tool joint wear.
- help to disturb cutting beds.

Drill Collars

Drill collars are used to apply weight to the bit. Their large wall thickness gives them a greater resistance to buckling than drill pipe. The neutral point should be located at the top of the drill collar section (never in the drill pipe section); 75 to 85% (maximum) of the drill collar section should be under compression (this will include HWDP if utilized).

The lower part of the drill collar section is under compression, therefore subject to buckling. This generates high stresses and potential fatigue failure, particularly at the connections.

The clearance between drill collars and the wellbore is smaller than with drill pipe, therefore increasing the possibility of differential sticking. If it is thought to be a potential problem, spiral drill collars can be used to reduce the contact area with the wellbore and consequently the chance of differential sticking.

In deviated holes, to avoid sticking of the drilling assembly, the minimum number of drill collars should be used (and these should be of the spiral type). Heavyweight drill pipe, in conjunction with undergauge stabilizers, should be used to substitute the drill collar weight.

Stabilization

Stabilizers

Full-gauge stabilizers provide a fixed standoff distance from the wall of the hole and keep the drill collars concentric with the wellbore, thus reducing the buckling and bending. Stabilizers may, however, increase torque and drag.

Recommended Type of Stabilizer

- The integral blade stabilizer is the preferred type.
- Although integral blade stabilizers are generally preferred, welded-blade stabilizers can be used for conductor and surface hole depending on the formation. Generally, soft formations and, in many cases, above the kickoff point for directional wells.
- Replaceable sleeve stabilizers are to be used only in areas of the world where logistics is a real problem (economical considerations). Their main disadvantage is that they restrict the flow circulation in smaller size holes.

The position, size (full, under, or adjustable gauge stabilizer), and number of the stabilizers in the bottomhole assembly are determined by the directional drilling requirement. In the vertical section their purpose is to maintain the drift angle as low as possible.

Note:

- The near-bit stabilizer may be replaced by a full-size roller reamer if excessive torque is experienced.
- Do not place a stabilizer at the transition from drill collars to HWDP.
- The use of stabilizers inside casing should be avoided as much as possible (or limited to a short period of time); e.g., while drilling out cement.

Roller Reamers

Roller reamers can be used for drill string stabilization where it is difficult to maintain hole gauge and, in hard, deep formations, where torque presents a problem.

Roller reamers do not stabilize as well as integral blade stabilizers. More walk is experienced when they are used, especially if a near-bit roller reamer is used. Used with a building assembly, they often increase the building rate.

The type of cutters will depend on the formation type. The same roller reamer body can be used for different applications.

Jars

Double-acting hydraulic jars are preferred. Jars are generally used from below conductors or surface casing.

The number of drilling hours and jarring hours should be recorded to enable replacement at the recommended time (this must be provided by the manufacturer). This varies, depending on the manufacturer, hole size, size of jar, and deviation.

Jar Position

Run a jar placement program, then optimize for position, considering all aspects of the BHA:

- The location of the neutral point in the drill string should be known and jars kept out of this area.
- When appropriate (see below), place jars in the drill collar section above the top stabilizer. Jars should not be run directly next to a stabilizer (minimum of one collar between them).
- Place a couple of drill collars above the jar for hammer weight where possible. HWDP is flexible and will not transmit a blow downward as well as drill collars.

- The anticipated problem can also influence where to locate the jar:
  1. If it is differential sticking or keyseating, then the jar should be run in the HWDP to avoid being stuck with the rest of the BHA.
  2. If it is stabilizer 'balling-up' or hole swelling, then the jars should be positioned in the collar above the top stabilizer.
  3. When drilling in a new area where common hole problems have not yet been identified, a good compromise is to run some smaller OD spiral drill collars above the jar.
- Jars have a pump open force which must be overcome when cocking the jar.

  pump open force = pressure drop below jars × washpipe area

  The washpipe area can be obtained from the manufacturer's data book.

Accelerators

Accelerators (also called jar boosters) are run in the string above the jars. They are used to increase the impact force exerted by a jar.

They consist of a slip joint that, as extension of the tool occurs, cause further compression of an inert gas (generally nitrogen) in a high-pressure chamber. Then, the gas under pressure forces the tool back to its original length. It allows the drill collars below the booster to move rapidly up the hole.

Accelerators are useful in a fishing string or drilling assembly, particularly in high angle holes where the string is in contact with the side of the hole and large amounts of friction may be developed.

Shock Subs

Shock subs are placed in the drill string, ideally directly above the bit to absorb vibration and shock loads.

They are useful, especially at shallow depth, when drilling hard rocks, broken formations, or intermittent hard and soft streaks to limit the wear and failure of the drill string components (MWD, bit, etc.).

Hole Openers and Underreamers

Hole openers and underreamers are used to enlarge holes. An underreamer is never as robust as a hole opener, but can pass through obstructions (e.g., casing string) of a smaller diameter than the hole it will drill.

Hole Openers

Applications

Use to enlarge a pilot hole, which may have been required for one of the following reasons:

- A core was required, standard coring equipment size start at 12½".
- High quality of wireline log was required which is not likely to be achieved in a big diameter hole.
- It is easier to control the trajectory of a smaller hole, especially in a very soft formation.
- Drilling through what may be a pressure transition zone or a gas pocket. In a small hole, circulation bottoms up take less time and kicks are easier to control due to the reduced volume.

A hole opener may also be required if the diameter of the hole has been reduced by the formation expanding into it so that the full-size bit can no longer pass. It may happen, in particular, in sections containing plastic shales or salt.

Guidelines for Use

A hole opener is run either with a pilot bit or with a bull nose which guides the hole opener along the pilot hole. Thus, there is no need to steer a hole opener and no risk to drill away from the pilot hole. The bull nose can be fitted directly to the hole opener or one joint below to give more flexibility.

As an alternative to a hole opener, especially in hole sizes less than 17½", a common bit may be used to enlarge the hole. This is not recommended in soft formations. In harder formations the bit is more likely to follow the path of least resistance, but it is necessary to measure the deviation of the well at frequent intervals to check that it is following the trajectory of the pilot hole.

The majority of hole openers still use roller cones with either steel teeth or tungsten carbide inserts depending on the formation. These are available from 8⅜" (6" pilot hole) to 48" (17½" pilot hole). The number of cones (from 3 to 8) is a function of the size of the hole.

Fixed-blade hole openers are available for smaller (less than 17½") hole sections. They remove the risk of cones falling off and can cut in an upward direction as well, should this become necessary ('squeezing formation').

While using a hole opener:

- Cutter selection will depend on the formation based on the same consideration as for bits.
- Soft formations will normally respond better to higher rpm and lower WOB, while hard formations require higher WOB and less rpm.
- If fractured formations are encountered, adjust drilling parameters to avoid bouncing.
- Use sufficient flow rate to obtain a good hole cleaning.
- Always stabilize the lower end of the hole opener to prevent it from rotating off center. A rock bit (i.e., if not anticipated to be clean) or a bull nose half an inch to an inch smaller than the pilot hole, should be efficient.

Underreamers

Typical applications include:

- Opening the hole below a casing shoe, to provide a larger annular space for cementing the next casing string. This permits, for example, the use of a larger intermediate casing string diameter than could be used otherwise.
- Overcome BOP or wellhead size diameter restriction.
- Enlarging the hole annulus within the producing zone for gravel pack completion.
- Opening a pocket to start a sidetrack.
- Reducing dogleg severity.
- Enlarging 'heaving areas' through problem fault zone.

Since the underreamer has to pass through a restricted bore, it incorporates expandable cutters which stay collapsed when the tool is *run-in-hole* (*RIH*). The cutters are then expanded into the formation by utilizing the differential pressure of the drilling fluid. Once the hole is underreamed to the desired depth, the pumps are turned off, allowing the arms to collapse back into the body for *pull-out-of-hole* (*POOH*).

Underreamers used to have rolling cones on extending arms but, nowadays, the tendency is to use extending arms fitted with PDC cutters. They can be run with a bull nose or a small drilling bit as for hole openers.

Should limited oversize be required, an alternative would be a bicentered bit (e.g., 8½" × 9⅞") which eliminates the risks associated with the underreamer.

DRILL STRING DESIGN

Objectives

The objective of drill string design is to:

- ensure that the maximum stress at any point in the drill string is less than the downrated yield strength.
- ensure that the components and configuration of the drill string minimize the effects of fatigue.
- provide equipment that is resistant to H_2S, if H_2S is anticipated.

Assumptions

The following assumptions are made:

- In low-angled holes, tension is approximated using the 'buoyed-weight' method. This ignores the effects of circulating pressure and hole angle on tension. Although not as exact as the 'pressure-area' method, any errors are compensated for by selecting an appropriate margin of overpull. *Extended-reach drilling* (*ERD*) and horizontal wells require computer modelling to evaluate torque and drag effects.
- In vertical holes, buckling is assumed to occur up to the point in the string where buoyed string weight equals weight on bit. This is incorrectly termed 'neutral point in tension'. In practice, if pressure-area forces are considered, the actual neutral point will always occur below this point unless either the drill pipe becomes stuck or bit pressure drop is increased with the bit on bottom.
- In inclined holes, buckling is assumed to occur when the compressive load in a component exceeds the component's critical buckling load.
- The tension calculations in the vertical and low-angle holes assume a vertical hanging string; i.e., a worst case with no hole support. If the hole is not vertical, then the design is a conservative one which is meant to offset the higher tensile drag as the hole angle and step-out increase.
- In the ERD designs, tensile drag is ignored for calculations in rotary drilling mode. Errors are small unless rotating very slowly with high penetration rates. Under normal drilling conditions, rotating speed will exceed axial speed.
- Drill string torsional load capacity is fixed at tool joint makeup torque.
- Material yield strength for all components is the specified minimum for the component being considered.
- Drill pipe tube wall thickness is the minimum for the stated drill pipe weight and class.
- Connection torsional strength and makeup torque are calculated using the A.P. Farr formula from API RP 7G.

Design Factors

Design factors are used to downrate the load capacities of components to provide additional margins for error caused by differences between the assumptions made in design and the real world.

Tension (DF_T)

This is used to reduce the drill pipe tensile capacity to establish the maximum allowable tensile load. DF_T is typically 1.15.

Margin of Overpull (MOP)

The desired excess tension above the normal hanging/working load to account for contingencies, such as hole drag, stuck pipe, etc. May be any positive amount but typically specified from 50,000 to 150,000 lb depending on hole conditions.

Excess BHA Weight (DF_{BHA})

Defines the amount of BHA weight in excess of bit weight that a given BHA will contain. This excess weight provides an extra margin to keep the neutral point below the top of the BHA. Recommended DF_{BHA} is 1.15.

Torsion

Applied torsion is limited to tool joint makeup torque. Standard makeup torque is 60% of tool joint torsional yield strength and standard tool joints are weaker in torsion than the tubes to which they are attached. Therefore, a design factor is not required.

Collapse Pressure (DF_C)

Collapse pressure capacities are first downrated to account for the effect of any tension and then further downrated by dividing by the collapse design factor. The DF_C is typically 1.1 to 1.15.

Burst pressure (DF_{BP})

This is used to reduce a component's burst pressure capacity to give a maximum permissible burst pressure that can be applied. Burst capacity is increased when tension is applied, but this is normally ignored.

Buckling (DF_B)

This is the high-angle well-safety factor equivalent to the excess BHA factor for vertical wells. Both serve to prevent drill pipe from buckling in rotary mode. The difference is that DF_{BHA} increases BHA length in vertical wells while DF_B decreases allowable bit weight in ERD and horizontal wells, where the traditional BHA is absent.

DESIGN FOR VERTICAL-TO-MODERATE ANGLE WELLBORES

Design Stages

Working from the bit to the surface:

- Choose drill collar size, connection, and connection features.
- Determine torsional strength of drill collar connections.
- Determine minimum lengths of drill collar and HWDP sections.
- Check slip-crushing forces.
- Set design factors and margin of overpull in tension.
- Calculate allowable and working tension loads.
- Calculate maximum permissible length of each drill pipe section.
- Calculate derated collapse pressure capacities of drill pipe tubes under tensile loading.

Drill Collar Size

- Unless mechanical hole sticking is a problem, the largest diameter BHA consistent with other needs should be used.
- The increased stiffness translates into better directional control.
- Presence of collars means fewer connections for a specified weight on bit.
- Larger collars mean reduced BHA length and, hence, reduced differential sticking risk.
- Larger collars have less lateral freedom of movement. This reduces the magnitude of the cyclic stresses generated by buckling and lateral vibration and thus increases connection fatigue life.
- Other considerations include:
  - ❖ ability to fish
  - ❖ effective range of pipe handling equipment
  - ❖ directional control requirements
  - ❖ hydraulics
  - ❖ desired features (spiral grooves, elevator groove, etc.)

BHA Connections/Features

The following points are applicable to all BHA components including crossovers, stabilizers, motors, LWD and MWD tools, hole openers, underreamers, jars, etc.

Bending Strength Ratio (BSR)

This is the ratio of the relative stiffness of the box to the pin for a given connection.

High BSRs can cause accelerated pin failure.

Low BSRs can cause box failures.

Field experience suggests that larger OD collars suffer predominantly from box fatigue cracks even when at or near the optimum BSR of 2.5. This suggests that higher BSRs might be a more appropriate guideline for large OD collars. Conversely,

4½" collars with BSRs of 1.8 are rarely found to exhibit box fatigue cracks. This serves to highlight the importance of field experience when choosing BSRs for particular collar sizes.

The recommended BSR for typical drill collar sizes is shown in the table below. These numbers should be adjusted as determined by local operating conditions.

RECOMMENDED BSR RANGES

Drill Collar OD	Traditional BSR Range	Recommended BSR Range
< 6"	2.25–2.75	1.8–2.5
6"–7⅞"	2.25–2.75	2.25–2.75
= or > 8"	2.25–2.75	2.5–3.2

Transitions between sections of different stiffness act as stress concentrators. This problem is worsened by short, straight crossovers. If a straight (nonbottleneck) crossover is used and its OD is larger than the HWDP tool joint OD, the resulting BSR of the upper crossover connection may be very high, resulting in accelerated pin fatigue. Bottleneck subs alleviate this problem by providing a smooth change in cross section.

The equations used in the calculation of BSR are given in Appendix 1.

BHA Connection Thread Form

Thread forms with full root radii should be used in all BHA connections to maximize fatigue resistance. API regular, NC, and full-hole connections all meet this requirement, although the API NC thread form (V-038R) is superior to the others. The H-90 thread form is also considered acceptable even though it does not have a full-root radius.

All connections that employ a 'standard' V-065 thread form, except PAC, are obsolete. The 'NC' thread form should be specified instead of the obsolete 'IF' or 'XH' names, as this will eliminate the possibility of receiving the fatigue prone V-065 thread form.

Stress Relief Features

Stress relief features should be specified on all BHA connections NC-38 and larger. These features include the 'stress relieved pin' and 'bore-back box'. Both extend connection fatigue life by eliminating disengaged thread roots which act as stress concentrators. Stress relief features are beneficial on all HWDP connections. Pin stress relief grooves are not recommended for connections smaller than NC-38 because they may weaken the connection's tensile and torsional strength and because fatigue is often less of a problem than noncyclic loads on small connections. Bore-back could be used on smaller connections without weakening them and should be considered if box fatigue is occurring.

Cold Rolling

Cold rolling BHA (and HWDP) thread roots and stress relief surfaces increase fatigue life by placing a residual compressive stress in the thread roots. It is not beneficial on normal weight drill pipe where fatigue is rarely a problem due to the relative stiffness of the tool joint compared to the tube.

BHA Connection Torsional Strength

Since torsion is transmitted from the top down, BHA connections are usually subjected to lower torsional loads than the connections above. However, if 'stick/slip' is occurring or a tapered assembly is being used, torsional strength should be checked to confirm that it is greater than the torsion expected within the operating BHA. Tool joint torsional strength tables cannot be used directly for this purpose because tool joint and drill collar materials have different yield strengths. Drill collar connection torsional strength can be calculated as follows:

$$T_s = \frac{MUT}{f}$$

where

T_S = DC connection torsional strength (ft lb)
MUT = DC makeup torque (ft/lb)
f = factor from table below.

Factors for Converting Drill Collar MUT to Torsional Strength		
Connection Type	OD < or = 6⅞"	OD > 6⅞"
PAC	f = 0.795	n/a
H-90	f = 0.511	f = 0.562
Other	f = 0.568	f = 0.625

Stabilizer and Jar Placement

Stabilizers

The number, size, and position of stabilizers is often determined by directional considerations. However, they also have an impact on other design aspects.

1. While rotary drilling vertical wells, the lower part of the BHA will suffer buckling and be supported by the sides of the hole. Stabilizers reduce connection stress/increase fatigue life by restricting the freedom of lateral drill collar movement.
2. If mechanical sticking is a concern, more or larger stabilizers may increase the likelihood of getting stuck. Conversely, when differential sticking is a concern, the presence of stabilizers can reduce this risk by keeping the collars off the sides of the wellbore.

Jars

Jar placement is dictated by the need to have maximum impact should the BHA become stuck while attempting to ensure that fatigue failure does not occur. Until recently, the rule of thumb was to run the jars in tension. More recently, in high-angle wells, it has become acceptable to run jars in compression. This has led to confusion regarding placement of jars, i.e., whether to run in tension or compression and whether buckled or not buckled. (Note: A rotating, buckled drill string element is always a concern in drilling and should be avoided.)

To clarify matters, the following rule is advocated:

"Do not run the jars buckled at any time."

This rule obviously precludes jars from being run in mechanical compression in vertical hole sections. However, in high-angle wells, it recognizes that a jar can be subjected to a large compressive load without buckling.

Length of BHA section

The length of the drill collar section will depend on the type of BHA being designed and whether or not HWDP is to be used for bit weight. Three types of BHA design are considered as illustrated in Figure 8.1.

- Type A : Full bit weight is supplied by drill collars. The HWDP is present to span the transition from DCs to DP.
- Type B : Sufficient DCs are used to achieve either directional control or other objectives except WOB. Bit weight derives from both DCs and HWDP. This BHA is easier to handle on the rig floor and appears to have reduced the incidence of drill collar failure.
- Type C : More than one size of drill collar is used but bit weight is still provided by both DCs and HWDP. As with Type B, the number of DCs is influenced by directional or other objectives (not WOB).

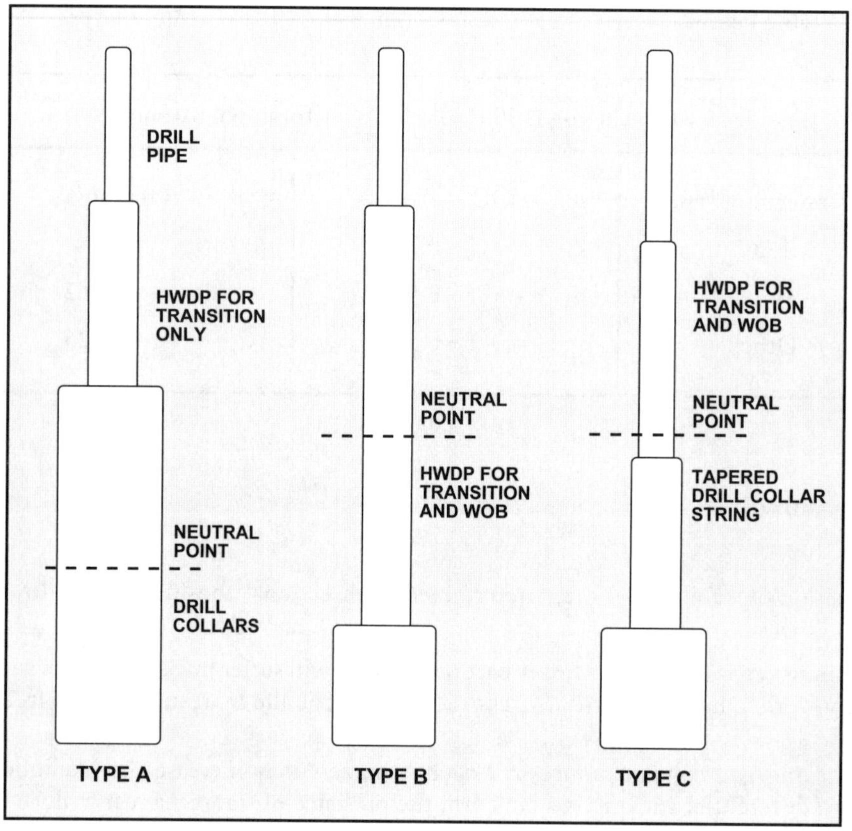

Figure 8.1 Three BHA configurations

Type A BHA

The minimum length of drill collars is calculated as follows:

$$L_{DC} = \frac{WOB \times DF_{BHA}}{W_{DC} \times K_B \times \cos\theta}$$

where

L_{DC} = minimum drill collar length (ft)
WOB = maximum weight on bit (lb)
DF_{BHA} = design factor for excess weight
K_B = buoyancy factor
θ = maximum hole angle at BHA (degrees)
W_{DC} = air weight of drill collars (lb/ft).

The design factor for excess BHA weight is chosen to ensure that the neutral point remains below the top of the BHA. This factor is typically assigned a value of 1.15 for most applications although hard drilling conditions may warrant a higher value. The minimum length of collars is often rounded up to the next stand of collars.

The amount of HWDP for transition should be determined by past experience and typically would be of the order 9–30 joints.

Type B and C BHAs

The amount of HWDP required to apply the necessary bit weight and keep the neutral point within the BHA can be determined using the following formula:

$$L_{HWDP} = \frac{\left[\dfrac{WOB \times DF_{BHA}}{K_B \times \cos\theta} - (W_{DC1} \times L_{DC1}) - (W_{DC2} \times L_{DC2})\right]}{W_{HWDP}}$$

where

L_{HWDP} = minimum HWDP length (ft)
WOB = maximum weight on bit (lb)
DF_{BHA} = design factor for excess weight
K_B = buoyancy factor
θ = maximum hole angle at BHA (degrees)
W_{DC1} = air weight of drill collars in first section (lb/ft)
L_{DC1} = length of first section of collars (ft)
W_{DC2} = air weight of drill collars in second section (lb/ft)
L_{DC2} = length of second section of collars (ft)
W_{HWDP} = air weight of HWDP (lb/ft).

As hole angle increases, a point will be reached where a heavy BHA would be more detrimental due to increased tensile and torsional drag than it would be beneficial for providing bit weight. The above two formulas will then cease to apply beyond that point; i.e., for higher hole angles.

For these higher hole angles, it is accepted practice to apply bit weight via normal weight drill pipe run in mechanical compression.

Tool Joint Torsional Capacity

To prevent downhole makeup and torsional failure, maximum operating torque should never exceed tool joint makeup torque.

Actual tool joint torque should be based on the actual OD and ID of the connection being used, rather than a general makeup torque for a given connection type.

When a high-operating torque is expected, makeup torque can be increased, but the effect of a possible reduction in the tensile capacity of the tool joint needs to be determined.

Connection dimensions and torque gauge accuracy should always be checked irrespective of whether high-operating torque is expected.

The equations used in the calculation of tool joint makeup torque are given in Appendix 1, at the end of this chapter.

Stiffness Ratio

The stiffness ratio (SR) of the sections above and below each transition must be compared to help in quantifying the abruptness of the section change and determine the need for transition pipe. This is achieved by dividing the section modulus (Z) of the lower section by the section modulus of the upper section.

$$SR = \frac{Z_{Lower}}{Z_{Upper}}$$

$$Z = \frac{\pi}{64} \times \frac{OD^4 - ID^4}{OD}.$$

As with BSR, the stiffness ratio is not a strictly quantitative performance limit and experience should be used to determine the optimum SR. If drill pipe failures are being experienced at the top of a BHA despite adequate drill collar weight for WOB, then transition pipe may be needed. The following guidelines have been found to be generally acceptable:

1. For routine drilling or very low failure rate experience, keep the SR below 5.5.
2. For severe drilling or for significant failure rate experience, keep the SR below 3.5.

Drill Pipe Tensile Design

The basis for selecting various pipe grades to make up a drill string is to always maintain the desired margin of overpull (MOP) at all points in the string (fig. 8.2). This is accomplished by adding the lowest pipe grade, one joint at a time, starting from the top of the BHA upward. Each joint must support the BHA weight plus the drill pipe below that joint. When the working load (P_W) is reached for that grade of drill pipe, the next higher grade should be used and the process repeated until the drill string is complete.

Note: The nominal weight of drill pipe is merely a descriptive term for identification purposes and refers to the line pipe. The actual weight (air weight or adjusted weight) which includes the tool joint weights should always be used in these calculations.

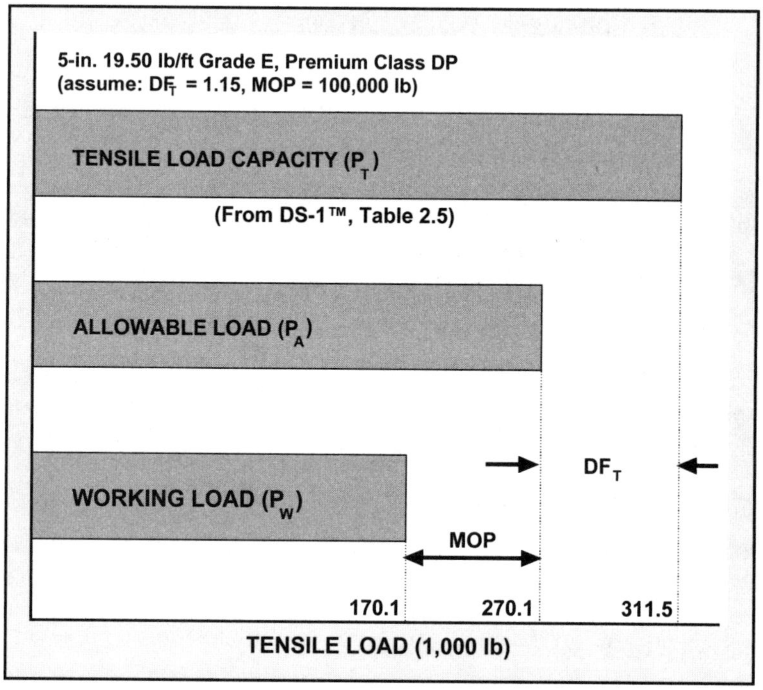

Figure 8.2 Tensile design nomenclature

Determine Tensile Load Capacity (P_T)

This is the calculated tensile pull to yield the pipe body. Values from tables, allowing for derating based on wall thickness or class of pipe.

Determine Design Factor in Tension (DF_T)

The factor used to derate the tensile load capacity to obtain the allowable load (P_A). Typically, a DF_T of 1.1 is used.

Calculate Allowable Load (P_A)

The maximum load placed on a pipe including contingency.

$$P_A = P_T/DF_T.$$

Set Margin of Overpull (MOP)

The design excess pull capacity above the working load (P_W) to compensate for expected drag, possible sticking, slip crushing and the effect of circulating pressure.

MOP values are typically 50,000–150,000 lb.

Calculate Working Load (P_W)

Working load is the maximum expected tension that will occur during normal operations.

$$P_W = P_A - MOP.$$

Calculate the Maximum Length of the First Drill Pipe Section

The maximum length of the lowest acceptable grade of drill pipe, in the first DP section above the BHA, can be determined using the following formula:

$$L_{DP1} = \frac{\frac{P_{W1}}{K_B} - (W_{DC1} \times L_{DC1}) - (W_{DC2} \times L_{DC2}) - (W_{HWDP} \times L_{HWDP})}{W_1}$$

where

L_{DP1} = maximum length of drill pipe in section 1 (ft)
P_{W1} = working load of drill pipe in section 1 (lb)
L_{HWDP} = length of HWDP (ft)
K_B = buoyancy factor
W_{DC1} = air weight of drill collars in first section (lb/ft)
L_{DC1} = length of first section of collars (ft)
W_{DC2} = air weight of drill collars in second section (lb/ft)
L_{DC2} = length of second section of collars (ft)
W_{HWDP} = air weight of HWDP (lb/ft)
W_1 = air weight of drill pipe in section 1 (lb/ft).

Calculate the Maximum Length of the Second Drill Pipe Section

To calculate the amount of drill pipe in the second section above the BHA (if required):

$$L_{DP2} = \frac{P_{W2} - P_{W2}}{W_2 \times K_B}$$

where

L_{DP2} = length of drill pipe in section 2 (ft)
P_{W2} = working load of drill pipe in section 2 (lb)
P_{W1} = working load of drill pipe in section 1 (lb)
W_2 = air weight of drill pipe in section 2 (lb/ft)
K_B = buoyancy factor.

Calculate the Maximum Length of the Third Drill Pipe Section

To calculate the amount of drill pipe in the third section above the BHA (if required):

$$L_{DP3} = \frac{P_{W3} - P_{W2}}{W_3 \times K_B}$$

where

L_{DP3} = length of drill pipe in section 3 (ft)

P_{W3} = working load of drill pipe in section 3 (lb)
P_{W2} = working load of drill pipe in section 2 (lb)
W_3 = air weight of drill pipe in section 3 (lb/ft)
K_B = buoyancy factor.

Burst Pressure

In general, drill pipe is not used in applications requiring high burst pressure loading if gas is the source of the pressure. Tubing with premium connections is often better suited.

Drill pipe should never be used to carry gas containing hydrogen sulfide at a partial pressure greater than 0.05 psia. Burst ratings can be found tabulated and assume minimum material properties and no axial load. While simultaneous tensile loading will increase burst pressure capacity, this is usually ignored and retained as an additional safety factor.

Collapse Pressure

The collapse pressure acting at any point within the drill string under static conditions is:

$$P_C = P_A - P_{DP} + D(G_A - G_{DP})$$

where

P_C = collapse pressure on drill pipe (psi)
P_A = surface annulus pressure (psi)
P_{DP} = surface drill pipe pressure (psi)
D = depth of interest (ft)
G_A = fluid gradient in annulus (psi/ft)
G_{DP} = fluid gradient in drill pipe (psi/ft).

Note: Simultaneous tension and annulus pressure will reduce collapse capacity which should be derated as below if tensile loads are anticipated.

Combined Loading

These cases refer to situations where there are several loads being exerted on a tubular at the same time.

1. Simultaneous tension reduces drill pipe collapse capacity and vice versa.
2. Simultaneous torsion reduces drill pipe tensile capacity and vice versa.
3. Connection makeup (torque) beyond a given point reduces connection tensile capacity.
4. Simultaneous tension reduces the torsional yield strength of pin-weak connections.

Derating Collapse Pressure Capacity of Drill Pipe Under Tension

The derating factor for the collapse capacity of pipe under tension can be calculated from equations given in API RP7G Appendix A. However, a quicker method is to use the graph in Figure 8.3 (also from API RP7G). The equations use an average yield strength while the graphical method uses the minimum yield strength.

Yield Strengths		
Pipe Grade	Minimum Yield Strength (psi)	Average Yield Strength (psi)
E	75,000	85,000
X	95,000	110,000
G	105,000	120,000
S	135,000	145,000

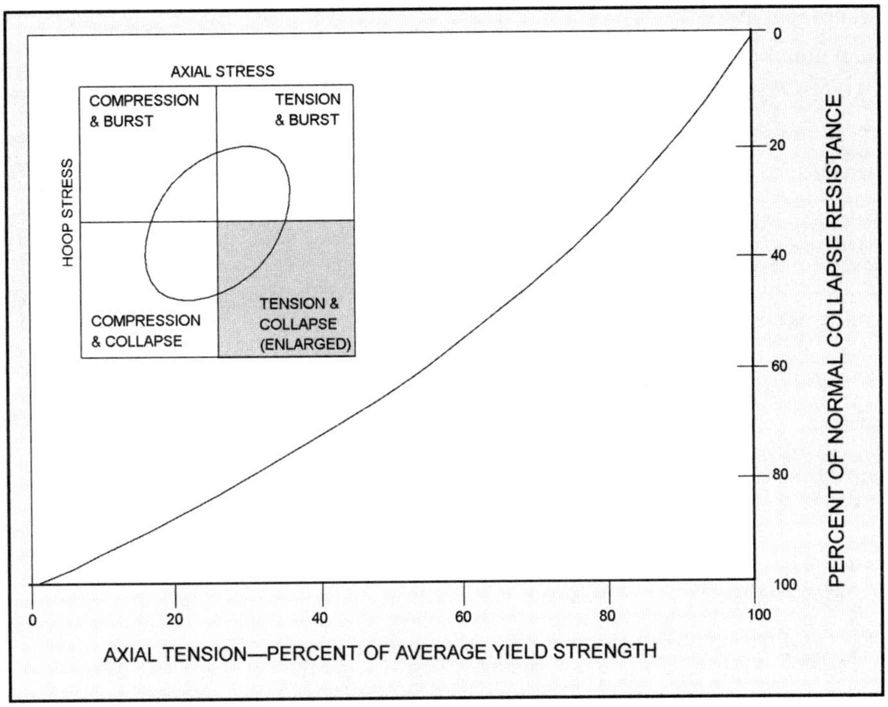

Figure 8.3 **Ellipse of biaxial yield stress**

Example

Determine the collapse pressure capacity of 5", 19.5 lb/ft grade E pipe under a tensile load of 50,000 lb.

1. From drill pipe tables, tensile capacity for this pipe is 311,535 lb and collapse capacity is 7,041 psi.
2. Express the axial load as a percentage of the minimum yield strength.
   % Min YS = (Axial load) x (100)/P_T
   = (50,000 × 100)/(311,535)
   = 16%
3. Plot 16% on horizontal axis of Figure 8.3. Take a perpendicular up to curve and then horizontal from curve to intersect y axis at 90% (a derating factor of 0.9).
4. Multiply the nominal collapse capacity by the derating factor.
   = 7,041 × 0.9
   = 6,337 psi
5. Assuming a design factor for collapse (DF_C) of 1.125, derated collapse capacity
   = 6,337/1.125
   = 5,632 psi

Derate Drill Pipe Tensile Load Capacity Under Combined Torsion and Tension

This situation could occur when back reaming, pulling on stuck pipe, or fishing. By inserting the applied tensile or torque load into the following equation, the maximum torque or tensile load respectively that can be applied simultaneously can be calculated.

$$Q_T = \frac{0.096167 \times J}{OD} \times \sqrt{Y_m^2 - \frac{P^2}{A^2}}$$

where

Q_T = minimum torsional yield under tension, lb/ft
J = polar moment of inertia
= $(\pi/32)(OD^4 - ID^4)$ for tubes

OD = outside diameter, in.
ID = inside diameter, in.
Y_m = minimum unit yield strength, psi
P = total load in tension, lb
A = cross section area
= $(\pi/4)(OD^2 - ID^2)$.

Example

What is the maximum torque that can be applied to stuck pipe under a tensile load as follows:

3½" OD 13.3 lb/ft Grade E drill pipe (new)
Tensile load 100,000 lb

ID (from tables) = 2.764"
Y_m = 75,000 psi

$$J = (\pi/32)(3.5^4 - 2.764^4) = 9$$

$$A = (\pi/4)(3.5^2 - 2.764^2) = 3.6209 \text{ in.}^2$$

$$Q_T = \frac{0.096167 \times 9}{3.5} \times \sqrt{75{,}000 - \frac{100{,}000^2}{3.6209^2}}$$

$$Q_T = 17{,}243 \text{ lb/ft.}$$

Tool Joint Tensile Load Capacity at Applied Makeup Torque

Makeup torque should not exceed the recommended value unless the impact of the excess torque on the tensile capacity of the string is first considered.

The tensile capacity of tubes is normally significantly lower than the tensile capacity of the tool joints and it is commonly assumed that a string's tensile capacity is limited by the tube.

Increased makeup torque places stress on the pin neck which is added to the tensile load in the string at each tool joint. Thus, a point will be reached at which the pin neck becomes the weakest part in the string.

For a particular tool joint it is possible to construct a tool joint combined loading curve, similar to that shown in Figure 8.4. The points *P1*, *T1*, *T2*, *T3*, and *T4* can all be derived from equations in API RP7G Appendix A.8.3.

Using these curves it is possible to determine the effect on increasing the makeup torque on the allowable tensile capacity of the tool joint. It is then necessary to determine if the tool joint or the drill pipe tube is the weaker part of the drill string.

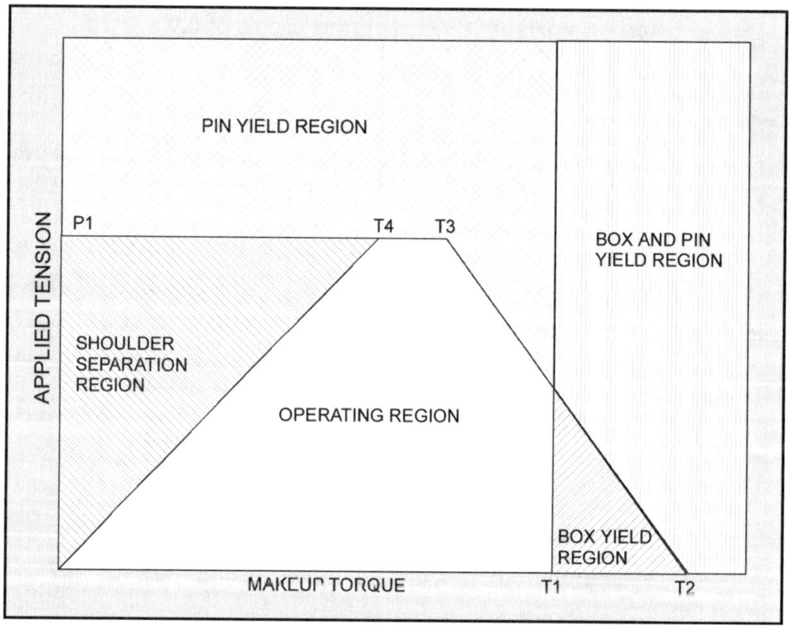

Figure 8.4 **Combined load curve for tool joints**

Stability Forces and Drill Pipe Buckling

Usually, full circulation is established while off bottom, thereby preventing any temporary buildup of pressure in the drill string. However, if bit pressure drop is increased while on bottom, the drill pipe can buckle above the BHA (fatigue damage) even though the neutral point is within the BHA. This occurs because the drill pipe is unable to stretch to compensate for the increase in internal pressure. However, any buckling that does occur is purely temporary and will be lost once sufficient new hole has been drilled to allow the pipe to stretch.

Whether or not drill pipe buckling occurs will depend on the conditions at the time. Typical conditions would include:

- Shallow drilling
- Thin-walled drill pipe in use
- Large changes in bit pressure drop
- High WOB for available BHA weight

Adherence to the following rule should help to avoid this situation occurring:

Any time an increase in pump rate occurs while the bit is on bottom, pick up the drill string until a gain in weight is first noticeable. This will allow the string to stretch and relieve the tendency to buckle.

Slip Crushing

Slips exert hoop compression on drill pipe and can deform it under certain conditions. A unit tensile stress (S_t) from hanging weight will result in a hoop stress (S_h) that is a function of many factors, such as slip length, coefficient of friction between slips and bowl, pipe diameter, etc. Slip-crushing constants (S_h/S_t) have been calculated for varying conditions, all of which assume a coefficient of friction between slip and bowl of 0.08.

Slip-Crushing Constants (S_h/S_t)		
	Slip Length (in.)	
Pipe Size (in.)	12	16
2⅜	1.25	1.18
2⅞	1.31	1.22
3½	1.39	1.28
4	1.45	1.32
4½	1.52	1.37
5	1.59	1.42
5½	1.66	1.47
6⅝	1.82	1.59

Assuming that the pipe is not stuck, the maximum tension carried by the slips will be the working load (P_W). To ensure that there is sufficient margin to allow for slip crushing, the following conditions must be obeyed:

$$P_W \times (S_h/S_t) < \text{or} = P_A$$

where

P_W = working load, lb
P_A = maximum allowable load, lb
S_h/S_t = slip-crushing constant.

Note: If the pipe is stuck and it is necessary to set the slips with additional tension, then the above calculation should include the extra string tension.

If the margin of overpull (MOP) is chosen to ensure that it always complies with the relationship below, then the slip-crushing effect will always be catered for.

$$\text{MOP} > \text{or} = P_W \times [(S_h/S_t) - 1].$$

The above assumes that the slips, slip element, and bowl are in good order and are regularly inspected.

Welded Components

Welded components should be avoided. Most components are made from relatively high carbon steels which are heat treated to achieve the required properties. Welding permanently alters these properties and, unless heat treated again, the component will be weakened and embrittled.

DESIGN FOR EXTENDED-REACH WELLBORES

General Design Considerations

In high-angle wells, traditional BHA components are often eliminated and bit weight is more likely to be applied by running normal weight drill pipe in compression (never considered in vertical wells).

For a given measured depth, surface tension load from hanging weight will decrease with increasing well angle due to increasing wall support. However, torque and drag will increase as hole angle increases. The load limit will more likely be its torsional capacity under these circumstances.

In vertical wells, frictional forces are either ignored or taken account of by utilizing design factors or margin of overpull. Highly deviated wells need to be able to account for frictional forces. Generally, this is an iterative process and is most efficiently tackled using computer simulation techniques as depicted in Figure 8.5.

Figure 8.5 outlines most of the design considerations for deviated wellbores but the single most important aspect will be availability. Can the well be drilled with the pipe that is on the selected rig? If so, this will inevitably be the most economical option.

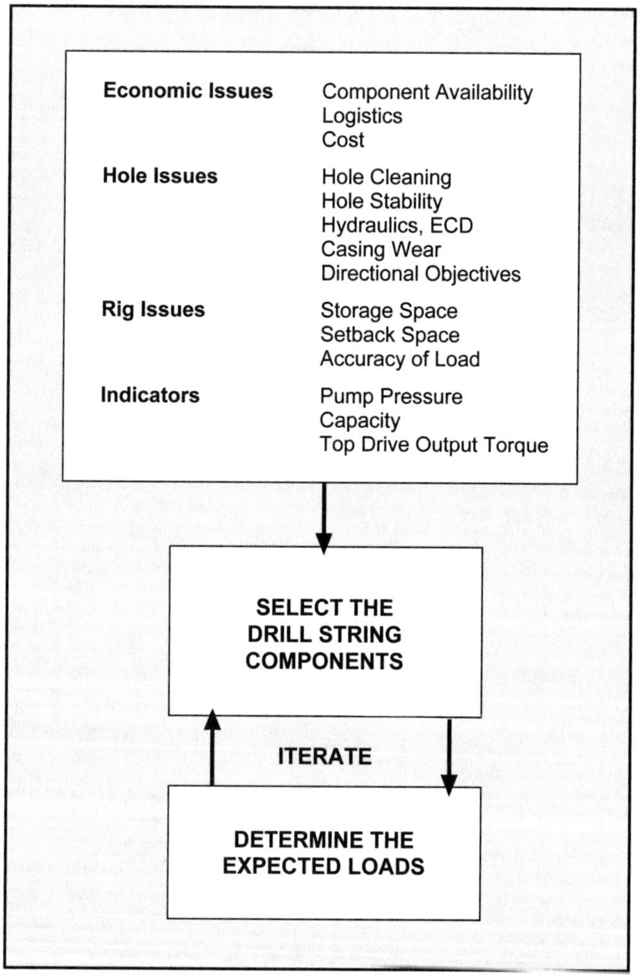

Figure 8.5 Designing a drill string for ER drilling may involve many considerations.

Estimating Drill String Loads

Hand calculating torque and drag is possible but not practical, bearing in mind the large number of calculations required. A computer program can complete the necessary iterations faster and more efficiently.

Most of these programs are based on the model by Johancik which analyses the drill string tensile and torsional loads in discrete sections and then sums the results over the entire hole.

Drag Coefficient

Torque and drag programs use an assumed friction coefficient which is initially based on area experience using whatever the proposed fluid system will be. Once drilling commences, the programs can be calibrated against actual loads to refine the model. However, torque and drag loads derive, not only from friction but also from the effects of hole tortuosity, cuttings buildup, swelling shale, differential sticking, etc. The friction coefficient would thus be more appropriately termed a drag coefficient. This coefficient will vary as both rotational and axial movement varies and as the direction of movement changes as well.

Critical Hole Angle

As hole angle increases, the BHA is less likely to slide down under its own weight. At the point at which it must be pushed to gain further advance, the wellbore angle is termed the critical angle.

$$\theta_{crit} = \arctan(1/f)$$

where

θ_{crit} = critical angle
f = coefficient of hole drag.

This is normally in the order of 70–80°, depending on the actual coefficient of drag.

Torsion

Tool joints are weaker than the tubes under torsion and the torsion capacity of a tool joint is taken to be its makeup torque. If predicted surface torque is likely to exceed makeup torque, then the string is torsion capacity limited and the following measures should be considered:

- The diameter and weight of drill string components in high angle sections should be minimized. This reduces the operating torque but must be balanced against the need to reduce internal pressure losses and maintain the stability of normal weight drill pipe.
- The makeup torque on the existing drill pipe can be increased as long as the tool joints are not overstressed and the pin neck tensile capacity is not reduced below that required by the operation.
- The drill pipe on the rig may be changed out for pipe with higher torsional capacity tool joints. This may be pipe with standard-sized tool joints that have higher ODs and smaller IDs if the hydraulics model permits.
- The operating torque can be reduced by mud selection, mud additives, or through the use of torque reducing equipment; e.g., torque reducing subs.

Nonstandard Joints

New and premium class tool joints are designed to be about 80% as strong in torsion as the tubes to which they are welded. Nonstandard tool joints do exist for various reasons, making it essential that all tool joints are measured to confirm actual dimensions. Nonstandard joints are acceptable but will have different makeup torque values.

Factors Affecting Torsion Capacity

All API tool joints are made from material having a minimum yield strength of 120,000 psi. Tool joint torsional capacity is determined only by connection type, pin ID, and box OD. The following table illustrates the impact of tool joint dimensions for 5" 19.5 ppf, S grade drill pipe along with the makeup torques assuming standard thread dope (friction factor = 1.0).

Properties of New Standard-Sized Tool Joints on 5" 19.5 ppf Drill Pipe

Grade	ID (in.)	OD (in.)	Makeup torque (ft-lb)
E	3¾	6⅝	22,840
X	3½	6⅝	27,080
G	3¼	6⅝	31,020
S	2¾	6⅝	38,040

Fatigue in ERD Wells

The prime cause of fatigue in high-angle wells:

1. Drill pipe buckling
2. Drill collar sag

Preventing buckling while rotating normal weight drill pipe and jars operating under mechanical compression will be a concern. Because of the high angle, it is often necessary to apply bit weight both with and through normal weight drill pipe in mechanical compression. As long as no buckling occurs, fatigue damage should be avoided. When sliding but not rotating, any buckling that occurs should not prompt fatigue damage. However, once rotation is applied, fatigue through buckling becomes an issue.

Drill collar sag relates to the low side sagging tendency of the BHA and its accompanying cyclic stress.

Drill Pipe Buckling

Buckling drill pipe while rotating can cause rapid fatigue failure. In high-angle wells, it is necessary to run drill pipe in mechanical compression to be able to apply bit weight. However, if the magnitude of mechanical compression does not exceed the critical buckling load, then the pipe will suffer little damage.

The maximum mechanical compression in a vertical string must remain below critical buckling load to ensure no fatigue damage.

In a deviated wellbore, the critical buckling load will be higher than for a vertical section due to the support provided by the inclined wellbore itself. The limiting factor will be the buckling load in the vertical sections measured as above. The problem is, though, determining where and at what bit weight buckling will commence. If the bit weight required to promote the onset of buckling can be determined, then, assuming that the well can be drilled with a lower bit weight, the buckling risk can be diminished.

Buckling initiation points are:

1. Lowermost joint in a tangent section.
2. In the straight hole section immediately at or above the kickoff point.
3. Over the entire length of a horizontal tangent section (initiated just below tangent section).

FATIGUE

- Fatigue is the progressive localized permanent structural damage that occurs when a material undergoes repeated stress cycles.
- Fatigue damage accumulates at high stress points and ultimately a fatigue crack forms. This can grow under continued cyclic loading until failure occurs.
- For a given material, the severity of fatigue attack is greater at higher cyclic stress amplitude and at higher average tensile stress.
- Failure will occur at points of highest stress on any component and failures are almost always near some stress concentrator such as a notch, pit, section change, or thread root.

Sources of Fatigue-Inducing Cyclic Stresses

- Rotating the string while part of it is bent or buckled
- Vibration

Mitigation

Actions that will minimize the occurrence of damaging cyclic stresses.

- Configure the bottomhole assemblies and limit bit weight so that simultaneous rotation and buckling does not occur in normal weight drill pipe or jars.
- Select products and components and configure string sections with smooth geometric transitions (sharp section changes magnify stress and accelerate fatigue).
- Reduce the degree of drill pipe bending (dogleg) and the degree of BHA bending and buckling to the lowest levels consistent with other objectives.
- Monitor and reduce vibration.

Critical Rotary Speeds

At certain speeds, defined as critical, the drill pipe experiences vibrations which can cause wear and deformation of the pipe body and lead to failure due to metal fatigue.

Critical speeds depend on the length and size of the drill pipe, drill collars, and hole size. A warning indication that the drill string may be working within the critical range of speed are high torque and, eventually, vibration at the rotary table.

To extend the life of the drilling equipment, rpm must be selected and monitored in order to avoid critical rotary speeds while drilling. Downhole vibration subs (MWD) or drill string vibration surface measurement techniques can be used.

Corrosion in Water-Based Drilling Fluid

In water-based drilling fluids, metallic corrosion reaction typically takes place due to the presence of three corrosion agents: gases (hydrogen sulphide, oxygen, and carbon dioxide), dissolved salts (sodium chloride, potassium chloride, calcium chloride, etc.), and acids (carbonic acid, formic acid, and acetic acid).

In order to limit corrosion in water-based drilling fluid, the following guidelines should be adhered to:

- If H_2S contamination is not anticipated, maintain a pH of the drilling fluid at 9.5 or higher. This will minimize the general corrosion and pitting corrosion that takes place due to the presence of dissolved oxygen.
- If H_2S contamination is anticipated, maintain the pH of the drilling fluid at 11 or greater through additions of caustic and/or lime.
- If H_2S is detected, scavenger should be used.

If the drilling fluid system requires that a low pH be maintained, treat the mud with a suitable scavenger and/or corrosion inhibitor. Concentrations should be specified only after pilot testing since overtreatment can actually increase the corrosion rate.

- If the drilling fluid becomes aerated, operate the degasser until the condition dissipates. If possible, premix additives in a mixing tank prior to addition to the active system. This will decrease entrained air entering the mud pump system and subsequently the drill string.
- Use corrosion rings for monitoring.

Drill String Operation Practices

The DP, HWDP, and DC are important parts of the drilling equipment cost, but the consequence of a downhole failure can be even greater.

Care must be taken while handling these tubulars, especially for the tool joint which is generally the 'weakest point'.

Recommendations for Handling New Tubulars

The break-in period of a tool joint's life is the most critical part, as newly machined surfaces are the most likely to gall. After some service, the surface undergoes changes which make them more resistant to galling.

As a consequence, the first couple of times tool joints are used, the following should be considered:

1. Make sure surface handling equipment is in good condition. Check slips and master bushing to prevent damage. Check that iron roughneck/tong dies are in good condition.
2. Make sure that the top drive saver sub is in good condition as it will mate with the majority of the joints.
3. Use thread protectors when picking up tubulars.
4. Thoroughly clean all pin and box threads to remove all grease, dirt, rust, preventive coating, or other foreign material.
5. Inspect for any damage on threads and for handling damage on threads and shoulder, such as scratches, gouges, and flat spots.
6. Thoroughly coat the shoulder and threads on both box and pin using a recommended tool joint compound.
7. Spin in new joints *slowly*. High speed rotation may cause galling. Make up joint with recommended torque.
8. Break out connection. Clean and inspect for damage; repair minor damage if possible.
9. Redope and remake to recommended torque.

General Recommendations

- Drill string component should be equipped with thread protectors when not in use and when being picked up or laid down.
- Ensure tool joint threads are clean and dry before doping them.
- Use tool joint dope specified. Do not use tubing or casing lubricants as they are too slick and can result in stretched and cracked pins. After break in, it is satisfactory to thoroughly dope the box threads and shoulder only.
- Lift sub pins should be cleaned, inspected, and lubricated on each trip. If damage to these pins goes unnoticed, they will eventually damage all the drill collar boxes.
- It is recommended practice to break a different joint on each trip, giving the crew an opportunity to look at each pin and box every third trip. This ensures that the connections are adequately doped at all times. Inspect the shoulders for sign of loose connections, galls, and possible washouts.
- Do not allow the pin end to be stabbed against the box shoulder. This can produce a low spot on the shoulder which will result in a washout.
- Do not stop downward movement of the drill string with slips. This can cause crushing or necking down of the drill pipe tube. Allowing slips to ride the pipe on trips out of the hole can also damage the pipe.
- Accidental catching of tool joints with slips will permanently damage the slips. This could lead to slips failing or damaging the pipe. In the event of such an accident, the slips should be inspected for deformations, breakage, or cracks.
- Make sure the setback areas are clean and the wood is in good condition. Use only tools designed to move joints on the setback. Sharp-edged tools can cause shoulder damage which will lead to washouts.
- Wash out drill string components when laying them out. Ensure thread protectors are installed.
- Inspect drill string components at regular intervals; e.g., at the end of every well, at six-month intervals, or as specified by operator contract. Inspection should be as per API RP7G or DS Hill Standard.

APPENDIX 1

CALCULATION OF BSR

BSR is calculated using the following equation

$$BSR = \frac{Z_b}{Z_p} = \frac{\left[\dfrac{D^4 - b^4}{D}\right]}{\left[\dfrac{R^4 - d^4}{R}\right]}$$

where

BSR = bending strength ratio
Z_b = box section modulus (in.^{3})
Z_p = pin section modulus (in.^{3})
D = outside diameter of box (in.)
d = inside diameter of pin (in.)
b = thread root diameter of box threads at end of pin (in.)
R = thread root diameter of pin threads ¾" from shoulder of pin (in.)

$$b = C - \left[\frac{tpr\,(L_{pc} - 0.625)}{12}\right] + (2 \times \text{dedendum})$$

where

C = pitch diameter (in.)
tpr = taper (inches per foot on diameter)
L_{pc} = length of pin (in.)

$$\text{dedendum} = (0.5\,H - f_m)$$

where

H = thread height not truncated (in.)
f_{rn} = taper (inches per foot on diameter).

$$R = C - (2 \times \text{dedendum}) - \left(\frac{tpr}{96}\right).$$

All of the variables in the above equations are specific to the various connection types and are available from various tables in API 7, *IADC Drilling Manual*, or manufacturer's catalogs. Details of common API and regular connections are shown in Appendix 2.

CALCULATION OF TOOL JOINT MAKEUP TORQUE

Recommended makeup torque is calculated using the following equation

$$T = \frac{S \times A}{12}\left[\frac{P}{2\pi} + \frac{Rf}{\cos\theta} + R_s f\right]$$

where

T = makeup torque (ft lb)
S = desired stress level from makeup (see the table on the next page).

Connection Type	Desired Stress (psi)
Used DP Tool Joints	72,000
New DP Tool Joints	60,000
PAC DC	87,500
H90 DC	56,200
Other DC	62,500

A = the smallest of the cross-sectional area ¾" from the pin shoulder or ⅜" from the box shoulder .

$$A_b = 0.25\,\pi\,[D^2 - (Q_c - E)^2]$$

Q_c = box counterbore (in.)

$$E = \frac{3\,tpr}{96}$$

$$A_p = 0.25\,\pi\,[(C - B)^2 - d^2]$$

$$B = (H - 2\,S_{rs}) + \left[\frac{tpr}{96}\right]$$

S_{rs} = root truncation (in.)
P = lead of threads (in.)
R_t = average mean radius of thread (in.)

$$R_t = 0.25\left(C + \left[C - (L_{pc} - 0.625) \times \frac{tpr}{12}\right]\right)$$

f = coefficient of friction (assume 0.08)
θ = ½ thread angle
R_s = mean shoulder radius (in.)

$$R_s = \frac{(Q_c + D)}{4}\ .$$

All of the variables in the above equations are specific to the various connection types and are available from various tables in API 7, *IADC Drilling Manual*, or manufacturer's catalogs.

APPENDIX 2

Rotary Shoulder Connection Dimensions—API Connections

Connection Type	Thread Form	Threads per in.	Taper	Angle	Thread Height, Not Truncated	Root Truncation	Dedendum	Pitch Diameter	Pin Length	Box Counterbore	Thread Root Diameter of Box Threads at End of Pin	Thread Root Diameter of Pin Threads ¾" from Shoulder of Pin
			tpr		H	f_m		C	L_{pc}	Q_c	b	R
			in./ft	degrees	in.	in.		in.	in.	in.	in.	in.
NC23	V-0.038R	4	2	30	0.216005	0.038000	0.07000	2.355	3	2.62500	2.09917	2.19416
NC26	V-0.038R	4	2	30	0.216005	0.038000	0.07000	2.668	3	2.93750	2.41217	2.50716
NC31	V-0.038R	4	2	30	0.216005	0.038000	0.07000	3.183	3.5	3.45313	2.84384	3.02216
NC35	V-0.038R	4	2	30	0.216005	0.038000	0.07000	3.531	3.75	3.81250	3.15017	3.37016
NC38	V-0.038R	4	2	30	0.216005	0.038000	0.07000	3.808	4	4.07813	3.38551	3.64716
NC40	V-0.038R	4	2	30	0.216005	0.038000	0.07000	4.072	4.5	4.34375	3.56617	3.91116
NC44	V-0.038R	4	2	30	0.216005	0.038000	0.07000	4.417	4.5	4.68750	3.91117	4.25616
NC46	V-0.038R	4	2	30	0.216005	0.038000	0.07000	4.626	4.5	4.90625	4.12017	4.46516
NC50	V-0.038R	4	2	30	0.216005	0.038000	0.07000	5.0417	4.5	5.31250	4.53587	4.88086
NC56	V-0.038R	4	3	30	0.215379	0.038000	0.06969	5.616	5	5.93750	4.66163	5.44537
NC61	V-0.038R	4	3	30	0.215379	0.038000	0.06969	6.178	5.5	6.50000	5.09863	6.00737
NC70	V-0.038R	4	3	30	0.215379	0.038000	0.06969	7.053	6	7.37500	5.84863	6.88237
NC77	V-0.038R	4	3	30	0.215379	0.038000	0.06969	7.741	6.5	8.06250	6.41163	7.57037

Rotary Shoulder Connection Dimensions—Regular Connections

Connection Type	Thread Form	Threads per in.	Taper	Angle	Thread Height, Not Truncated	Root Truncation	Dedendum	Pitch Diameter	Pin Length	Box Counterbore	Thread Root Diameter of Box Threads at End of Pin	Thread Root Diameter of Pin Threads ¾" from Shoulder of Pin
			tpr		H	f_{rn}		C	L_{pc}	Q_c	b	R
			in./ft	degrees	in.	in.		in.	in.	in.	in.	in.
2⅜ REG	V-0.040	5	3	30	0.172303	0.020000	0.06615	2.36537	3	2.68750	1.90392	2.20182
2⅞ REG	V-0.040	5	3	30	0.172303	0.020000	0.06615	2.74037	3.5	3.06250	2.15392	2.57682
3½ REG	V-0.040	5	3	30	0.172303	0.020000	0.06615	3.23987	3.75	3.56250	2.59092	3.07632
4½ REG	V-0.040	5	3	30	0.172303	0.020000	0.06615	4.36487	4.25	4.68750	3.59092	4.20132
5½ REG	V-0.050	4	3	30	0.215379	0.025000	0.08269	5.23402	4.75	5.57813	4.36815	5.03739
6⅝ REG	V-0.050	4	2	30	0.216005	0.025000	0.08300	5.7578	5	6.06250	5.19464	5.57096
7½ REG	V-0.050	4	3	30	0.215379	0.025000	0.08269	6.71453	5.25	7.09375	5.72366	6.51790
8½ REG	V-0.050	4	3	30	0.215379	0.025000	0.08269	7.66658	5.375	8.04688	6.64446	7.46995

CHAPTER 9
Surveying and Directional Drilling

SURVEYING

Why Survey?

Accurate data about the position of a borehole is required in order to monitor and control where a borehole is and where it is going for the following reasons:

- To hit geological targets
- To provide a better definition of geological and reservoir data to allow for production optimization
- To avoid collision with other wells
- To define the target of a relief well for blowout contingency planning
- To provide accurate vertical depths for the purpose of well control
- To provide data for operational activities such as running and cementing casing
- To fulfill the requirements of local legislation

Models of the Earth

The earth is commonly described as a spherical object but it has a very irregular surface and carries mountain chains and deep-sea canyons in excess of 5 miles above and below mean sea level. The problem confronting surveyors is how to represent any point on the earth's surface on a flat sheet. Small areas of the earth may appear to have a flat surface but, by and large, this is not the case. This has rendered it necessary to look more closely at the shape of the earth so that a method of representing this shape on a flat surface can be used (fig. 9.1).

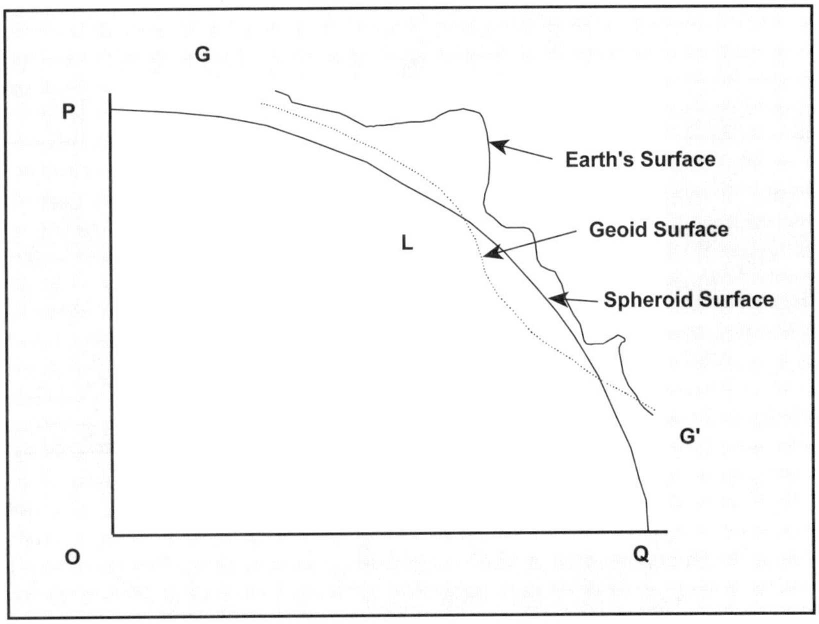

Figure 9.1 Shapes representing the earth

The Geoid

A smooth surface representing the earth's surface and referred to as the geoid can be produced, but it is impossible to describe any point on this surface mathematically. The geoid effectively smoothes out the irregularities of the earth's surface, but in so doing, creates an irregular shaped object itself. If mean sea level could be established everywhere, then this would be the surface of the geoid. All astronomical observations are made relative to the geoid and astronomical latitudes and longitudes are positions on the geoid.

The Spheroid

The earth can be more accurately represented in shape by that of an oblate spheroid flattened at the poles by approximately one part in three hundred due to rotation. This can be described mathematically by an algebraic equation, which can then be used as the basis for calculations. Over a dozen different ellipsoid shapes describing the earth mathematically have been generated and are in use today.

In 1924, an official ellipsoid was defined (based on the existing Hayford Ellipsoid of 1909) and called the International Ellipsoid. This had a flattening factor of 1:297, a polar radius of 6,356,911.9 m and an equatorial one of 6,378,388 m.

Many countries did not adopt this and chose, instead, to define their own because of irregularities in the spheroid's shape over different parts of the globe. The current range of spheroids used is tabulated below.

Name	Date	Equatorial Radius, m	Polar Radius, m	Flattening Factor	Use
GRS	1980	6,378,136	6,356,752.3	1/298.257	Newly adopted
WGS 72°	1972	6,378,135	6,356,750.5	1/298.6	NASA
Australia	1965	6,378,160	6,356,774.7	1/298.25	Australia
Krasovsky	1940	6,378,245	6,356,863.0	1/298.25	Soviet Union
International	1924	6,378,388	6,356,911.9	1/297	Rest of world
Hayford	1909	6,378,388	6,356,911.9	1/297	Rest of world
Clarke	1880	6,378,249.1	6,356,514.9	1/293.46	Africa, France
Clarke	1886	6,378,206.4	6,356,583.9	1/294.98	North America, Philippines
Airy	1849	6,377,563.4	6,356,256.9	1/299.32	Great Britain
Bessel	1841	6,377,397.2	6,356,079.0	1/299.15	Central Europe, Chile, Indonesia
Everset	1830	6,377,276.3	6,356,075.4	1/300.8	India, Burma, Pakistan, Afghan, Thailand

Figure 9.2 represents a meridian section of a spheroid through its polar axis OP. OQ represents the plane of the equator. The figure is an ellipse defined by the lengths of OQ and OP.

To calculate connecting points on this surface, it is first necessary to develop formulas linking the curvature of this surface with the elements *a*, *b*, and *f* as defined below.

$$f = (a - b)/a$$

where

f = flattening factor

a = length of OQ (semimajor axis) or the equatorial radius

b = length of OP (semiminor axis) or the polar radius.

Thus, we now have three shapes depicting the earth—its actual irregular shape, the geoid (smooth earth), and an ellipsoid (mathematically defined shape that most closely fits the geoid). Measurements made on the earth can be transferred to the geoid surface with minimal error. The spheroid that is the closest fit to the geoid can be defined by observing its meridian arcs (see fig. 9.3).

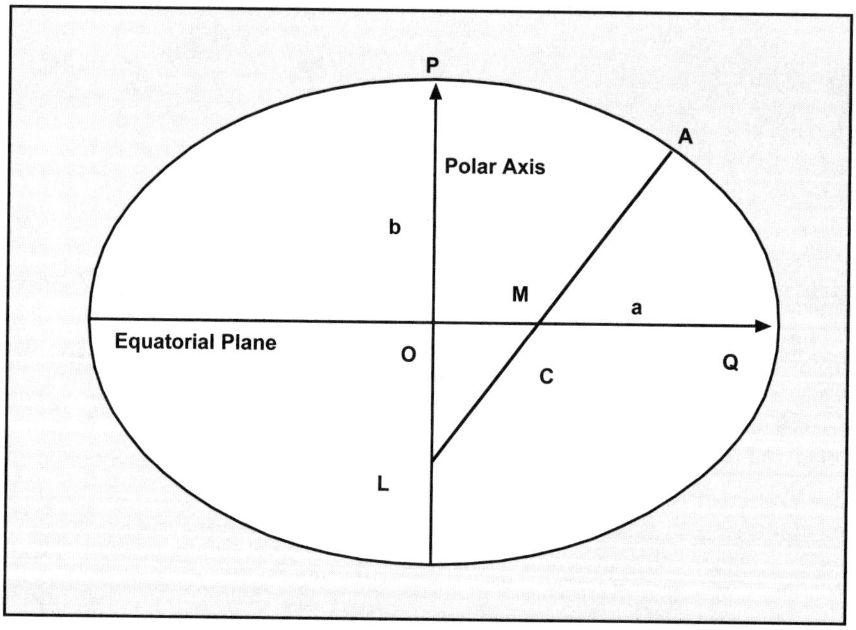

Figure 9.2 Meridian section through a spheroid through its polar axis

Figure 9.3 represents a meridian section through a spheroid whose shape is to be compared with that of the geoid. The shape of the ellipse *PAQ* can be calculated if lengths *OQ* and *OP* are known. The latitudes at *A* and *B* can be determined astronomically. The position of *B* can also be determined mathematically by analyzing the spheroid surface specified by known lengths *OP* and *OQ* and by using the initial fixed point at *A*. It can also be determined by triangulation. The base line at *A* (spheroid surface) is dropped to sea level (geoid surface). Careful triangulation will enable the position of *B* to be determined. The difference in the position of *B* between the two methods above and that determined astronomically is a measure of how closely the spheroid matches the geoid. The above should be repeated for several meridians at differing latitudes to confirm the match; i.e., repeat with points *C* and *D*. Gradually, the lengths of *OP* and *OQ* are varied until a close fit between the geoid and spheroid is obtained.

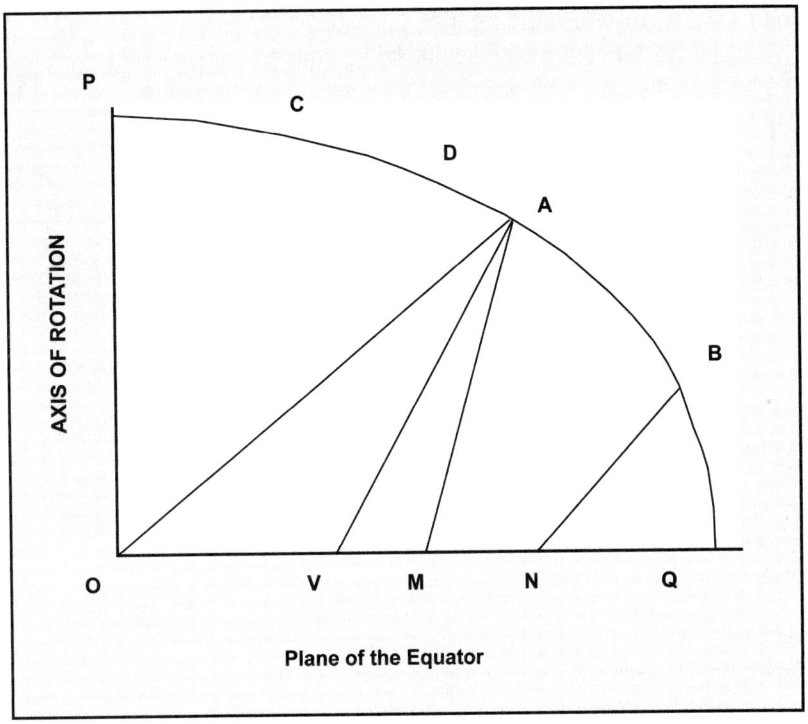

Figure 9.3 Meridian section through a spheroid whose shape is to be compared with that of the geoid.

Grid Systems

Grid systems are lines running east–west and north–south to generate a pattern of squares. In attempting to superimpose a grid system over a map of the earth, problems arise in trying to represent the spherical surface of the earth in two dimensions without incurring too much distortion. The cartographer must decide which characteristics he wishes to display precisely at the expense of others. Different grid systems are in use today to project part of the earth onto a flat surface and are generally classified according to their method of construction with regard to the shape of the developed surface used; i.e., cylindrical (UTM) and conical (Lambert).

Universal Transverse Mercator Grid (UTM)

This is the most common worldwide grid system used today and is based on the Cylindrical Mercator Conformal Projection developed by Johannes Lambert in 1772. In this projection, the spheroid representing the earth is wrapped in a cylinder, which touches the surface of the spheroid along a chosen line of longitude (fig. 9.4).

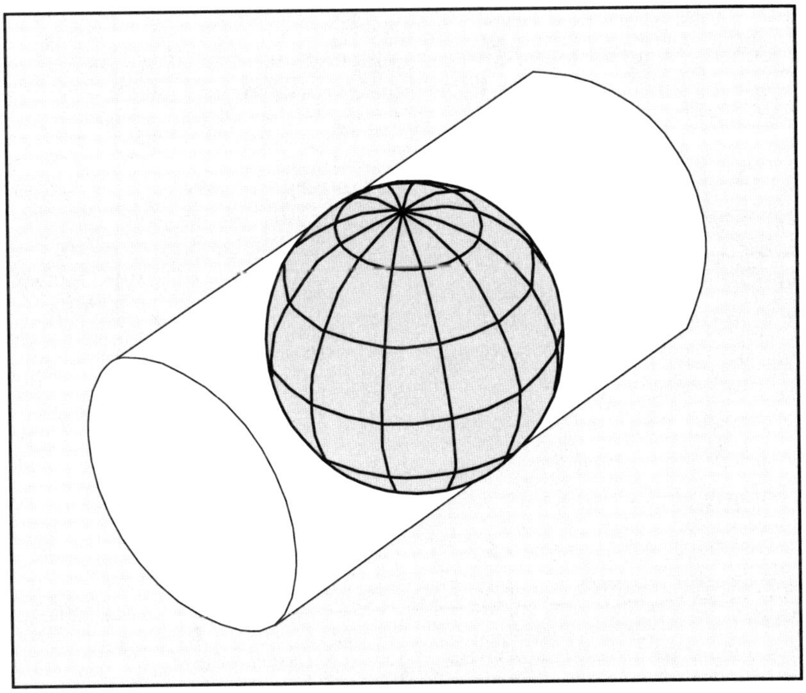

Figure 9.4 Transverse cylindrical map projection

The UTM grid (fig. 9.5) divides the world into 60 equal zones between 80°N and 80°S and each are 6° wide being numbered 1 to 60 beginning at the International date line (180°W) and reading eastward around the globe. Each zone is flattened and a square grid superimposed on it. Any point in the zone can be referred to by its zone number, its distance in metres from the equator (northing), and its distance in metres from the north–south reference line (easting).

To avoid negative numbers, an arbitrary value of 500,000 metres easting is assigned to the central meridian in each zone. Easting values typically range from 200,000 metres to 800,000 metres at the equator (spanning the 3° either side of the central meridian in any zone).

Northings for a point north of the equator begin with a value of zero at the equator increasing northwards. For points to the south of the equator, the equator is assigned an arbitrary value of 10,000,000 metres and values decrease southwards.

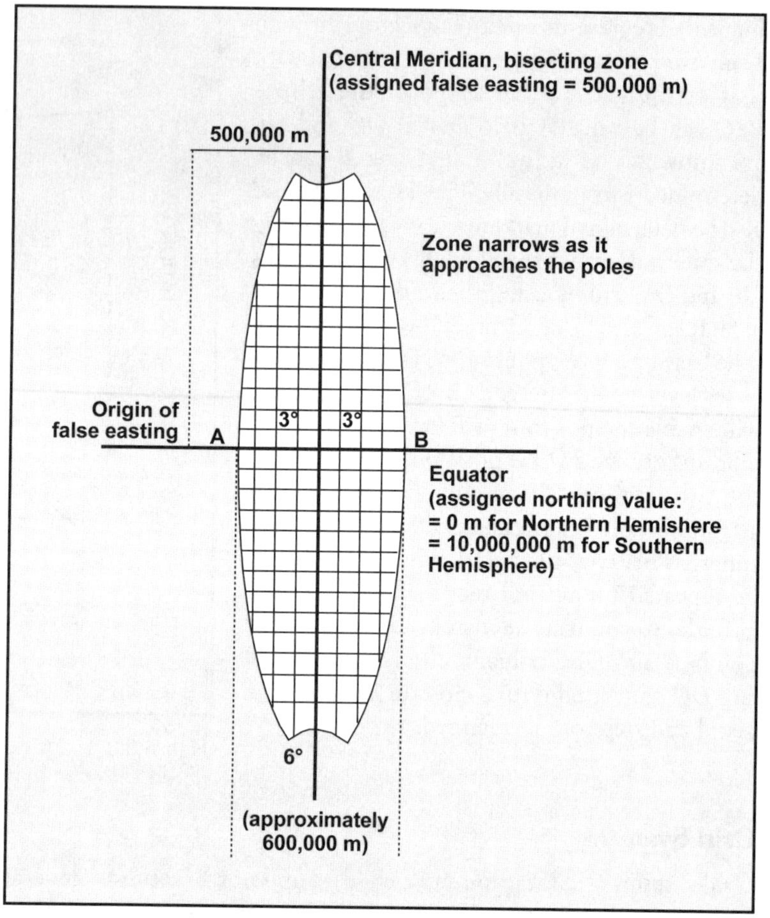

Figure 9.5 UTM grid zone

Lambert Conformal Conic Projection

This projection was first described by Lambert in 1772 but was little used until the First World War when France made use of it in drawing up battle maps. This system uses a cone as opposed to a cylinder to cover the spheroid being considered. This produces a representation with lines of longitude appearing as convergent lines and lines of latitude as arcs of circles (fig. 9.6).

The scale along the parallels is usually undistorted and a pair of parallels suitably separated would be selected for a particular area of interest. This system is utilized by 48 states in the USA (including most of the oil-producing states). The grids across each state are measured in feet with the east–west axis assigned a value of zero while the north–south axis is assigned a value of 2,000,000 ft. The example in Figure 9.7 is for the South Louisiana Lambert System.

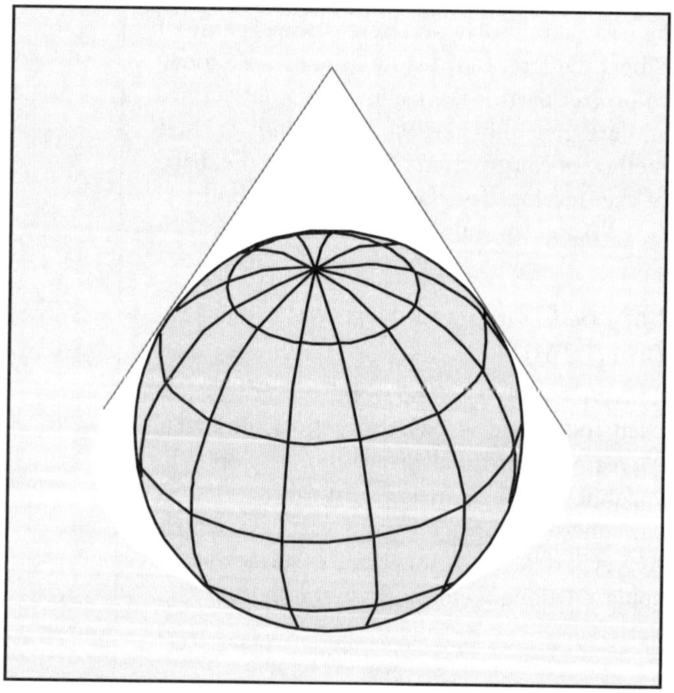

Figure 9.6 Regular conic map projection

Other Grids

There are other grid systems in use, such as:

Universal Polar Stereographic Grid (UPS) used in the polar regions.

Transverse Mercator (Gauss-Kriiger) with 6° zones used in the FSU, China, and Eastern Block.

Magnetic Declination

The earth possesses a magnetic field due to its relatively iron-rich core. The lines of force associated with this field are horizontal at the equator while, at the poles, they are represented by vertical lines. The dip angle is the angle between the horizontal and magnetic lines of force. The poles of this magnetic field 'wander' with time and the difference between their position; i.e., magnetic north, and the geographical poles; i.e., true north (the axis about which rotation of the earth occurs) is known as the magnetic declination. The actual measured distance is about 1,000 miles. The angle of declination is taken to be the angle between the horizontal component of the earth's magnetic field and the lines of longitude.

The magnetic field strength is measured in micro Tesla (mT) and it varies from 30 mT at the equator to 60 mT at the poles (fig. 9.8).

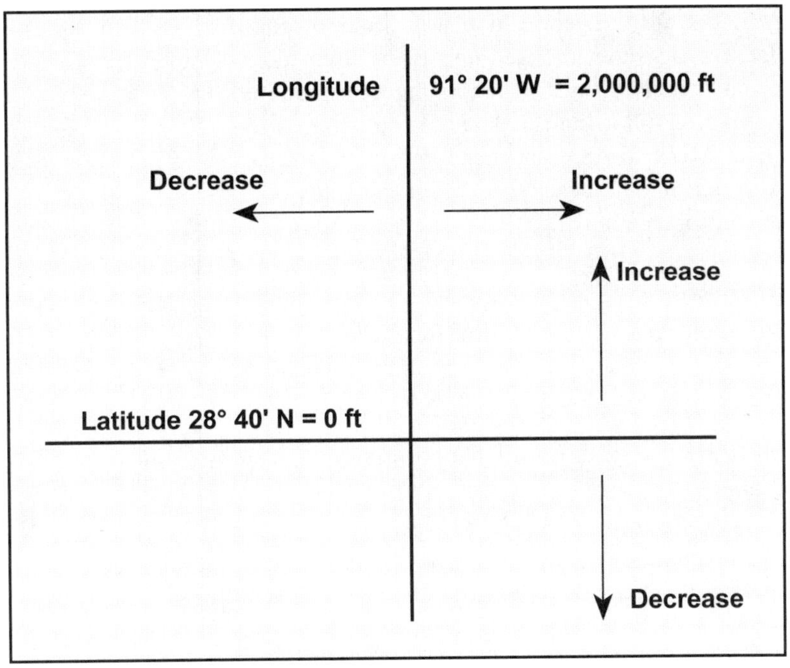

Figure 9.7 South Louisiana Lambert System

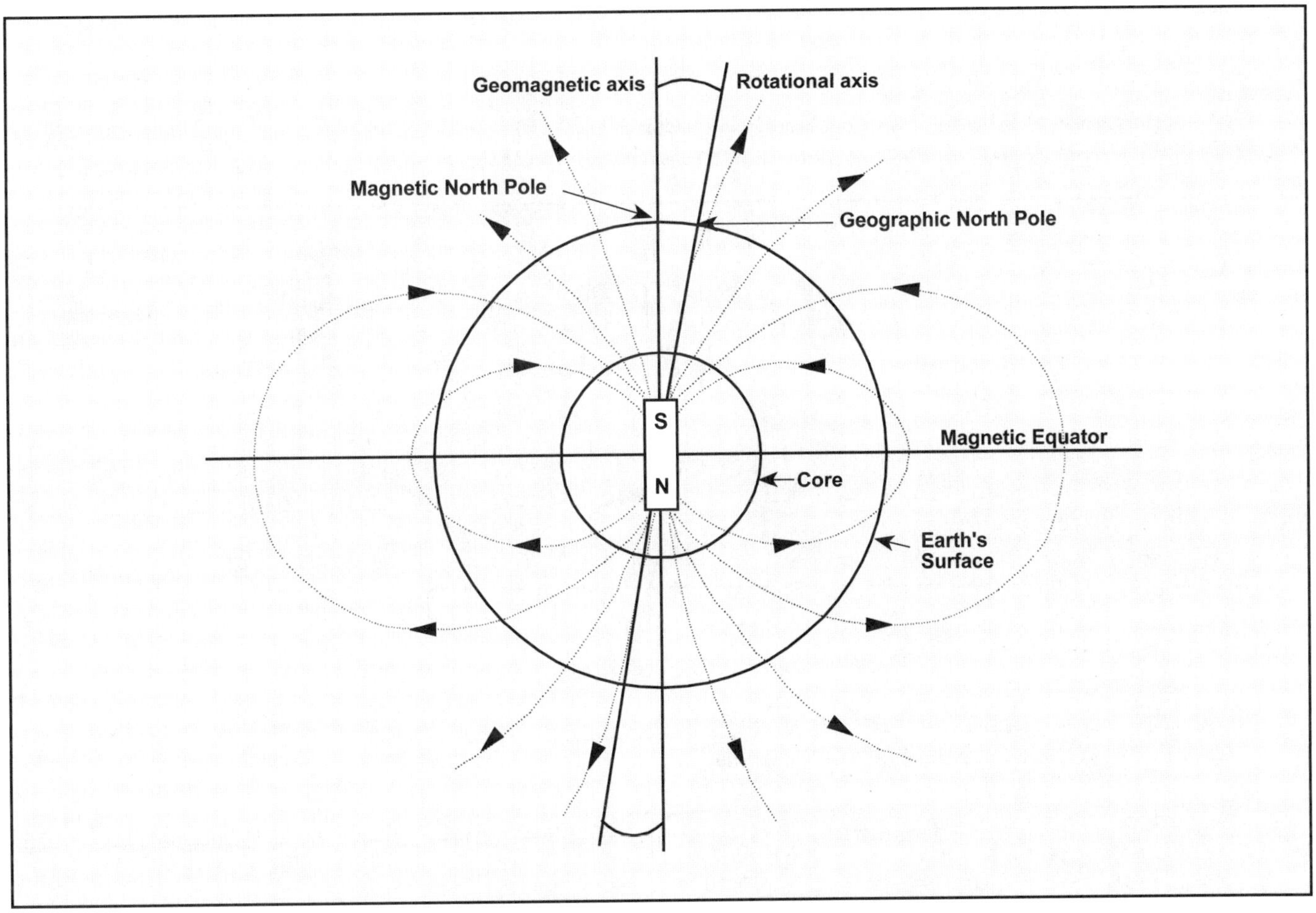

Figure 9.8 Magnetic surveys follow the flux lines of the earth.

An isogonic chart is a map with lines of equal declination superimposed and is usually accompanied with an annual correction figure that should be applied following the date of production of the chart. Declination can either be west or east; i.e., west declination means that the magnetic pole is to the left of true north.

Example

Magnetic declination of Houston from the 1985 isogonic chart was 7° east. The annual change is about 8' per year West (note 1° = 60').

In 1991, the magnetic declination should have changed:

8' per year × 6 years = 48' west

The new magnetic declination in 1991 = 7° 0'–0° 48' = 6° 12' east.

To convert from magnetic north to true north is a matter of simply considering the positions of the two poles in relation to the magnetic reading.

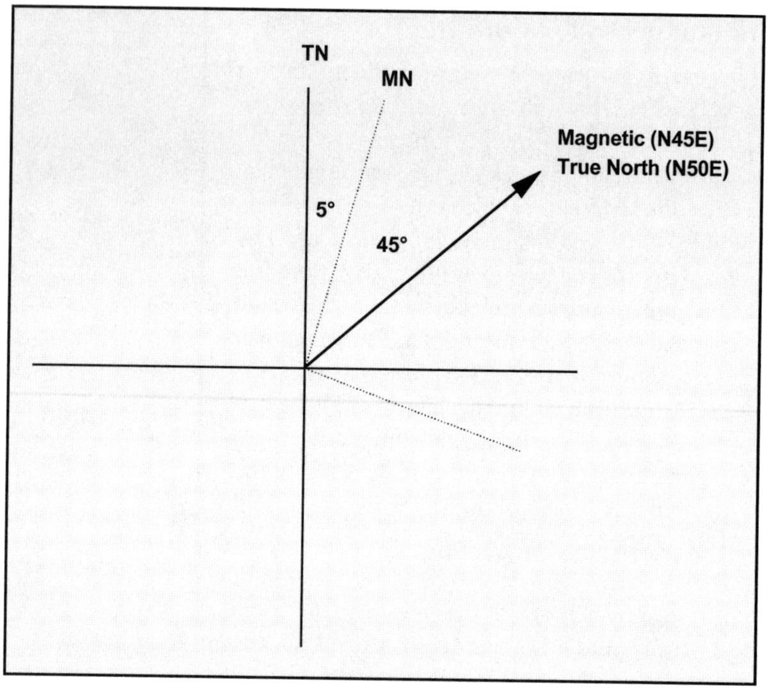

Figure 9.9A Magnetic north readings

Example

Convert the following magnetic north readings to true north (figs. 9.9A and 9.9B).

1. N45E (Azimuth 45) with 5° east declination. Answer: N50E (50)
2. S80W (Azimuth 260) with 5° west declination. Answer: S75W (255)

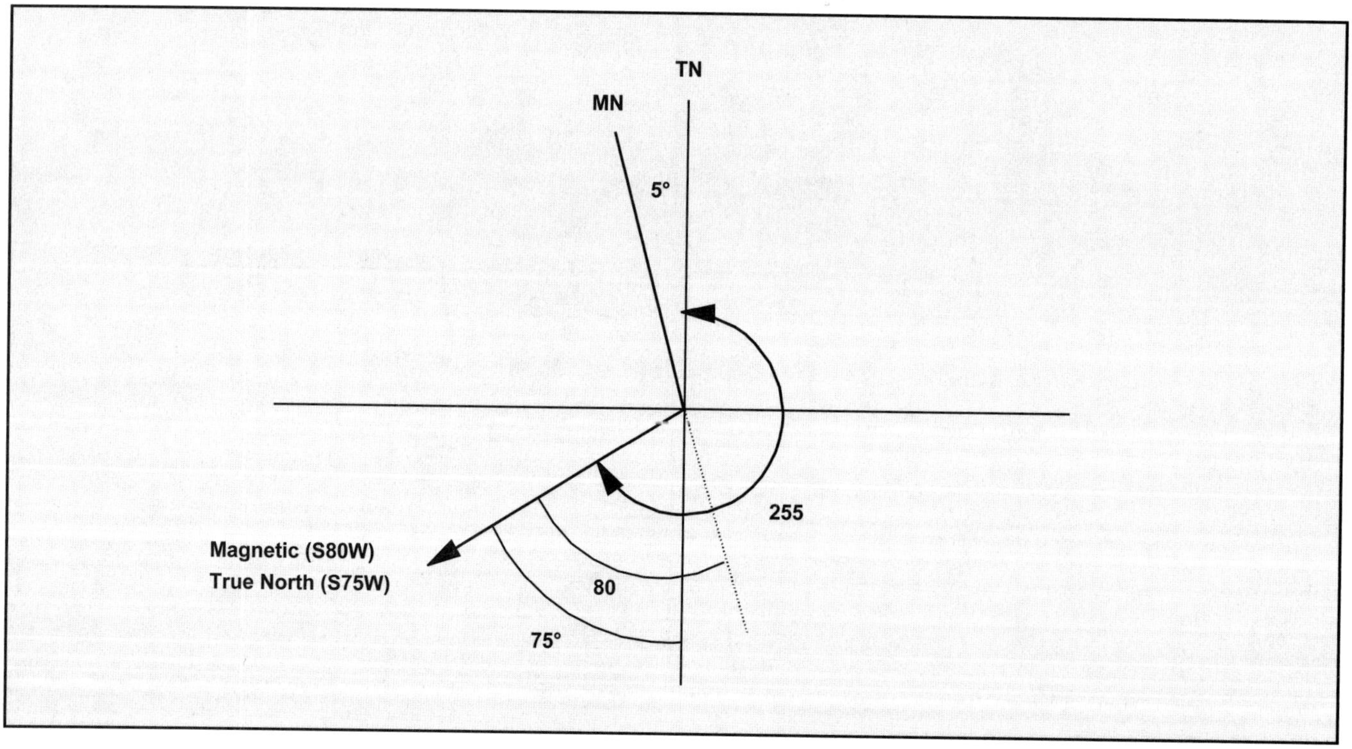

Figure 9.9B Magnetic north readings

Mapping Grids

On land, surveys are corrected to true north while offshore, they are corrected to the grid north standard. A grid system is merely a rectangular coordinate system charted on a map. An arbitrary latitude and longitude has usually been selected and, in this instance, grid north coincides with true north (fig. 9.10).

Grid declination—angular correction converting readings of magnetic north to grid north.

Grid convergence—angle between true north and grid north.

Misunderstandings surrounding the relationships between these various references have prompted serious survey errors in the past. Polar diagrams, as shown in Figure 9.11, should be included with all survey data to clarify which reference system is being used.

Summary

Using the previous information, it is now possible to calculate with reasonable precision where we are in relation to other points on the earth's surface. However, it is imperative that one knows which model and grid system are being used. These will vary from area to area. UK Ordnance Survey maps are derived from the Airy Spheroid based on measurements of the Greenwich Meridian and use the British National Grid. For North Sea offshore positioning and survey work, normal practice is to work with the International Spheroid based on the Central Meridian (3° east) on a UTM grid projection. Near shore areas of the North Sea can thus be very confusing regarding which systems are being used. The moral is to ensure that the systems being referenced are clearly stated on the survey data before such data is used.

SURVEYING TOOLS

Surveying is the science of accurately locating a point in space. In well construction this means accurately locating a point in the wellbore. It enables the bottomhole location to be determined relative to a surface location at a given vertical depth. It also provides information on wellpath irregularities (doglegs), drift during drilling, and the orientation of deflecting tools.

Current systems include telemetry, electronic measurement packages, photographic-based systems, and continuous surface readout systems.

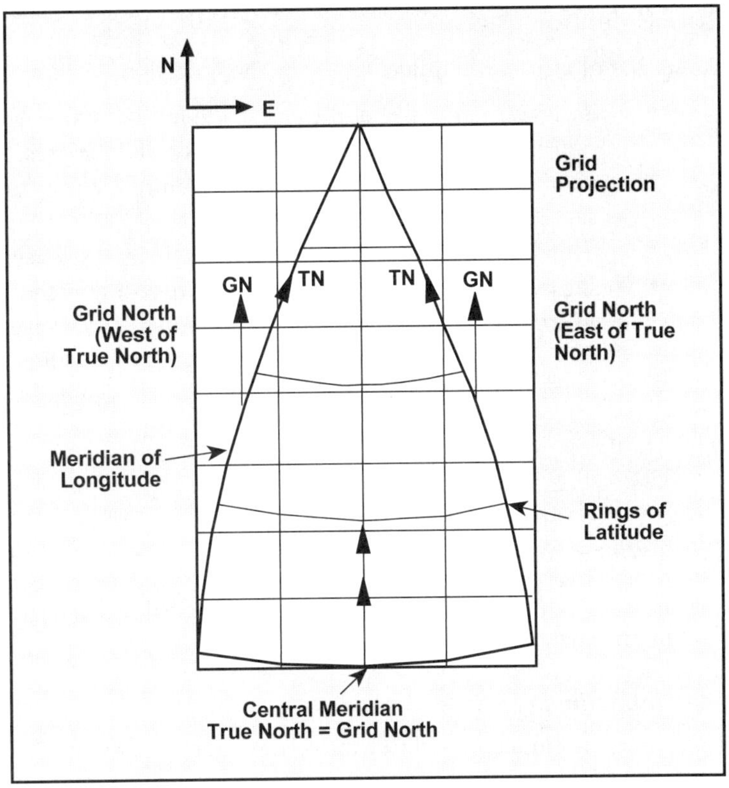

Figure 9.10 Grid north vs. geographical north

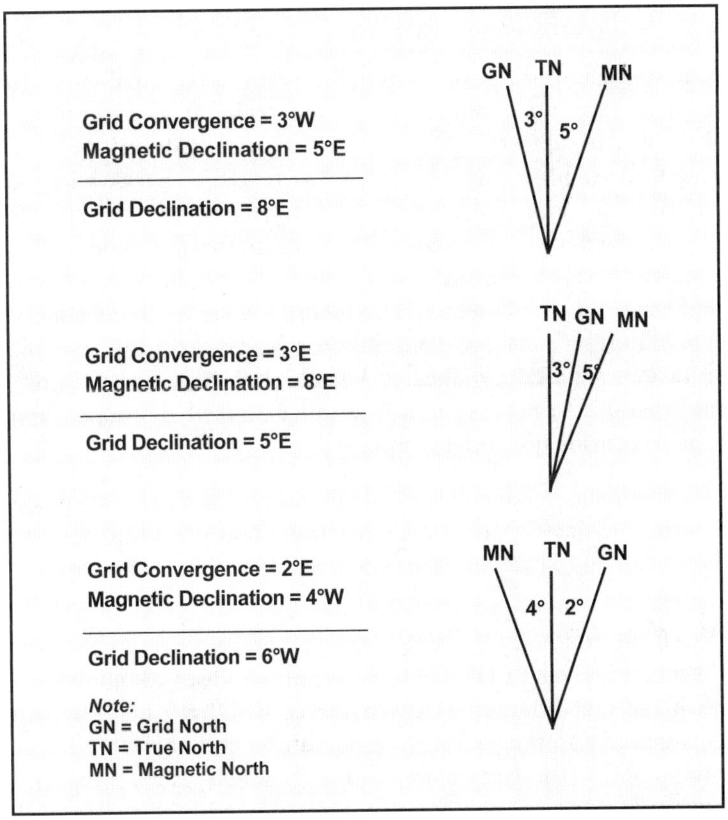

Figure 9.11 Relationship between grid, true, and magnetic north

Tool Selection Factors

A number of factors will influence survey tool selection:

- *Target size*—the size of the target will influence how accurate the survey must be.
- *Latitude of the well*—the greater the latitude, the greater the effect on both magnetic and rate gyroscopic tools.
- *Target directions*—east/west surveys require special procedures for both magnetic and north-seeking gyroscopic sensors.
- *Type of installation*—presence of magnetic interference on multiwell platforms.
- *Rig costs*—MWD system might be more cost effective on high day-rate rig.
- *Maximum inclination proposed*—some survey tools have hole angle limitations.
- *Hole conditions*—high temperature, open, or small hole size may limit the use of some tools.
- *Survey depth*—the accuracy of a survey is dependent on the depth.
- *Open* or *cased hole*—impact on magnetic tools.

Magnetic Tools

Photographic Magnetic Survey Tools

All these tools must be run inside a nonmagnetic drill collar (manufactured from a nickel alloy) to eliminate any magnetic interference from the drill string.

Magnetic Single Shot

Used to record simultaneously the magnetic direction of an uncased borehole and its inclination from the vertical. Component parts comprise:

- Timing device or a motion sensor
- Camera
- Angle indicating unit

Due to the uncertainties regarding the time taken for the tool to drop from the surface down to the survey depth, a motion sensor was developed to supersede the timer. After the tool has been *armed* and launched in a protective barrel (1.75" or 1.375") either on wireline or free fall, once it becomes stationary on bottom, electronic circuitry within the tool senses this and activates the camera. The prefocused camera records both a magnetic compass orientation and the position of a plumb bob giving the angle of inclination. The film discs with angle indicating scales on them are available in various ranges depending on the hole angle; e.g., 0–10°, 0–20°, and 15–90°. Back on surface, the disc is extracted, developed, and read to provide hole orientation.

The single shot is normally run on wireline during the drilling phase to provide a single measurement for current bottomhole orientation or it can be launched free fall onto a dull bit prior to tripping.

Magnetic Drop-Type Survey

This tool is similar to that above except that it uses 10 mm film to enable several different surveys to be recorded (fig. 9.12). The tool is launched and eventually seats in a catcher just above the bit. A mechanical timer, synchronized with a stopwatch at surface, enables a series of surveys to be stored as the drill string is recovered giving a very simple view of the open-hole orientation.

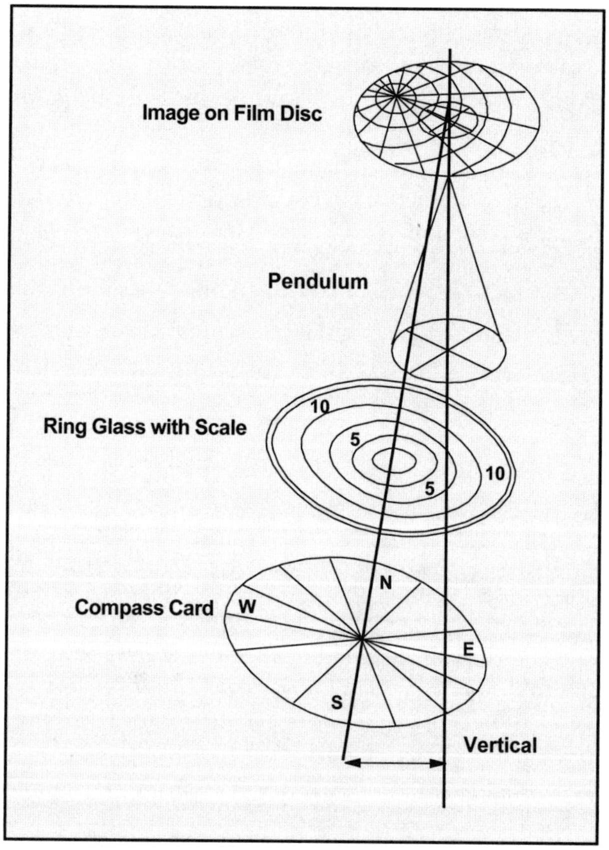

Figure 9.12 Magnetic drop-type survey

Magnetic Multishot (MMS)

This is normally run at the end of a section and is used to record the orientation of a wellbore. The tool is a battery-powered unit comprising a timing device, a camera loaded with 16 mm film, and a compass/pendulum assembly, and is normally encased in a 1.75" barrel. A mini multishot version in a 1.375" barrel is also available. Essentially, the tool takes a photograph of the compass/pendulum periodically as the drill string is pulled out. Once back on surface, the film can be developed, correlated against depth, and the open-hole section orientation plotted out.

Electronic Magnetic Multishot (EMS)

This is the latest technology and comprises both triaxial accelerometers and magnetometers for taking measurements of hole angle and direction. It also calculates magnetic dip angle and field strength at each survey station. These are used to calculate downhole magnetic interference giving greater confidence in the survey data. The tool also measures downhole temperature over the range 0–125°C (32–257°F). After arming at surface, the tool is run in a similar manner to the simpler multishot and can be programmed to operate in either single shot, multishot, or core orientation mode. Survey data for up to 1,023 data points can be stored.

Following the survey, the tool is reconnected to the system computer and the information processed using a reference tie-in point; i.e., previous casing shoe. As an option to running the required length of NMDCs to avoid interference, it is possible to shorten this and apply a correction to eliminate any additional interference.

Gyroscopic Survey Tools

Where magnetic interference cannot be averted, gyroscopic systems are used. There are two types of gyro used—those that drift with time and those that sense the rotation of the earth; i.e., 'rate gyros'. These can be run without nonmagnetic drill collars since the magnetic compass is replaced by a gyro compass disc controlled by a high rpm electric motor.

Gyroscopic Single Shot

Used to orient tools in areas of high magnetic interference; e.g., kicking off on a platform well with other conductors around. The tool consists of a timing device, a camera section, and a very sensitive gyro compass. The gyro is first oriented at surface to a known direction and then lowered to the survey depth in a protective casing. After recording the survey, the tool is recovered and the film unloaded and developed.

Gyroscopic Multishot

This is used to record the orientation of a cased or uncased wellbore. The tool is battery-powered similar to the magnetic multishot. The gyro unit is initially aligned with a known direction and the timing device synchronized with a surface watch. It is then run in on wireline and surveys are taken as the tool is going down. This reduces the error due to gyro drift, which increases nonuniformly with time. To correct for this, several drift checks are made both running in and pulling out. At these, the tool is held stationary while a number of surveys are taken at the same location. These can then be compared afterwards to enable a drift correction to be applied.

Surface Read-Out Cyro (SRC)

Example—*Seeker from BHI, Keeper from Sperry Sun*

Used for drilling and orienting in high cost areas where magnetic interference is a problem; e.g., kicking off platform wells adjacent to other conductors. The tool is connected to a direct surface readout system via a hard-wired cable and side entry sub. Once the desired orientation has been achieved, the tool is recovered to allow the side entry sub to be removed. Later variants now come equipped with a wet connect that enables the electrical lead to be severed and rotary drilling to continue for a period without pulling the assembly.

The gyro system used in the SRG comprises an orthogonal axis (all at right angles to each other) accelerometer and magnetometer. The accelerometer measures the gravitational vector relative to the tool axis from which the tool face and inclination can be determined. The magnetometer measures the components of the earth's magnetic field relative to the tool axis which, when combined with the accelerometer readings, determines the azimuth.

Ring Laser Inertial Guidance Surveyor (RIGS)

This is a high-accuracy, high-speed surveying system, which gathers survey data as the tool is run through the borehole. It comprises a three axis inertial navigation system and is accurate to within 1–2 ft/1,000 ft of hole surveyed with a horizontal accuracy of 2.6 ft/1,000 ft. These results are typically three times more accurate and completed in half the time compared with using a rate gyro.

At the beginning of the survey, the tool is aligned to derive a true north reference by measuring the earth's rotation. As it is then lowered into the well, the inertial navigation system measures the changes in three-dimensional space generating coordinates for north/south, east/west and down along the wellbore. This continuous monitoring system eliminates the errors generated using a point-to-point wellbore geometry survey calculation method as used with rate gyros and conventional survey systems.

A precision wireline depth measurement system and a casing collar locator are used to verify sensor depth and restrict errors to less than 0.5 ft/1,000 ft while roller centralizers keep the tool centralized within the wellbore. The tool can only be used in hole/casing sizes down to 7" due to its physical size.

A typical survey using the RIGS system would comprise:

- Align the sensor at surface—12 mins
- Perform a drift check—3 mins
- Run the tool into the hole at 300 ft/min
- At final depth, perform inertial drift check—3 mins
- Retrieve the tool while completing second survey at 300 ft/min
- At surface, complete a final drift check—3 mins

Total time to survey a simple 10,000 ft well = 88 mins.

MWD Survey Measurement Systems

The MWD systems of today evolved from the hard-wired surface read-out systems. Power is generated either downhole via a mud turbine or from batteries. Data transfer is by pressure pulse via the drilling fluid column (either negative or positive pulse).

These systems were developed in the 1970s with Teleco being the first company to provide one commercially in 1978. These became commonplace during the 1980s and have continued to be developed with additional sensors now being the norm; e.g., gamma ray and resistivity sensors. These systems are, however, still magnetic measurement based and must therefore be compensated or protected from magnetic interference.

Inclination only MWD

Example—*Teledrift or Anderdrift*

These systems comprise a mechanical signalling instrument capable of detecting hole inclinations up to 10.5° (Anderdrift 5°, Teledrift 10.5°). Signal transmission is by way of a series of mud pulses, which are detected by a transducer usually sited on the standpipe.

In principle, the downhole tool contains a pendulum at the bottom of the tool that is able to move along a series of shoulders and a signal plunger at the top capable of traversing a series of annular restrictions. The latter creates the mud pulses. In the Teledrift system, there is a patented coding system so that the deviation from vertical increases the number of pulses issued. A maximum of seven pulses can be generated (0.5°/pulse) meaning that the tool can operate over a 3.5° range between 0 and 10.5°. Adjustment is easily achievable in the field to ensure that the tool can respond to any hole angle up to 10.5°. Several options also exist that enable the tool to detect a greater range of hole angle. This is done by increasing the angle measured per pulse such that it can either report 1° or 1.5° per pulse. This gives a maximum measurable hole inclination of 15°.

The Anderdrift system generates similar mud pulses but there is a different relationship between the pendulum position and the plunger. The Anderdrift can generate up to eleven pulses, again each representing 0.5°, but these function in a reverse manner. Thus, one pulse will indicate a hole angle of 5° or more while eleven pulses indicate a hole angle of zero degrees.

Readings are generally taken at a connection. The pumps are shut down for about a minute to allow the tool to measure the drift angle and prepare for signalling. The pipe must remain stationary during this period. The pumps are then restarted and brought up to a maximum of 360 gpm and held at that. With the Teledrift tool, the first pulse should arrive after about 10–15 secs and subsequent readings should arrive after a similar period. Pumping should continue until about a minute after the last recorded pulse. The pulse period for the Anderdrift tool is about 5 seconds.

Magnetic Interference

During the drilling process, the steel components of the drill string become magnetized due to the earth's magnetic flux lines. This induced magnetism influences the magnetic survey tools and so they must be protected by being sheathed in nonmagnetic collars. The length of nonmagnetic spacing required is dependent on the following factors:

- Magnetic pole field strength of the magnetized steel drill string above and below the sensing device
- Borehole direction as related to magnetic north or south
- Borehole inclination
- Geographic location (related to dip angle)

Magnetic Pole Strength

As the drill string becomes magnetized, the two ends become the magnetic poles of the string. In the northern hemisphere, the north pole end of the string points down the hole. The magnetic interference varies with the inverse square of the distance between the source and the compass/sensor.

$$\text{Interfering Force} = (\text{Pole Strength})/(\text{Distance})^2$$

Thus, if a force F_1 is found to result from a separation of 4 ft between magnetized steel collar and the sensor, then, at a separation of 12 ft, the interfering force F_2 would be:

$$F_1 \times (4)^2 = F_2 \times (12)^2$$

$$F_2 = F_1/9.$$

Since the drill string is much longer than its diameter, it can be analyzed as though it comprised discrete point sources of magnetism located at the extreme ends of the hole. The strength of the magnetic poles depends on the individual steel components which make up the drill string. Grindrod and Wolff in 1983 presented pole magnetic strength values for four different bottomhole assemblies. These generalized values are:

Component	Electromagnetic Units
Upper drill string	300 EMU
Bit, mud motor, and bent sub	2500 EMU
Bit, NB stab, short collar (packed assy)	200 EMU
Bit and NB stabilizer	500 EMU
Crossovers	250 EMU

The approximate pole strength is selected for the appropriate pole—either the upper pole (UP) or the lower pole (LP). If the pole strength is measured in microwebers (μWb), then the distance is in metres; if the pole strength is in electromagnetic units (EMU), then the distance is in feet.

The upper distance (UD) and lower distance (LD) are merely the respective distances from the sensor to the nearest magnetized section of the drill string. The interfering force (IF) can now be calculated:

IF (upper) = UP/(UD)² and
IF (lower) = LP/(LD)².

Thus

IF (total) = IF (upper) + IF (lower) or
= IF (upper 1) + IF (upper 2) + IF (lower).

These two cases are represented schematically in Figures 9.13A and 9.13B.

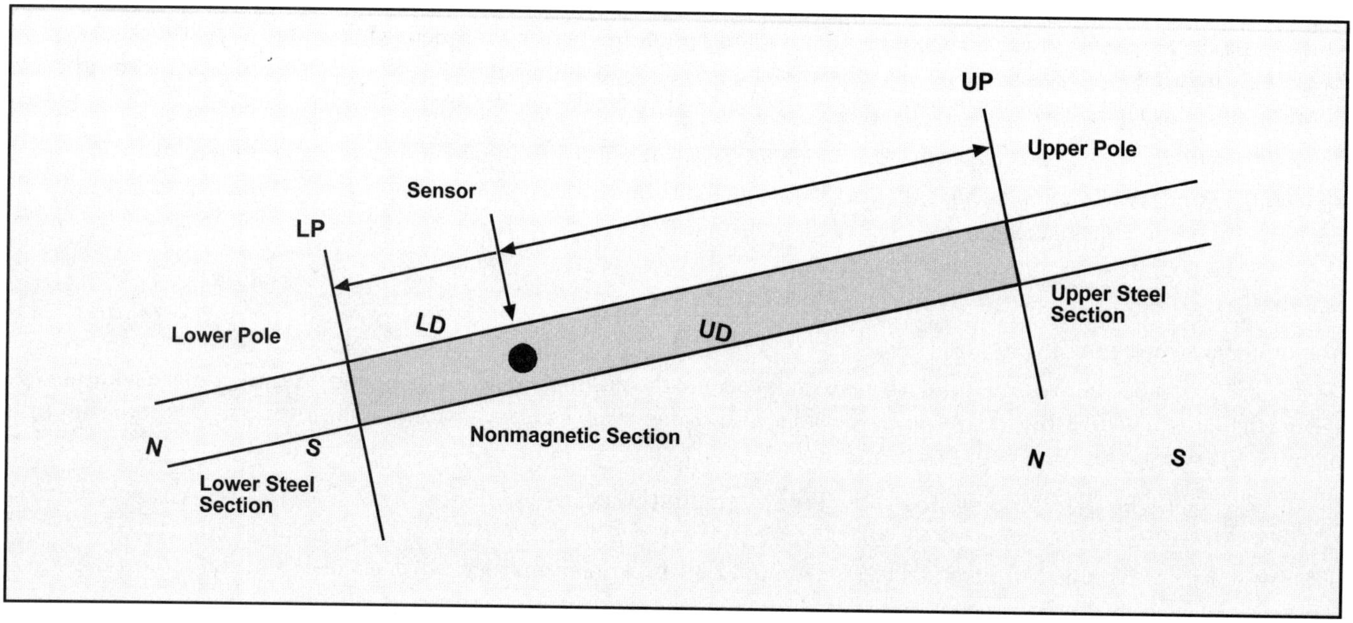

Figure 9.13A Both lower and upper steel sections have individual poles

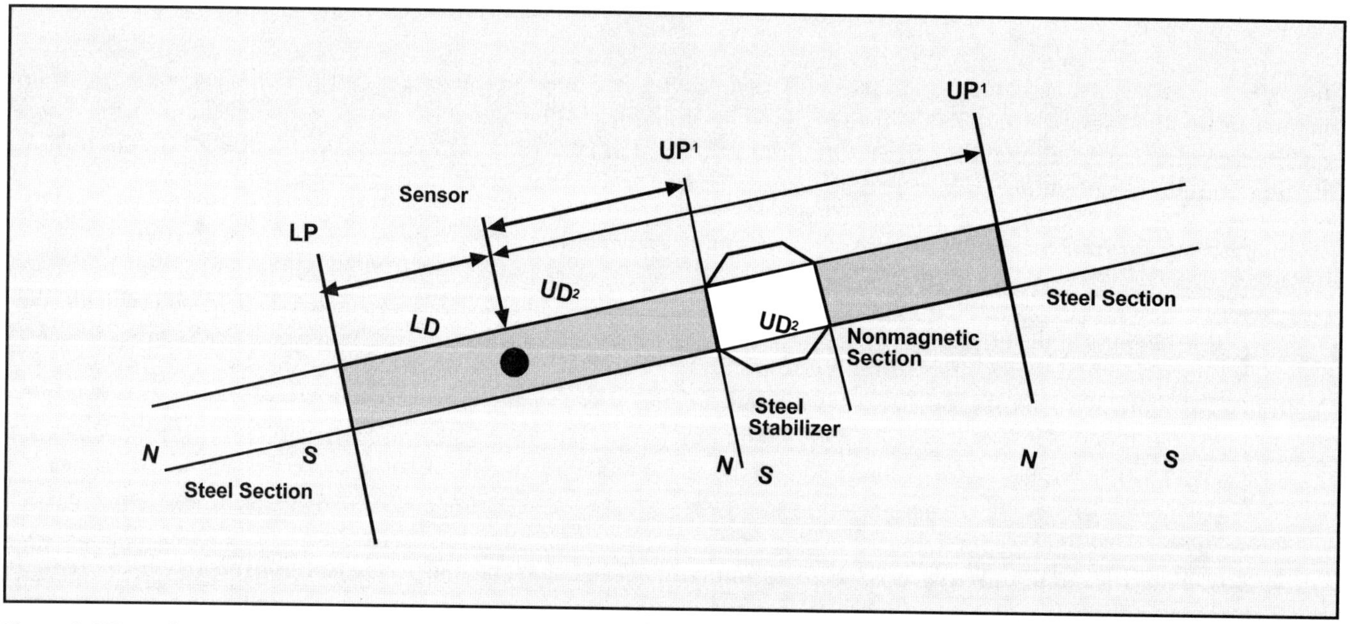

Figure 9.13B The steel stabilizer breaks the (upper) distance into two segments

Borehole Inclination

A levelled magnetic sensing device (compass or magnetometer) only relies on the horizontal component of the earth's magnetic field. Similarly, it will be influenced by the horizontal component of any induced magnetism in the drill string. As the drill string approaches greater angles in a deviated well, then the interfering force will be greater.

Borehole Direction

The direction (azimuth) in which the drill string is positioned is also a factor in determining the effect of the interfering force on the magnetic sensor (fig. 9.14). As the direction approaches due east or west, the effect is greatest.

For a well being drilled in a due easterly direction, the magnetic sensor will read an azimuth that is smaller than the actual azimuth.

For a well being drilled in a due westerly direction, the magnetic sensor will read an azimuth that is greater than the actual azimuth.

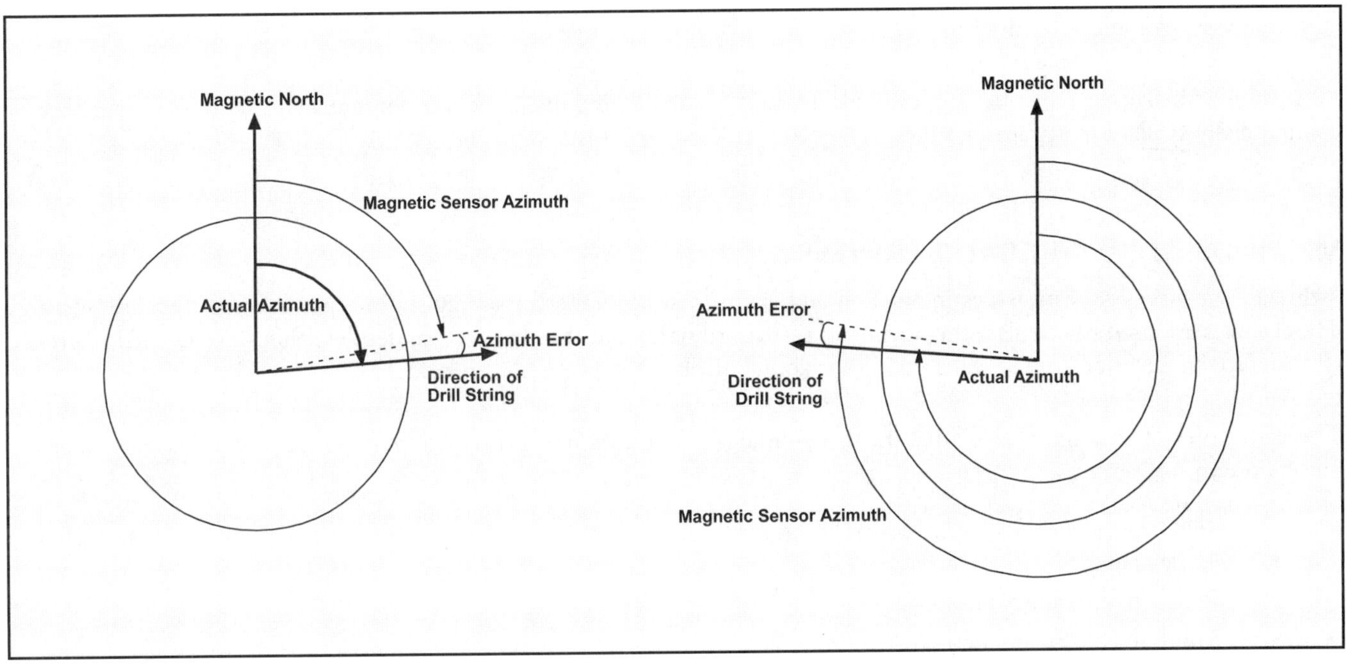

Figure 9.14 Effect of azimuth on survey readings

Geographical Location

Just as a levelled sensing device will be affected by the horizontal component of the earth's magnetic field, so it will also be affected by the amplitude of this horizontal component. This varies with geographical location (see fig. 9.8) and is greatest along the equator. Moving either north or south from the equator, the angle of dip of the earth's magnetic field increases. The resultant effect of this is that the vertical component of the earth's magnetic field increases while the horizontal component decreases. Thus, any tool that is dependent on the horizontal component of the earth's magnetic field is more likely to be affected by other horizontal field effects the farther away from the equator that it is used.

A magnetic sensor will therefore sense magnetic north more strongly at the equator than at the poles. If there is an interfering force present, this will be seen to have more influence on directional readings nearer the poles than at the equator even though the interfering force itself remains the same. In effect, any lessening of the earth's natural field allows the drill string field to have more impact.

Nonmagnetic Collar Selection

The magnetic sensors must be housed in nonmagnetic drill collars to minimize the interference effects derived from induced magnetism in the drill string and to prevent any distortion of the earth's magnetic field in the vicinity of the sensors. The composition of the nonmagnetic collars is usually one of the following:

1. K Monel, an alloy containing 30% copper and 65% nickel
2. Chrome/nickel steels comprising 18% chrome and 13% nickel
3. Austenitic steels based on chromium and manganese (>18% Mn)
4. Copper beryllium bronzes

- K Monel and copper beryllium are expensive but are corrosion resistant.
- Austenitic steel is most common but is susceptible to stress corrosion in high salt environments.
- Chrome/nickel steels tend to gall, causing thread damage.

As described above, the number of nonmagnetic drill collars and the position of the sensor/compass within the nonmagnetic collars will depend on the borehole inclination, azimuth, and geographical location. Charts are available from most of the surveying contractors that indicate the number of nonmagnetic drill collars required at different inclinations and azimuths in different geographical locations.

Alternatively the survey data obtained can be corrected for drill string magnetic interference by using one of a number of commercially available software packages.

METHODS OF SURVEY CALCULATIONS

A directional survey tool measures inclination and azimuth at a number of survey stations at specified measured depths. These values are used to calculate the north and east coordinates and the true vertical depth within the specified reference system. Dogleg severity and vertical section (horizontal displacement) can also be calculated.

There are several methods of calculating the three-dimensional location of a survey station. These methods are listed below in the order of increasing accuracy:

- Tangential method
- Average angle method
- Radius of curvature method
- Minimum curvature method

Tangential Method

This was the earliest used survey method. Calculations are based on the drift angle (inclination) and direction of drift angle (azimuth) at the lower of two survey points. The course length (distance between two survey points) is taken to be a straight line. This line is assumed to have the same drift angle and direction of drift angle as the lower survey point. The compound errors with this method can become very significant and hence the method is inaccurate.

For two given survey stations, S_1 and S_2, where the measured depth, MD_1 and MD_2, the inclination, I_1 and I_2, and the azimuth, A_1 and A_2, are known, then the north, east, and TVD can be calculated at S_2 as follows:

$$\text{North}_2 = \text{North}_1 + [(MD_2 - MD_1) \times \sin I_2 \times \cos A_2]$$

$$\text{East}_2 = \text{East}_1 + [(MD_2 - MD_1) \times \sin I_2 \times \sin A_2]$$

$$TVD_2 = TVD_1 + [(MD_2 - MD_1) \times \cos I_2].$$

Example

Measured Depth	Inclination	Azimuth	North	East	TVD
0	0	0	0	0	0
500	1	N28E	7.70	4.10	499.92
750	1.5	N10E	14.15	5.23	749.84
1,000	2	N35E	21.30	10.24	999.69
1,250	4	N25E	37.10	17.61	1249.08
1,500	8	N30E	67.23	35.00	1496.64
1,750	12	N35E	109.81	64.82	1741.18
2,000	15	N40E	159.38	106.41	1982.66
2,250	17	N43E	212.83	156.26	2221.74
2,500	18	N40E	272.02	205.92	2459.50
2,750	19	N37E	337.02	254.90	2695.88
3,000	20	N38E	404.40	307.54	2930.81

Average Angle Method

This method uses the average drift angle (inclination) and average direction of drift angle (azimuth) between two survey points. This significantly reduces the errors generated using the previous method and assumes that the course length lies in a straight line between survey points. The dogleg between two survey points is assumed to be small. If the distance between the two points is small, then the straight-line path assumed approximates very closely with the original wellbore.

It is simple to hand calculate in the field. For two given survey stations, S_1 and S_2, where the measured depth, MD_1 and MD_2, the inclination, I_1 and I_2, and the azimuth, A_1 and A_2, are known, then the north, east, and TVD can be calculated at S_2 as follows

$$\text{North}_2 = \text{North}_1 + \left[(MD_2 - MD_1) \times \sin\left(\frac{I_1 + I_2}{2}\right) \times \cos\left(\frac{A_1 + A_2}{2}\right)\right]$$

$$\text{East}_2 = \text{East}_1 + \left[(MD_2 - MD_1) \times \sin\left(\frac{I_1 + I_2}{2}\right) \times \cos\left(\frac{A_1 + A_2}{2}\right)\right]$$

$$\text{TVD}_2 = \text{TVD}_1 + \left[(MD_2 - MD_1) \times \cos\left(\frac{I_1 + I_2}{2}\right)\right].$$

Example

Measured Depth	Inclination	Azimuth	North	East	TVD
0	0	0	0	0	0
500	1	N28E	4.23	1.06	499.98
750	1.5	N10E	9.39	2.83	749.92
1,000	2	N35E	16.44	5.75	999.80
1,250	4	N25E	27.77	12.29	1249.46
1,500	8	N30E	50.95	24.36	1498.09
1,750	12	N35E	87.57	47.69	1744.29
2,000	15	N40E	133.87	83.21	1987.39
2,250	17	N43E	185.48	128.88	2227.70
2,500	18	N40E	241.78	178.69	2466.13
2,750	19	N37E	303.86	228.07	2703.21
3,000	20	N38E	370.07	278.87	2938.87

Radius of Curvature Method

This method assumes that the borehole is a smooth arc between surveys. Requires some complex calculations, which are best suited to a programmable calculator or computer. The calculation method is unaffected by long course lengths. It assumes wellpath has a constant radius of curvature.

For two given survey stations, S_1 and S_2, where the measured depth, MD_1 and MD_2, the inclination, I_1 and I_2, and the azimuth, A_1 and A_2, are known, then the north, east, and TVD can be calculated at S_2 as follows

$$North_2 = North_1 + \left[\frac{(MD_2 - MD_1) \times (\cos I_1 + \cos I_2) \times (\sin A_2 + \sin A_1)}{(I_2 - I_1) \times (A_2 - A_1)}\right]$$

$$East_2 = East_1 + \left[\frac{(MD_2 - MD_1) \times (\cos I_1 + \cos I_2) \times (\cos A_1 + \cos A_2)}{(I_2 - I_1) \times (A_2 - A_1)}\right]$$

$$TVD_2 = TVD_1 + \left[\frac{(MD_2 - MD_1) \times (\sin I_1 + \sin I_2)}{(I_2 - I_1)}\right]$$

Example

Measured Depth	Inclination	Azimuth	North	East	TVD
0	0	0	0	0	0
500	1	N28E	4.19	1.05	499.97
750	1.5	N10E	9.33	2.81	749.91
1,000	2	N35E	16.32	5.71	999.80
1,250	4	N25E	27.64	12.25	1249.44
1,500	8	N30E	50.81	24.31	1498.02
1,750	12	N35E	87.40	47.62	1744.17
2,000	15	N40E	133.68	83.13	1987.24
2,250	17	N43E	185.29	128.78	2227.54
2,500	18	N40E	241.58	178.59	2465.97
2,750	19	N37E	303.66	227.97	2703.05
3,000	20	N38E	369.86	278.77	2938.70

Minimum Curvature Method

This method takes the inclination and azimuth information from each survey point and creates a smooth curve. This is a complex method not suitable for hand calculations. This method is widely used by operators and directional drilling companies.

For two given survey stations, S_1 and S_2, where the measured depth, MD_1 and MD_2, the inclination, I_1 and I_2, and the azimuth, A_1 and A_2, are known, then the north, east, and TVD can be calculated at S_2 as follows

$$\text{North}_2 = \text{North}_1 + \left[\left(\frac{(MD_2 - MD_1)}{2}\right) \times [(\sin I_1 \cos A_1) + (\sin I_2 \cos A_2)] \times RF\right]$$

$$\text{East}_2 = \text{East}_1 + \left[\left(\frac{(MD_2 - MD_1)}{2}\right) \times [(\sin I_1 \sin A_1) + (\sin I_2 \sin A_2)] \times RF\right]$$

$$\text{TVD}_2 = \text{TVD}_1 + \left[\left(\frac{(MD_2 - MD_1)}{2}\right) \times (\cos I_1 + \cos I_2) \times RF\right]$$

where

$$\cos DL = \cos(I_2 - I_1) - [\sin I_1 \times \sin I_2 \times (1 - \cos(A_2 - A_1))]$$

$$RF = \frac{180}{\pi} \times \frac{2}{DL} \times \tan\frac{DL}{2}$$

$$DL = 0$$

if $RF = 1$

and

$$\text{dogleg severity (DLS)} = \frac{DL}{MD_2 - MD_1} \times 100\ (°/100\ \text{ft}).$$

Example

Measured Depth	Inclination	Azimuth	North	East	TVD	DLS °/100 ft
0	0	0	0	0	0	
500	1	N28E	3.85	2.05	499.97	0.20
750	1.5	N10E	9.00	3.64	749.92	0.25
1,000	2	N35E	15.80	6.71	999.80	0.36
1,250	4	N25E	27.27	12.90	1249.45	0.82
1,500	8	N30E	50.25	25.29	1498.03	1.61
1,750	12	N35E	86.62	48.90	1744.19	1.64
2,000	15	N40E	132.71	84.61	1987.26	1.29
2,250	17	N43E	184.23	130.34	2227.57	0.87
2,500	18	N40E	240.55	180.10	2466.00	0.54
2,750	19	N37E	302.64	229.42	2703.08	0.55
3,000	20	N38E	368.83	280.23	2938.74	0.42

Survey Uncertainty

Survey uncertainty is the 3D range that the actual position of a survey station may be compared to its calculated position. It is the result of combining the uncertainties of all the individual data measurements for the survey. The magnitude of the uncertainty is determined by the tool type, quality, and uncertainty model of the survey tool combined with the uncertainties of the technique used to run the tool and the environment in which it is run.

Uncertainties occur due to:

- statistical uncertainties in the sensor readings from which inclination and azimuth values are derived.
- systematic uncertainties due to variations in calibration of linearity, sensitivity, bias, and drift.

For normal borehole surveys, which have a significant length and significant number of survey stations, the systematic uncertainty will dominate the statistical uncertainty. Therefore, Wolff and de Wardt in 1981 introduced a systematic tool error model. This approach to calculation of the borehole position uncertainty has become the industry standard.

The effect of these differences is an uncertainty in the measured positions, resulting in an uncertainty around the calculated borehole position. Typically, the uncertainty in the lateral direction is larger than the uncertainties in upward and measured depth. Consequently, the resulting uncertainty at a survey station has an ellipsoidal shape. The uncertainty accumulates along the well trajectory, and will be largest at the last survey station.

The FINDS tool (Ferranti Inertial Directional Tool) was the most accurate survey tool used to date at 0.5 ft in any direction. However, its size (10.625") effectively restricted it to 13.375" casing runs. Current wireline gyros are accurate to 1.5°/1,000 ft up to 60° rising to 3°/1,000 ft in horizontal sections. The accuracy of MWD tools is quoted in degrees variation for the following:

Inclination 0.1 – 0.25°
Azimuth 1.0 – 1.5°
Tool face 1.5 – 3.0°

In holes less than 5°, accuracy diminishes and an alternative system should be used.

DIRECTIONAL DRILLING

Why Directionally Drill?

Directional drilling is the science of directing a wellbore along a predetermined path to a designated subsurface target.

The most common applications of directional drilling are:

- Drilling multiple wells from offshore structures
- Controlling inclination in vertical wells
- Sidetracking
- Drilling relief wells
- Drilling horizontal or multilateral wells to expose more producing formation

Deflection Techniques

The main techniques for deflecting a well are:

- Whipstocks
- Jetting
- Rotary drilling
- Motors

Whipstocks

This was the main method of deflecting a well from 1930–1950. It was superseded by the introduction of mud motors. It has recently seen a revival due to multilateral and reentry drilling.

There are two variants of this tool, the retrievable and the permanent whipstock. Both provide a means of orienting a steel, concave wedge, which is used to deflect the drill string. Depending on the style of whipstock used, the number of trips to initiate a deflected wellbore can be a single trip to multiple trips. A lot depends on how the whipstock is set and oriented in the hole and how the starting mills perform.

Jetting

Jetting can be used to steer in soft formations and is typically used in top hole. The assembly consists of a modified tricone bit with either one jet significantly larger than the other two or with one open and two blanked jets. Essentially, the bit is oriented and washed down at maximum pump rate for 5–10 ft, rotary drilled for the remainder of the single, and then, a survey is taken with a surface readout gyro. This procedure can be repeated until the desired angle and deflection is obtained. This technique can be used to build angles up to 15° and create doglegs of 3°/100 ft.

Jetting is economic in enabling a hole to be drilled quickly without having to resort to assembly changes. It enables a full-gauge hole to be drilled with gentle changes in direction in soft formations with reduced rotating hours, and provides a useful means of steering safely through a top hole where there are numerous·conductors from adjacent wells.

Rotary Drilling

Historically, it has always been possible to control the inclination of directional wells during rotary drilling by correct assembly design and the use of appropriate drilling parameters. Azimuth control, however, has always been difficult.

The factors affecting the behavior of rotary assemblies will be discussed later.

Motors

Motors (either positive displacement motors or turbines), equipped with a bent sub or bent housing, allow the bit to be orientated and drilled in a preferred direction without any drill string rotation. This allows full control over azimuth and inclination.

Other factors affecting the behavior of these steerable systems will be discussed later.

Directional Control with Rotary Systems

Directional trends are related to the direction of the resultant force at the bit. In this respect, bit tilt angle (angle between bit axis and hole axis) is believed to be influential. This is because a drill bit is designed to drill parallel to its axis. In rotary assemblies with a near-bit stabilizer, the bit tilt angle is small and the magnitude of the side force is the key factor.

Trajectory is affected by the following parameters:

- Gauge and placement of stabilizers
- Diameter and length of drill collars
- Weight on bit
- Rotary speed
- Bit type
- Formation anisotropy (properties vary horizontally/vertically) and dip angle of the bedding planes
- Formation hardness
- Flow rate
- Rate of penetration

Gauge and Placement of Stabilizers

The gauge and placement of stabilizers, combined with drilling parameters, has a marked effect on the ability of a rotary assembly to build, drop, or hold inclination. There are three fundamental principles:

- Fulcrum principle
- Stabilization principle
- Pendulum principle

Fulcrum Principle

An assembly with a full-gauge near-bit stabilizer and 40 ft–120 ft of drill collars before the next stabilizer will build angle when weight is applied. Applying weight will cause the drill collars to flex and the near-bit stabilizer to become a fulcrum or pivot point. This creates an opposing force at the bit (fig. 9.15), which creates an upward curving hole until bit weight is reduced.

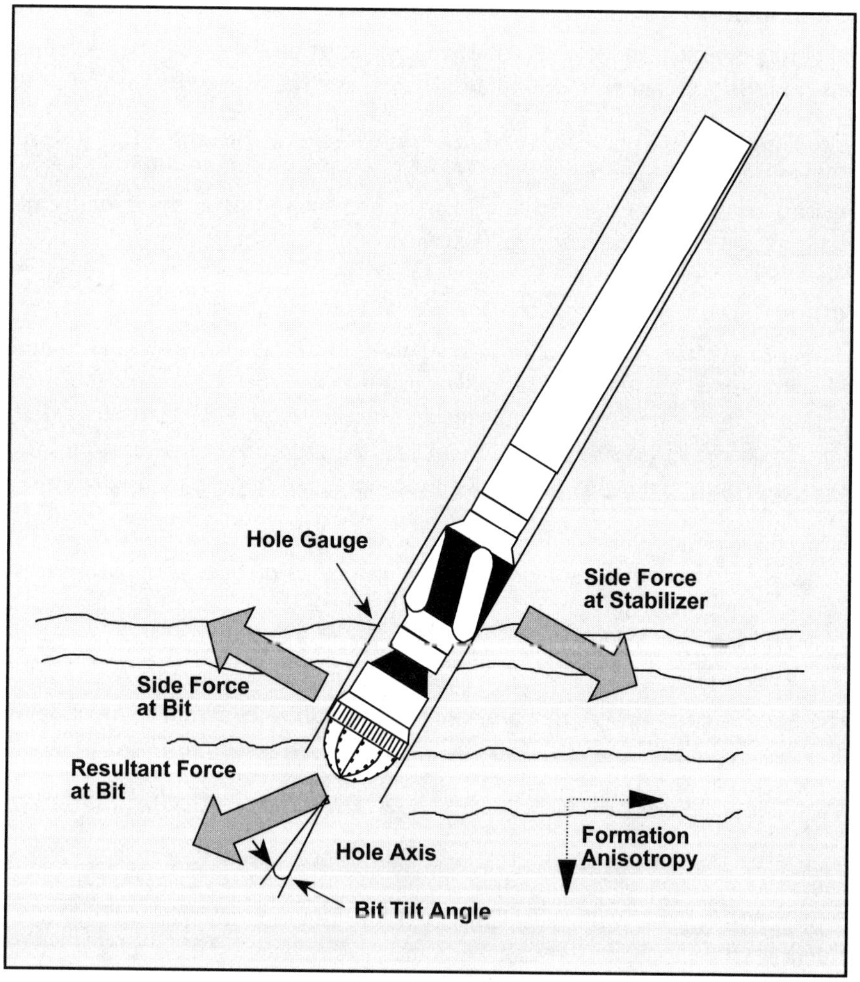

Figure 9.15 Fulcrum principle

The build rate is *increased* by:

- *An increase in distance between near-bit and first-string stabilizer*—this is the most important design element in a fulcrum assembly for building angle. As this distance is increased, the capacity to build angle increases due to the greater leverage exerted at the pivot point (near-bit stabilizer). Once the first-string stabilizer is about 120' above the near bit, no further effect will be felt since the collars will be contacting the wellbore.

- *Increase in hole inclination*—as this increases, the drill collars have a greater component at right angles to the hole axis and hence exert a greater force about the fulcrum.
- *Reduction of drill collar diameter*—the stiffness of a drill collar is proportional to the fourth power of the diameter. Thus, using a slightly smaller collar will greatly increase the flexibility of an assembly. A smaller collar will also increase the annular distance and allow the collar to flex further before contacting the wellbore.
- *Increase in weight on bit*—as this increases, it bends the collars between the two stabilizers, which increases the side force at the bit.
- *Reduction in rotary speed*—a higher rotary speed will tend to straighten the drill string. Hence, low speeds (70–100 rpm) tend to be used with these assemblies.
- *Reduction in flow rate*—in soft formations, a high pump rate can wash out the hole ahead and reduce build rates.

Stabilization Principle

The principle used is that three or more full-gauge stabilizers each separated by a stiff collar, including the near-bit one, will resist any bending effects and will prefer to follow a straight trajectory. These assemblies are termed packed assemblies and are typically used on tangent sections in conjunction with high rotary speeds (120–160 rpm).

Pendulum Principle

This was the first directional principle developed and comprises either no near-bit stabilizer or one that is undergauge. The bit experiences a low side force due to gravity. Reducing the weight on bit and reaming each stand to help in bedding it can increase the effect. If the bit to the first-string stabilizer is too great, then the collars may contact the wall reducing the effectiveness and could turn the bit to the high side. Maintaining a high rotary (120–160 rpm+) plus low weight on bit initially will help to initiate a drop. Once the trajectory has been started, more weight can be introduced to speed up the process.

Guidelines:

- The sections between the bit and first-string stabilizer, and first- and second-string stabilizer, should be as stiff as possible.
- No near-bit stabilizer required with a PDC bit or when there are no azimuth problems. Using a tricone bit, insert an undergauge stabilizer (¼" or ½").
- There should be two string stabilizers with no more than 30 ft between them.
- Use a low bit weight initially until a dropping tendency has been established and then increase gradually.
- Use high rotary speed (dependent on bit type).

Variable Gauge Stabilizers

These tools enable a rotary assembly to behave in a manner dependent on the stabilizer gauge status; i.e., whether full or under gauge.

The Andergauge tool comprises a stabilizer with adjustable blades created by cam activated cylinders, which can be extended or withdrawn to vary the gauge of the tool. This is affected by setting down weight on the bit with the pumps off to lock the pistons in place, and then repeating the procedure to retract the pistons; i.e., undergauge status. If appropriately located in a bottomhole assembly, then the assembly can be made to behave according to the gauge setting and as discussed above for traditional rotary assemblies.

The adjustable blades on the Sperry-Sun AGS™ adjustable stabilizer also comprise several rows of cylinders that can be extended and retracted by using a 'pumps on/pumps off' procedure. This cycling of the pumps allows the cylinders to fully retract each time and then settle in alternating gauge and undergauge positions. The tool position can be easily monitored by observing the standpipe pressure (fig. 9.16) as there is a 150-250 psi pressure differential between positions.

Diameter of Drill Collars

Drill collars add stiffness as well as weight to a drilling assembly and can be likened to thick-walled cylinders for modelling purposes. Their stiffness depends on their axial moment of inertia and modulus of elasticity (see Chapter 8, Drill String Design).

With a fulcrum-type assembly, reducing collar OD will increase the build tendency because the collars are more flexible and they can flex further before contacting the wellbore. Once this situation occurs, increasing the weight on bit will have little added effect.

With a packed-type assembly, reducing collar OD may create a slight build tendency as the collars are more flexible.

With a pendulum-type assembly, this should be as stiff as possible and so the largest collars practicable should be used. Reducing collar OD will increase the annular clearance (distance from collar to borehole wall) which will enable the collars to bend further toward the low side before making contact. This will reduce the drop rate as it turns the bit toward the high side. Using smaller OD collars also reduces the BHA weight and hence reduces the pendulum effect due to gravity.

Bit Type

Roller cone bits have no effect on whether an assembly will build, hold, or drop angle. They do, however, have an impact on azimuth and will tend to 'walk' to the right. This effect is emphasised with long-toothed rock bits in soft formations, partly due to the enhanced penetration rate and partly due to the greater cone offset. Conversely, short-toothed bits in hard formations will have a much lesser tendency to walk to the right.

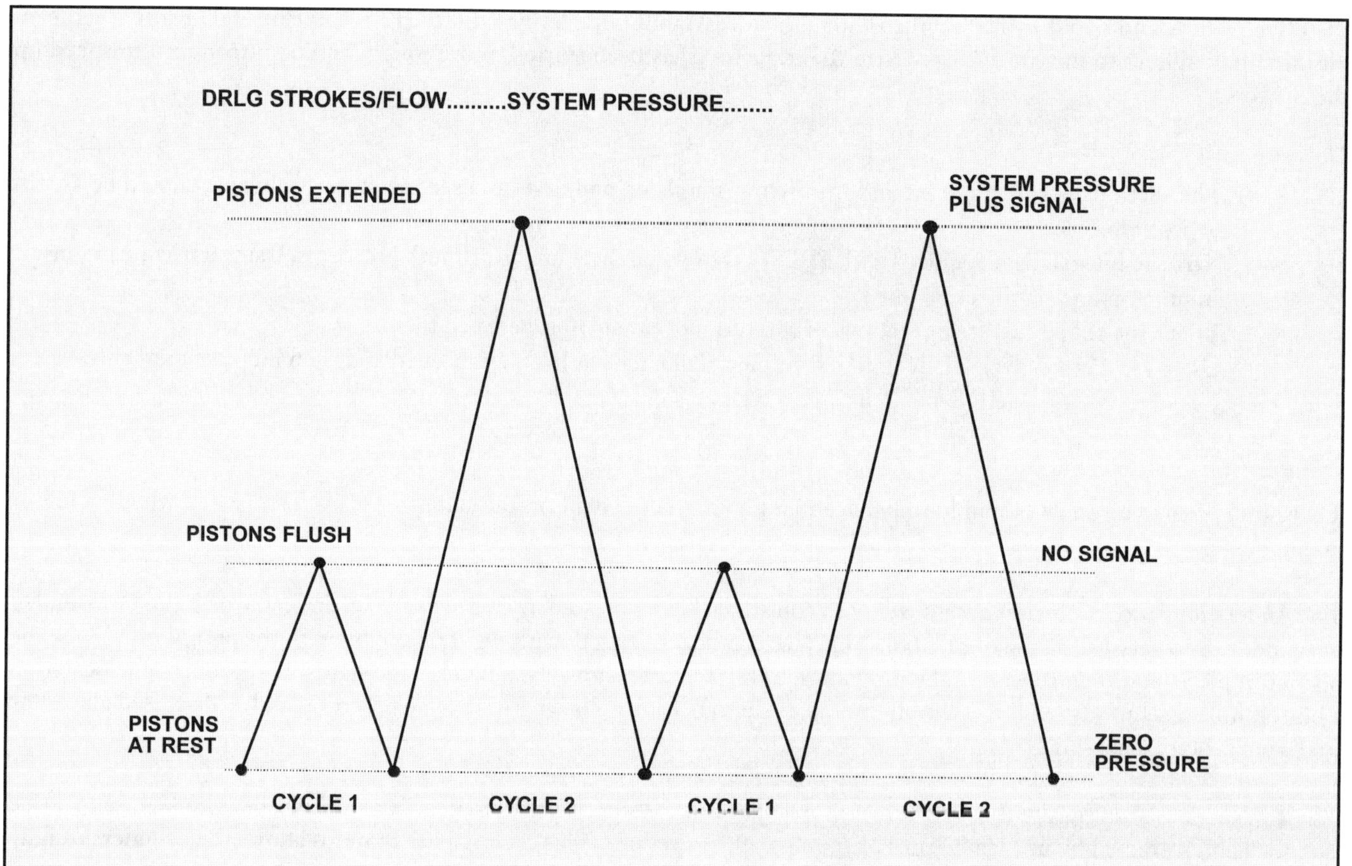

Figure 9.16 AGS status diagram

PDC bits using low weight on bit and high rotary speeds have been shown to produce little 'walk' tendency. However, they do have an effect on hole angle. A short-gauge bit will have a greater impact on hole-angle adjustment than a long-gauge bit, which will have a tendency to maintain trajectory. The gauge of the bit acts as a stabilizer and the longer the gauge, the greater the stabilizer effect.

Formation Anisotropy

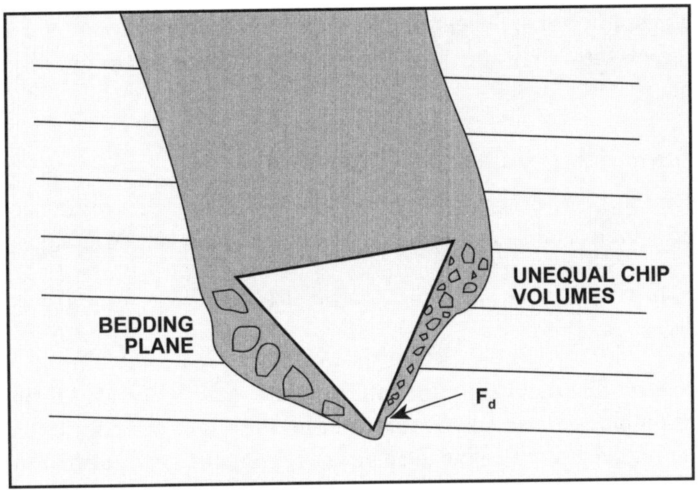

Figure 9.17 Preferential chip formation

Formation anisotropy (the variation in formation properties along different directions in the rock) does certainly affect directional response. Most oilfield drilling has been conducted in sedimentary rocks which, by definition, were usually deposited in layers and thus exhibit anisotropy; e.g., bedding planes, graded bedding, fine interbedding. Drilling experience has demonstrated that the bit will be deflected in a manner dependent on the incident angle of the bit with the dipping beds, and different theories have been propounded to describe the reaction of a drill bit under these conditions. The 'preferential chip formation theory' suggests that a bit tooth generates compressive stresses perpendicular to the tooth face and that shear failure will occur more readily along the bedding planes. Thus, chip size will vary around a tooth depending on tooth/bedding plane angle. As depicted in Figure 9.17, the bit is preferentially directed towards the area of maximum chip removal by the resultant deviation force (F_d) and the graph in Figure 9.18, derived experimentally, can be used as a guide to indicate deflection tendencies.

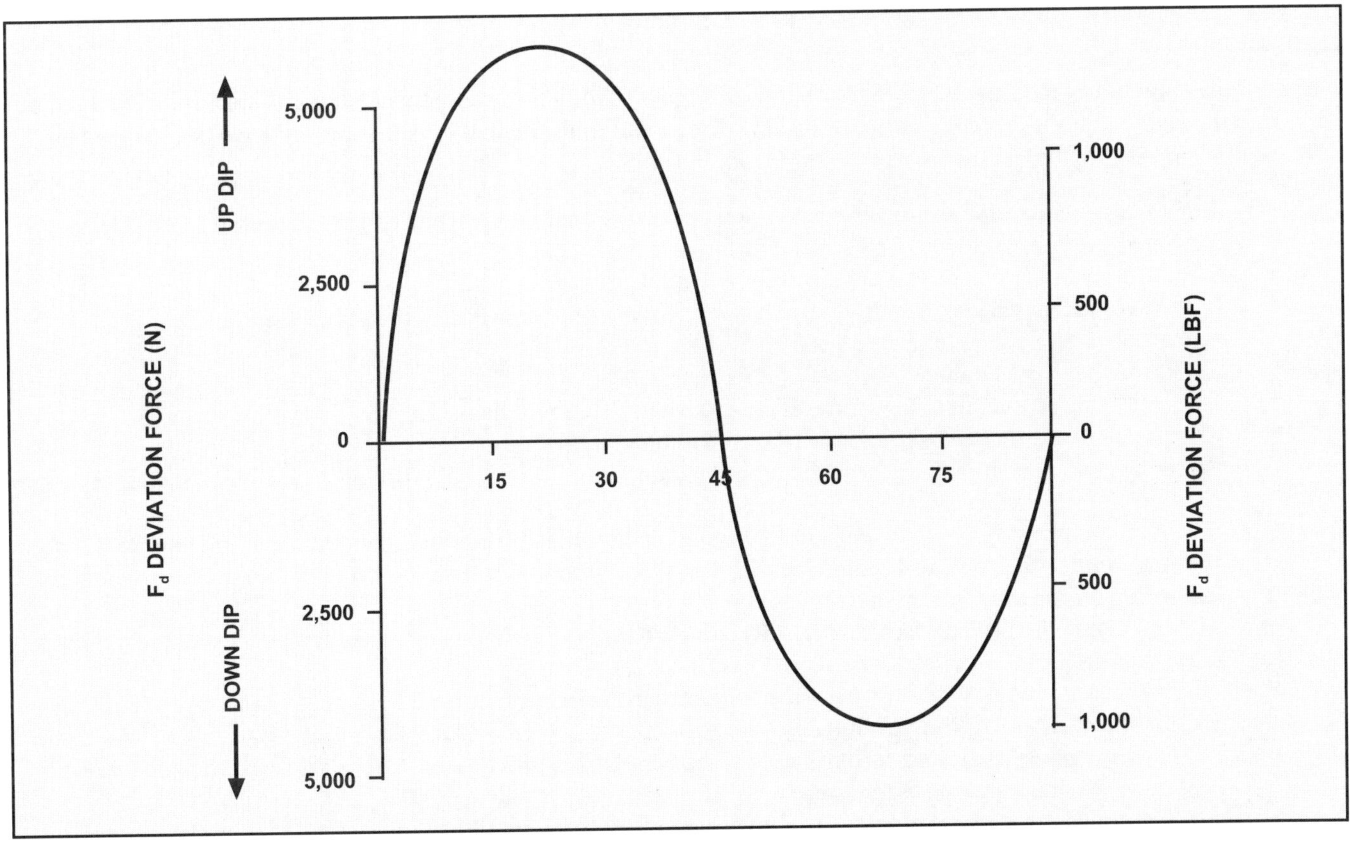

Figure 9.18 Maximum deviation force as a function of formation dip

These unwanted deflection trends can be reduced through the use of packed assemblies. A full-gauge, near-bit stabilizer will reduce bit walk while more serious deflection trends might require countertrends being designed into the drilling assembly beforehand (figs. 9.19 and 9.20).

Formation Hardness

Very soft formations may be washed away by the mere jetting action of circulating mud. Under these conditions, it can prove difficult to build any angle. If anticipated, the bit can be equipped with large nozzles to reduce nozzle velocity. Reduced pump rates can be used except just prior to connections when they should be raised just to ensure that the bit/BHA are not fouled with settled cuttings during the connection. Excessive washing can produce a dropping effect even with a fairly stiff packed assembly.

In harder formations, BHAs behave more predictably due to the hole being more usually in gauge and able to take some contact force, enabling the drill collars to flex and create bit tilt angle. The major problem in hard formations is getting an assembly to drop. Reduced weight on bit might be required to maximize the drop rate, but this conflicts with perhaps an increased weight on bit to achieve a reasonable penetration rate.

Directional Control with Downhole Motors

There are two distinctive classes of motor—turbines and positive displacement motors (PDMs). They offer the following advantages regarding direct provision of motive power to the bit rather than transmitting it from the surface.

- Elimination of lateral vibration
- String and casing wear reduced
- Lower torque in string, especially in deviated holes
- Reduced fatigue loads on drill pipe
- Can be run with light weight at continuous speeds
- The ability to orient and drill ahead

Turbines

Turbine development began in about 1924 in both the USSR and USA. The development has continued in Russia up to the present to the extent that these are the norm for drilling directional wells there. This situation will have been helped by the lack of sufficient quantity of fatigue resistant pipe for rotary drilling.

Elsewhere, PDMs have superseded the turbine as the premier directional motor while turbines are used more selectively.

- Turbines can only be powered by a liquid drilling fluid.
- A turbine consists of blades and stators mounted at right angles to fluid flow.
- The rotors are attached to the drive shaft while the stators are attached to the outer casing.
- Each rotor/stator pair is known as a stage.
- Typical turbines are equipped with 75–250 stages.
- The stators direct the fluid flow onto the rotors driving the shaft clockwise.
- The power of a turbodrill is proportional to the number of stages.
- Power output from any stage is dependent on the number of blades, blade spacing, blade angle, blade shape, and flow area.
- Side tracking turbines are short units (30 ft) while straight hole turbines are much longer and comprise several sections—circulating sub, top motor section, 1–2 power sections, a bearing section, and a drive shaft section incorporating the near-bit stabilizer.
- Turbines are high-speed devices (500–1,000 rpm). As they have an open fluid path, applying excessive weight will cause stalling.
- Turbines usually operate best in hard-to-medium-hard formations where high speeds and low weight on bit provide the optimum drilling parameters.
- For hard abrasive formations, turbines used to be run with diamond bits. Nowadays, with the advent of PDC bits, these would tend to be used instead.
- The continued development of positive displacement motor/bit combinations has led to these largely replacing turbines since the late 1980s.

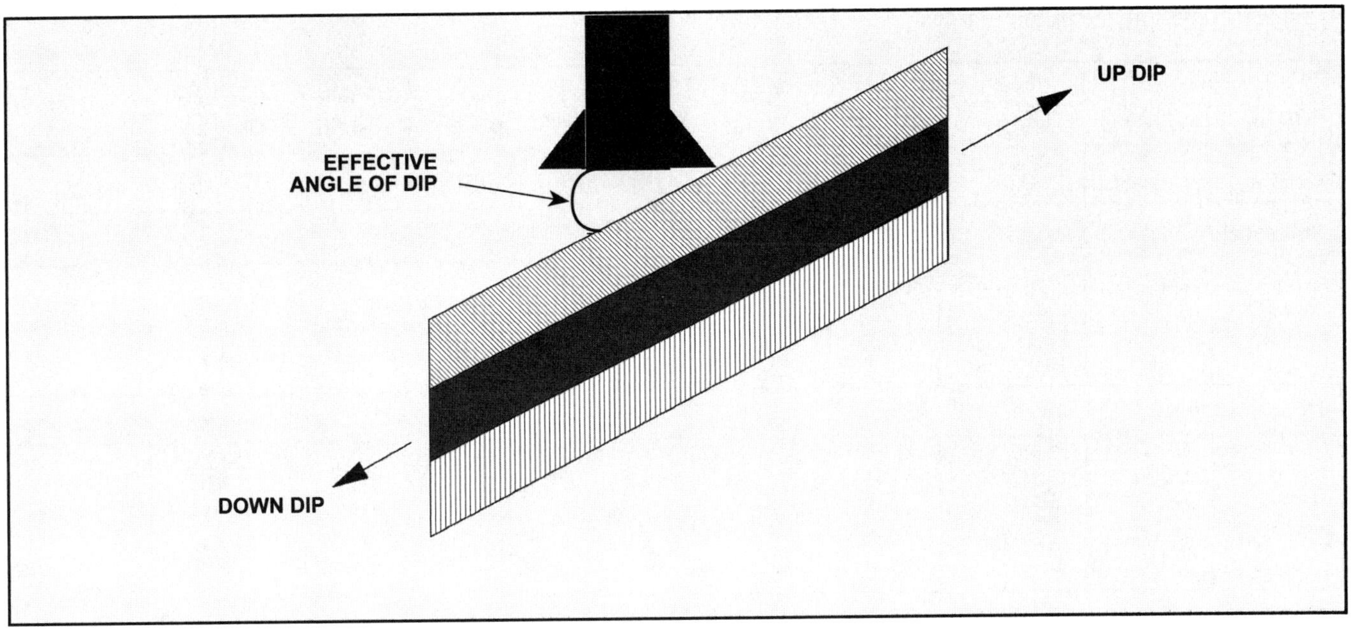

Figure 9.19 Meaning of up dip and down dip

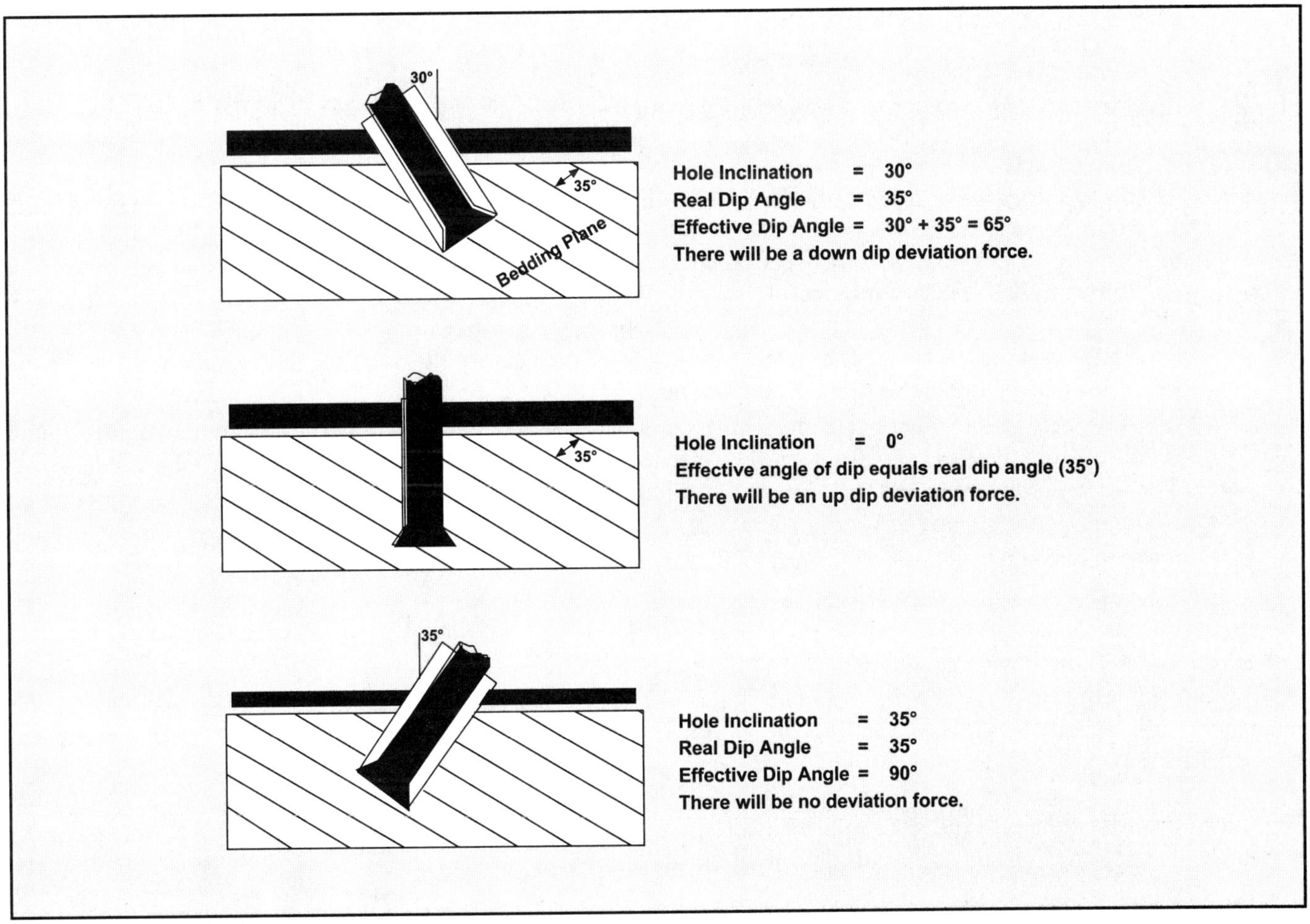

Figure 9.20 Deviation responses to various angles of dip

Typical Turbine Specifications—Drilex

OD in.	Stages	Flow gpm	Flow	Speed rpm	Torque ft/lb	Power hp	Hydraulic Thrust (lb)	Pressure Drop psi	Length ft
			Nominal performance with 10 ppg mud						
Straight Hole Tools									
5	150	100–200	150	975	375	70	10,480	1,095	34
7, 7¼	150	375–500	425	910	1,410	244	26,345	1,430	38
7¼	120	250–500	475	1,010	1,430	275	25,574	1,450	47
Directional tools									
5	90	150–250	150	975	225	42	6,298	656	22
7, 7¼	70	375–500	425	910	660	115	12,380	670	21
7, 7¼	100	375–500	475	1,010	1,192	229	21,312	1,208	28

Positive Displacement Motors

- The original PDM concepts were developed in 1956 based on the Moineau pump principle in reverse; i.e., the fluid flow drives the pump shaft.
- The pump can be powered by drilling fluid, air or gas, and comprises four major sections:
  1. *Dump valve*—a bypass valve allowing the drill string to fill or empty when tripping.
  2. *Motor assembly*—comprises a rubber-lined stator with a spirally-shaped cavity of elliptical cross section and a spiral, solid steel shaft rotor running throughout the length of the cavity. The top end is free while the lower end is fixed to a connecting rod.
  3. *Connecting rod*—equipped with a universal joint at each end to accommodate the eccentric rotation of the rotor and transfer this rotation to the drive shaft.
  4. *Bearing and drive shaft assembly*—consists of thrust bearings and a radial bearing to allow smooth rotation of the drive shaft. The bearings are lubricated by the mud. The drive shaft is then connected to a bit sub, which is the only external rotating part of the mud motor.
- The motor is designed such that the rotor is forced to turn clockwise when drilling fluid is pumped through the motor into the cavities between the rotor and stator.
- Motors are defined by the ratio of the number of lobes in section of rotor to stator. There is always one more stator lobe and these can vary from 2 to 11 for stator lobes with corresponding 1 to 10 rotor lobes.
- The torque produced by PDMs is proportional to the pressure differential across the motor. When weight on bit is applied, the circulating pressure must increase. As the bit drills off, the pressure decreases. Thus, the pump pressure can be used as both a bit weight and torque indicator.

Bit Tilt

Directional control with downhole motors is based on tilting the axis of the bit with respect to the axis of the hole and/or creating a side force at the bit. If the drill string and hence the body of the motor is rotated from the surface, then the bit will tend to drill straight ahead. However, if the drill string is not rotated, then the bit will drill a curved path determined by the orientation of the side force or the tilt of the bit axis.

The tilt of the axis of the bit is caused either by incorporating a bent sub above the motor or by using a motor with an adjustable bent housing. The tilt angle is dependent on the directional requirements.

- A high tilt angle is recommended for kicking off and sidetracking. It should produce a greater dogleg than that programmed to enable a portion of the drilling to be completed in rotary mode, which should produce a higher penetration rate. Thus, in a typical deviated section, a single will be drilled by oriented sliding for the initial 10 ft and the remaining 20 ft drilled in rotary mode. As a rule of thumb, the tilt angle selected should produce a dogleg of 1.25 times the maximum dogleg severity required by the well plan.

- With a choice available, a tool with a higher dogleg capability should be selected as this will enable a greater portion of the drilling to be completed in rotary mode.
- When tangent or straight-hole drilling, a lower tilt angle is recommended to reduce bit wear.

Bent Sub

This is a short length of drill collar-sized pipe placed directly above the motor, with its lower section offset from the vertical. The amount of offset varies with from 1 to 3 degrees being common. The inside of the bend is scribed to indicate the direction in which the bit will drill (tool face) and an orienting tool (MWD tool) is usually positioned above the bent sub. This type of assembly offers control over direction, smoother deviation, and can be used to build or drop angle.

Adjustable Bent Housing

An adjustable bent element is positioned in the motor housing which also holds the flexible coupling/connecting rod. This places it below the motor assembly and nearer the bit, giving greater directional control (fig. 9.21). The angle of the bent housing ranges in increments from 0 to 3 degrees and is easily reset at the site.

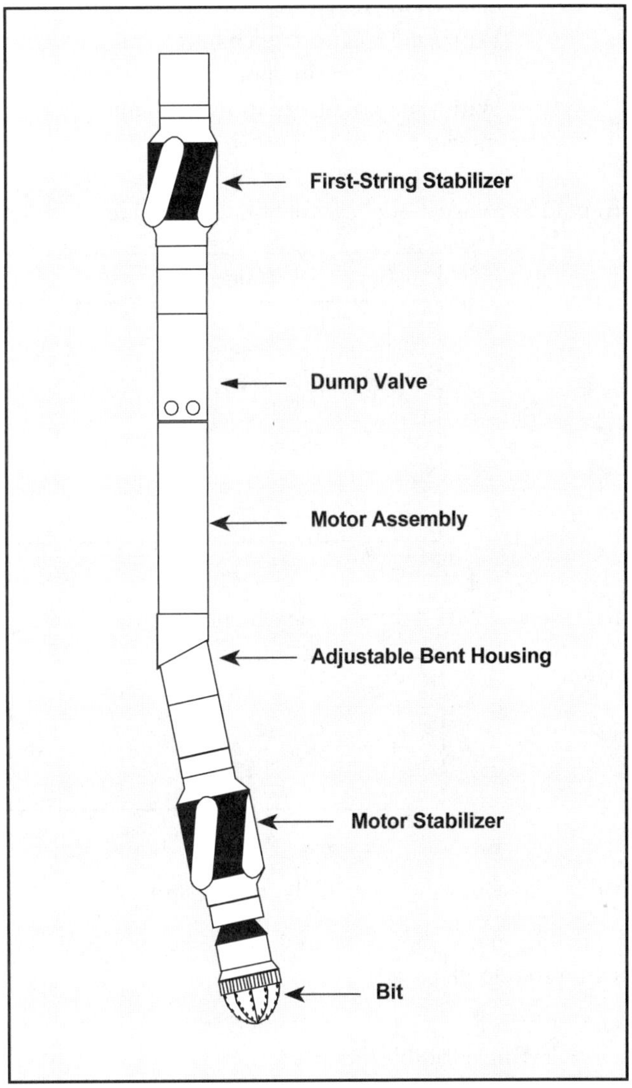

Figure 9.21 Steerable system—motor/bent housing

Reactive Torque

Reactive torque is the tendency for the drill string to turn in the opposite direction from the bit. As the rotor turns to the right, the stator is subjected to a left-turning force. Depending on the type of formation and length of drill string, the pipe will twist causing the bit to drill to the left. This left-hand torque increases as more weight on bit and pump pressure are applied. Local experience can be used to counter this effect by adjusting the bent sub orientation.

Stabilizer Size and Placement

Most bent housing/motor combinations can be used as a

- fully stabilized system with bearing and top (first-string) stabilizers.
- partially stabilized system with bearing housing stabilizer only.
- slick system; i.e., no stabilization.

The degree of stabilization is determined by the desired directional response.

The dogleg severity that can be generated is a function of the

- tilt angle (as discussed above).
- size of the motor stabilizer and the first-string stabilizer above the bit.
- distances between the bit, motor stabilizer, and first-string stabilizer (this is a combination of the fulcrum and stabilization principle for rotary assemblies).

Motor Stabilizer

The motor stabilizer centralizes the bit and motor in the hole.

- Is usually a screw-on or clamp-on design
- Is always undergauge
- Typically, ⅛" undergauge for hole sizes up to 17½" and ¼" for larger hole sizes
- Gauge length should not exceed bit gauge length

First-String Stabilizer

This is normally run either directly above the motor or with a pony drill collar in between.

- It should be no bigger than the motor stabilizer and be of a similar design.
- The build or drop capacity of the BHA is increased as the degree undergauge of the first-string stabilizer increases.
- With long motors (which have a build tendency in rotary), the first-string stabilizer should always have the same gauge as the bearing housing stabilizer.
- With standard length motors, an undergauge first-string stabilizer is required to produce a holding tendency in rotary.

To hold in rotary mode, use the following guide:

Hole Size	First-String Stabilizer Gauge
8½"	8" to 8⅜"
9⅞"	9⅛" to 9⅝"
12¼"	11¾" to 12"
14¾"	14⅛" to 14½"
17½"	16" to 17"

To build at 0.25°/100 ft in rotary mode, use the following guide:

8½"	⅛"
12¼"	¼"
17½"	⅜"

Note: The above does not apply to long motors.

Note: The above should be used as a guide only. Actual results can be affected by the formations drilled and will vary from region to region.

Variable-Gauge Stabilizers

Variable-gauge stabilizers are often run as the first-string stabilizer to give a greater degree of control when drilling in rotary mode. A number of manufacturers are also designing variable-gauge motor housing stabilizers.

Amount of Slide Drilling

An estimate of the proportion of slide drilling that will be required can be determined by the following equation:

$$\% \text{ Footage in Sliding} = \frac{(DL - DLR)}{(DLO - DLR)} \times 100$$

where

DL = required dogleg
DLO = actual dogleg when oriented
DLR = actual dogleg when rotary drilling.

It must be remembered that sliding is detrimental to hole cleaning, especially on long tangent sections at relatively high angle. So a balance must be struck between hole cleaning and the amount of slide drilling.

TYPICAL BOTTOMHOLE ASSEMBLIES

Rotary Assemblies

BHA for Holding

Depending on formation characteristics, WOB, rpm, bit type, etc., the following assembly will generally hold or show a slight build or drop tendency at 0.1–0.5°/100 ft.

Bit
Full-gauge near-bit stabilizer
Short drill collar
Full-gauge string stabilizer
Drill collar
Full-gauge string stabilizer
Drill collars as required for WOB
Jar
Drill collars as required to operate jar
HWDP as required

BHA for Building

Depending on formation characteristics, WOB, rpm, bit type, etc., the following assembly will generally build at 1.5–2.5°/100 ft:

Bit
Full-gauge near-bit stabilizer
2 × drill collars
Full-gauge string stabilizer
Drill collar
Full-gauge string stabilizer
Drill collars as required for WOB
Jar
Drill collars as required to operate jar
HWDP as required

This assembly is often referred to as a 60-ft building assembly. Increasing the spacing between the near bit and the first stabilizer will increase the build rate. Decreasing the spacing between the near bit and the first stabilizer will decrease the build rate.

BHA for Dropping

Depending on formation characteristics, WOB, rpm, bit type, etc., the following assembly will generally drop at 1.5–2.0°/100 ft:

Bit
Drill collar
Full-gauge string stabilizer
Drill collar
Full-gauge string stabilizer
Drill collars as required for WOB
Jar
Drill collars as required to operate jar
HWDP as required

This assembly is often referred to as a 30-ft pendulum assembly. Increasing the spacing between the bit and the first stabilizer will increase the drop rate. Decreasing the spacing between the bit and first stabilizer will decrease the drop rate.

A 60-ft pendulum assembly is sometimes used to drill vertical wells through soft-to-medium-hard formations, although care must be taken with how much WOB is applied, because this could easily become a building assembly.

Steerable Assemblies

BHA for 17½" Hole—Holding Tendency in Rotary

17½" Rock bit
11¼" Steerable motor with 17¼" motor stabilizer
16½" First-string stabilizer
Float sub
9½" NMDC
9½" MWD
16½" NM stabilizer
2 × 9½" NMDC
Crossover
2 × 8" Steel DCs (increase or decrease if required)
Jars
8" Steel DC
Crossover
HWDP (sufficient to provide weight on bit)

BHA for 17½" Hole—Building Tendency in Rotary

17½" PDC bit
11¼" Steerable motor with 17¼" motor stabilizer
17¼" First-string stabilizer
Float sub
9½" NMDC
9½" MWD
16½" NM stabilizer
2 × 9½" NMDC

Crossover
2 × 8" Steel DCs (increase or decrease if required)
Jars
8" Steel DC
Crossover
HWDP (sufficient to provide weight on bit)

BHA for 12¼" Hole—Holding Tendency in Rotary

12¼" Bit
9½" Steerable motor with 12⅛" motor stabilizer
12" First-string stabilizer
8" NMDC
8" MWD
12" NM stabilizer
2 × 8" NMDC
Jars
8" Steel DC
Crossover
HWDP (sufficient to provide weight on bit)

BHA for 12¼" Hole—Building Tendency in Rotary

12¼" Bit
9½" Steerable motor with 12⅛" motor stabilizer
11" First-string stabilizer
8" NMDC
8" MWD
11¾" NM stabilizer
2 × 8" NMDC
Jars
8" steel DC
Crossover
HWDP (sufficient to provide weight on bit)

Rotary Steerable Assemblies

Rotary steerable assemblies, which allow inclination and azimuth control while rotary drilling, will be discussed in Chapter 13, Advances In Technology.

CHAPTER 10
Formation Evaluation

INTRODUCTION

A wide variety of information is available from the well that can be used by the geologist and petrophysicist to refine the geological and petrophysical models and to gain a better understanding of the reservoir, assess how large the reservoir is, and how it will perform if placed on production.

Information can be obtained from the following sources:

- *Data collected while drilling*
  - Penetration rate
  - Cuttings analysis
  - Mud losses/gains
  - Shows of gas/oil/water
- *Core analysis*
  - Lithology
  - Presence of shows
  - Porosity
  - Permeability
  - Special core analysis
- *Log analysis* (wireline and MWD/LWD)
  - Electrical logs
  - Acoustic logs
  - Radioactivity logs
  - Pressure measurements
  - Special logs
- *Productivity tests*
  - Formation tester
  - Drill stem test
  - Production test

Mud Logging

Data collected while drilling is usually incorporated as part of the mud logging service.

The mud log provides a record of the penetration rate, lithology (inferred from cuttings analysis), and cuttings description on a depth basis together with general comments on the drilling parameters, mud type and properties, hydrocarbon shows, logs run, cores cut, etc.

Coring

Cores provide more accurate information than cuttings. However, unless special circumstances dictate, it is usually only cost effective to core the reservoir section of a well.

A core allows a detailed lithological description of the reservoir to be made. Additional tests can be performed in the laboratory to establish the porosity and permeability of the rock, which can then be used to calibrate the response from logging tools.

Log Analysis

Logs can be obtained by running specialist tools on wireline or, as is becoming more common, by including a LWD tool as part of the MWD tool string. Note that not all wireline logs are available as LWD logs.

A log is the recording of the physical properties of the formations drilled on a depth basis.

Log Types

Logs can be classified into three families

1. Reservoir thickness (gamma ray and spontaneous potential). Discriminate between reservoir and nonreservoir rocks.
2. Porosity (density, neutron, sonic). Used to calculate porosity, identify lithologies, and differentiate oil from gas.
3. Resistivity (laterolog, induction, microresistivity). Together with porosity logs, they are used to calculate hydrocarbon saturations.

Additional logs can also be run to measure specific properties.

Formation tester—measures formation pressures and can retrieve small samples
Dipmeter—measures the formation dip and azimuth
VSP—used to calibrate seismic surveys
Caliper—used to correct other log measurements, cement volumes, etc.

Response from the tools varies under differing conditions of pressure, temperature, fluid environment, and borehole size. Calibration is normally required to ensure quality and accuracy of the log.

Recording speed is a function of the type of sonde, the quality of the data required, and of the information to be collected.

Detailed log interpretation is performed by powerful workstations, often installed in the logging units. Quick-look analysis can still be performed; however, it is not the intent of this section to cover this subject.

Mud Invasion

The majority of wells are drilled with a mud hydrostatic that is greater than the formation pressure. In other words, there is a differential pressure attempting to push the mud in the well into the formation. Across shales, nothing happens as the shale is not permeable and there is nowhere for the drilling mud to go.

Across a porous and permeable layer (a reservoir), this pressure differential pushes mud filtrate into the reservoir rock displacing the reservoir fluid. The depth of this invasion varies from centimetres to metres depending on the differential pressure, the porosity and permeability of the formation, and the ability of the drilling fluid to form a filter cake, which helps reduce the amount of fluid invasion. The depth of this invasion or flushed zone can be important as, if it is too deep, then some tools will not be able to read past it.

RESERVOIR DEFINITION

A reservoir is a porous and permeable formation saturated with one or more fluids. The reservoir is composed of a rock matrix and a certain number of voids that contain the fluids. It can be highlighted by the following parameters

Porosity

The volume occupied by all the fluids of the formation. It is the ratio of the pore volume to the total volume.

$$\phi = \frac{V_p}{V_b}$$

where

ϕ = porosity
V_p = pore volume
V_b = bulk volume.

Permeability

The ability with which a fluid can flow through the formation. Permeability is not quantitatively and directly obtainable from logs. Expressed using the Darcy equation

$$Q = \frac{K \times A \times \Delta P}{\mu \times L}$$

where

Q = flow rate
A = cross-sectional area
ΔP = pressure differential
μ = viscosity of the flowing fluid
L = length
K = permeability.

Permeability is measured in Darcies where 1 Darcy = 0.9869 x 10^{-12} m^2.

Fluid Saturation

The ratio of the volume of fluid present in the pores to the total pore volume

$$S_f = \frac{V_f}{V_p}.$$

LOG PRESENTATION

Log measurements are presented graphically on a standard API grid, which consists of three data tracks and a depth column.

Track 1, on the left-hand side of the depth column, always has a linear scale and usually presents lithological data.

Tracks 2 and 3, on the right-hand side of the depth column, are often combined and may be either on a linear or logarithmic scale or a combination of the two. These tracks present formation properties such as resistivity and porosity data.

Several parameters may be recorded on the same track.

The vertical depth scale of the log may be 1:1000 or 1:500 for correlation purposes, 1:200 for petrophysical evaluation, or 1:40 for correlation with cores. In the USA, vertical depth scales are often 1:1200, 1:600, and 1:240.

The log header also contains useful information about the well and the drilling fluid in use at the time that the log was recorded.

WIRELINE LOGS

Caliper

Principle

Caliper tools measure the diameter and shape of the borehole using two or more arms symmetrically placed on the logging tool. Variations in the borehole diameter cause the caliper arms to open or close.

Log Presentation

The caliper log is always recorded on track 1, on the left of the depth column, on a linear scale, graduated in inches. A vertical line indicated on the same scale represents the bit size and the caliper fluctuates around this value.

Three typical cases can be observed:

- Caliper value = bit size
  Hole is considered to be in gauge—typical of a tight formation
- Caliper value > bit size
  Hole is overgauge caused by a weak or unconsolidated formation or the formation has been partially dissolved by the drilling fluid
- Caliper value < bit size
  Hole is considered to be undergauge caused by swelling formation or the buildup of a thick filter cake

Application

Caliper measurements can be used for:

- measurement of hole volumes for cement calculations.
- selection of suitable points to take formation pressure tests.
- selection of suitable packer seats for open-hole well testing.
- correction of log readings for the effect of hole size and thickness of filter cake to generate a more quantitative interpretation.

Gamma Ray

Principle

The gamma ray log measures the natural radioactivity of the formations. The most common elements that emit naturally occurring radiation are potassium (K), thorium (Th), and uranium (U). These three elements continuously emit gamma rays, which are picked up by a radiation detector mounted on the gamma ray tool.

Most reservoir rocks (sandstone, limestone, dolomite) contain none or very small amounts of these three elements and therefore have a low gamma ray radiation level. Other rocks types (e.g., shales) have a large amount of potassium and thorium atoms, resulting in a high gamma ray radiation level. This contrast in radiation levels can be used to differentiate between different formations.

Log Presentation

The GR log is normally presented on track 1. It is recorded on a linear scale, graduated in API unity from 0 to 100 or 0 to 150 API, increasing toward the right.

While the GR log is a general indicator of shaly formations, care must be taken as other formations can also exhibit high GR values.

Radioactivity In Shales or Clays

Potassium and thorium exist with shales and clays and are very good indicators of shales.

Radioactivity In Clastic Formations

Generally, clastic formations exhibit weak radioactive emissions which give low GR values. However, in certain circumstances, these formations are associated with detrital elements, which can contain potassium and thus exhibit higher levels of natural radioactivity. An example of this is the presence of mica within the Brent sandstone sequence in the North Sea.

Radioactivity In Carbonate Formations

Generally, carbonates are not radioactive. However, under certain conditions, they can contain organic matter which gives a radioactive response due to the presence of uranium 238. There are also shaly carbonates with significant radioactivity according to the type of shale.

Radioactivity In Evaporite Formations

Evaporites such as halite and anhydrite have extremely low values of GR (< 5 API).

Application

The GR has three main uses:

- Discrimination between reservoir and nonreservoir formations
- Correlation tool
- Calculation of clay or shale content

Correlation

Due to its ability to repeat, the GR is an extremely useful tool to correlate not only other logs within the same well but also to correlate information between different wells.

Calculation of Clay Content

An approximation of the clay or shale content of a formation can be made using the following formula:

$$V_{sh} = \frac{GR_{read} - GR_{min}}{GR_{max} - GR_{min}}$$

where

V_{sh} = clay or shale content
GR_{read} = GR reading at the point of interest
GR_{min} = GR reading from a clean formation
GR_{max} = GR reading in 100% shale formation.

Limitations

For quantitative use, the logging speed should be around 1,200 ft/hr.

For qualitative use, higher logging speeds can be used but a smoothed curve will result.

Potassium muds will increase the readings.

Density Log

Principle

A strong gamma ray source bombards the rock with medium energy gamma rays. These gamma rays collide with electrons in the formation. In the process, the gamma rays lose some of their energy and are scattered in all directions. The count rate of these scattered gamma rays reaching the tool detector gives an indication of the formation bulk density.

From the bulk density an assessment of the formation porosity can be made.

Log Presentation

Density is generally recorded on track 2 or 3 often together with the neutron log, on a linear scale.

Application

From the bulk density recorded by the tool, porosity can be calculated using the following equation:

$$\phi = \frac{\rho_m - \rho_b}{\rho_m - \rho_f}$$

where

ϕ = porosity
ρ_m = density of the rock matrix
ρ_b = bulk density from the log
ρ_f = density of the mud filtrate.

Limitations

The density tool has a depth of investigation of 10 to 15 cm so only records in the flushed zone.

Neutron Log

Principle

A neutron source bombards the formation with high energy neutrons. As the neutrons collide with atoms present in the formation, they slow down. Hydrogen atoms slow them down especially quickly. The tool measures the returning neutrons at two distances from the source and, from this, computes the amount of hydrogen atoms in the formation. The tool assumes that all hydrogen atoms are only present in the pore space of the formation and, from this, computes a porosity value.

The main tools run are:

- Sidewall neutron porosity (SNP), which is a pad-type tool
- Compensated neutron log (CNL), which is a mandrel-type tool

Log Presentation

The neutron log is generally recorded on track 2 or 3 together with the density curve. The curve is scaled in porosity units (pu). As the pu reference is water-filled limestone, a correction is required for different formations.

Application

Log is used to calculate porosity values for the formation.

Limitations

In gas-bearing formations, the low density of hydrogen atoms causes a low porosity reading. When combined with the density log, a marked 'gas effect' is noticeable where the density log kicks to the left (increasing porosity/decreasing density) and the neutron log kicks to the right (decreasing porosity).

Shales contain clay-bound water. The neutron tool interprets this water as porosity, when in reality there is no effective porosity present.

The neutron tool has a depth of investigation of 15 to 20 cm so only records in the flushed zone.

Sonic Log

Principle

The sonic tool measures the time it takes for an acoustic or sound wave to travel through the mud and formation.

Tool most commonly run today is the borehole compensated (BHC) sonic which automatically corrects the readings for any borehole effects.

Tool generally needs to be well centralized in the hole.

Log Presentation

The sonic log is usually recorded on track 3 on a linear scale. The travel time is recorded in microseconds/ft on a scale ranging from 40 to 140 increasing to the left.

Application

Sonic log can be used as follows:

- Calculate porosity (not as effectively as density or neutron logs)
- Lithology identification and correlation
- Overpressure detection

Resistivity Log

Principle

The resistivity tool measures the resistivity of the formation to an electrical current.

Most rocks are resistant to an electrical current compared to fluids which are generally good conductors, except hydrocarbons.

In the case of a porous formation, and as a result of fluid invasion, the mud filtrate displaces the formation fluids and creates three different zones with different resistivities.

These zones are known as:

- Flushed zone, close to the borehole wall, with a resistivity of Rxo
- Transitional zone, with a resistivity of Ri
- Uninvaded or virgin zone, with a resistivity of Rt

According to the Archie, in a water-bearing formation, the ratio of Rt to Rxo is a constant, while in a hydrocarbon-bearing reservoir, the ratio is not constant and Rt increases more than Rxo.

Using this information, different readings can be taken in the uninvaded zone and the flushed zone.

In nonreservoir rock, there is no filtrate invasion, so Rxo = Ri = Rt.

Large spacing sondes, such as the laterolog (LL), dual laterolog (DLL), and spherically focused log (SFL), read in the uninvaded zone. They have a large spacing between the transmitter and the receiver (24"–64") and can be influenced by mud type.

Small spacing pad-type sondes, such as the microlog (ML), microlaterolog (MLL), and micro spherically focused log (MSFL), read in the flushed zone. They have a short spacing of a few inches between the transmitter and the receiver.

If nonconductive muds are being used, then induction logs can be used to measure the resistivity of the formation.

Log Presentation

Resistivity is recorded on track 2 on a log scale from 0.2 to 2,000 ohm m.

Application

Used to differentiate between hydrocarbon- and water-bearing intervals.

Used to quantify the water saturation in hydrocarbon-bearing intervals.

Formation Tester

The formation tester (RFT—repeat formation tester, MDT—modular dynamics tester, or FMT—formation multiple tester being the more common varieties of this tool) is run to obtain formation pressures, pressure gradients, permeability estimates, and formation samples.

The tool has a pad containing a seal packer opposite an arm. At the depth of interest, the arm is opened, forcing the pad and seal packer into contact with formation. Pretest and pressure buildup information is obtained. From this information, the formation permeability and formation pressure can be determined. By reviewing this information at different depths, pressure gradients can also be determined.

Depending on the formation permeability, it might take some time to obtain a valid set of readings. In high overbalance wells, the risk of the tool becoming differentially stuck can be high (the RFT is not known as the repeat fishing tool for nothing!). Contingency plans for the recovery of stuck formation tester logs should be established prior to running this tool (or any other log for that matter).

Care should be taken with the MDT tool as this tool has the ability to pump small volumes of formation fluid into the wellbore. In gas or condensate wells, small volumes at depth can easily become large volumes at surface, so it is always wise to follow the MDT program carefully and not assume that it is just another logging run.

PIPE CONVEYED LOGGING

In highly deviated wells (angle greater than 60 degrees) and in other wells with poor hole condition, it is common to run wireline logs on the end of drill pipe, using the drill pipe to push the logs to the required depth.

A typical procedure for such an operation is as follows:

- Make up wireline tool string
- Make up wet connect sub to drill pipe
- RIH drill pipe to required depth, breaking circulation on a regular basis to ensure that the ports in the wet connect sub are not blocked
- Make up side entry sub
- Rig up wireline sheaves
- Install wet connect and pump down
- Latch wet connect and test. Secure side entry sub pack off.
- RIH to required depth
- POOH logging as required
- Once logging is complete, POOH to side entry sub.
- Recover wet connect and rig down wireline
- POOH

Prior to commencing a pipe-conveyed logging operation, a detailed planning meeting is required that needs to focus on:

- communication between the driller and the wireline winch operator.
- connection procedure after running the side entry sub to minimize the risk of crimping the cable.
- working with sheaves and cable in the derrick while running the blocks up and down.

- Contingency measures in the event of
  - ❖ stuck pipe.
  - ❖ well control incident (note that most annular preventers will NOT seal against drill pipe and wireline cable).

LWD

Apart from inclination and azimuth measurements (the traditional MWD tools), the following data can also be measured and transmitted to surface in real time or recorded for later analysis:

- Gamma ray
- Resistivity
- Density
- Neutron porosity
- Sonic
- Drill mechanics (vibration, downhole weight on bit, downhole torque)
- Pressure while drilling
- Acoustic caliper

Apart from proving information on the formations, as with wireline logs, the formation evaluation logs (gamma, resistivity, density, and neutron porosity) can also be used to geosteer horizontal wells, ensuring that the well trajectory remains within a predefined formation, or at least minimizes the amount of nonreservoir formation that is drilled.

The drilling mechanics and pressure-while-drilling logs provide information that can be used to optimize drilling performance, reduce problems associated with hole cleaning, and minimize problems associated with drill string failure by alerting the driller and directional driller to downhole vibration and drill string compression.

The formation evaluation logs work on the same principle and have the same application as wireline logs.

Telemetry

Log data acquired while drilling can either be recorded in memory mode or transmitted real time back to surface. The three techniques currently available are:

- Negative pulse
- Positive pulse
- Electromagnetic

Negative Pulse

The negative pulse tool uses a sliding gate and seat valve to momentarily vent fluid from the inside of the drill string to the annulus, creating a pressure drop that propagates to surface, where it can be recorded by a pressure sensor on the standpipe manifold.

The amplitude of the negative pulse signal is a function of the differential pressure between the bore of the drill string and the annulus. Most tools specify a minimum pressure drop of at least 350 psi.

Most tools have a mass flow rate limit to prevent excessive erosion of the gate and seat components.

The pulser valve is powered by lithium batteries.

If a float sub is being run in the string, always ensure that it is run above the pulser sub to prevent backflow through the valve when tripping in to prevent damage.

Positive Pulse

The positive pulse system is usually powered by a mud turbine that fits within the pulser sub. The turbine powers an electrical generator and a hydraulic pump. The generator provides electrical power for both the pulser control circuitry and the sensors, while the hydraulic pump supplies power for the hydraulic solenoid, which activates a poppet valve to induce a pressure pulse.

To transmit data, the poppet valve is extended into a flow orifice, restricting the flow of mud inside the drill string and creating a momentary pressure increase that propagates to surface, where it can be recorded by a pressure sensor on the standpipe manifold.

Electromagnetic

Electromagnetic telemetry systems can only be used on land wells at present. Data is transmitted to surface using electromagnetic waves, generated by a downhole sub, that travel through the formations to surface. At surface, the electromagnetic wave is received as a voltage potential between the conductive drill string and a ground electrode. The ground electrode is usually a metal stake placed in the ground a few hundred feet from the well.

This system does not require a drilling fluid to transmit data, so is ideally suited to underbalanced drilling.

The tool is normally battery powered.

RIG-SITE SAFETY WITH DENSITY AND NEUTRON LOGS

As density and neutron logs require the use of radioactive sources, a safety meeting should be held prior to picking up or running these tools. Only approved personnel are to handle radioactive materials.

CHAPTER 11
Rig Equipment and Sizing

GENERATIONS OF OFFSHORE DRILLING UNITS

Generations are used to differentiate semisubmersible hulls. This section describes the term *Generations.*

Year of Construction

Generation is traditionally based on age. Semisubmersibles are built to satisfy demand and construction dates coincide with peaks in oil price and increased demand. Generations are based on the following construction dates.

Generations	Year of Construction
1st	1962 to 1969
2nd	1970 to 1981
3rd	1982 to 1986
4th	1987 to 1998
5th	1999 onwards (?)

Technical Capability

Generation is based on the technology of equipment installed on the rig. When rigs are built, they generally reflect the technology available at the time. As technology develops, more complex work can be carried out and, over the past thirty years, semisubmersibles have moved into deeper water to drill deeper more complex wells.

Generations	Examples of Development of Technology
1st	800 ft water depth, 2 × 1,250 hp mud pumps, kelly, 1,450 ton variable deck load (VDL), manual derrick
2nd	1,500 ft water depth, 2 × 1,600 hp mud pumps, kelly, 3,000 ton VDL, manual derrick
3rd	2,500 ft water depth, 2 × 1,600 hp mud pumps, kelly, 3,800 ton VDL, automatic pipe handling
4th	3,500 ft water depth, 3 × 1,600 hp mud pumps, TDS3 top drive, 4,300 ton VDL, automatic pipe handling
5th	8,000 ft water depth, 5 × 2,200 hp mud pumps, TDS8 top drive, 5,000 ton VDL, dual activity

Generations of Semisubmersibles

Generations of semisubmersibles are traditionally based on year of construction; however, if a rig's equipment is upgraded to a more modern level of technology, then it would effectively be a newer generation rig.

Example

The Ocean Victory was built in 1972 and should be classed as a 1st generation semi. In 1996 the rig was upgraded to 5,000 ft water depth, 5,600 ton VDL, 3 × 1,600 hp mud pumps. Based on technology of rig equipment, the Ocean Victory would be classed as a 4th generation semi.

Time Line of Generations

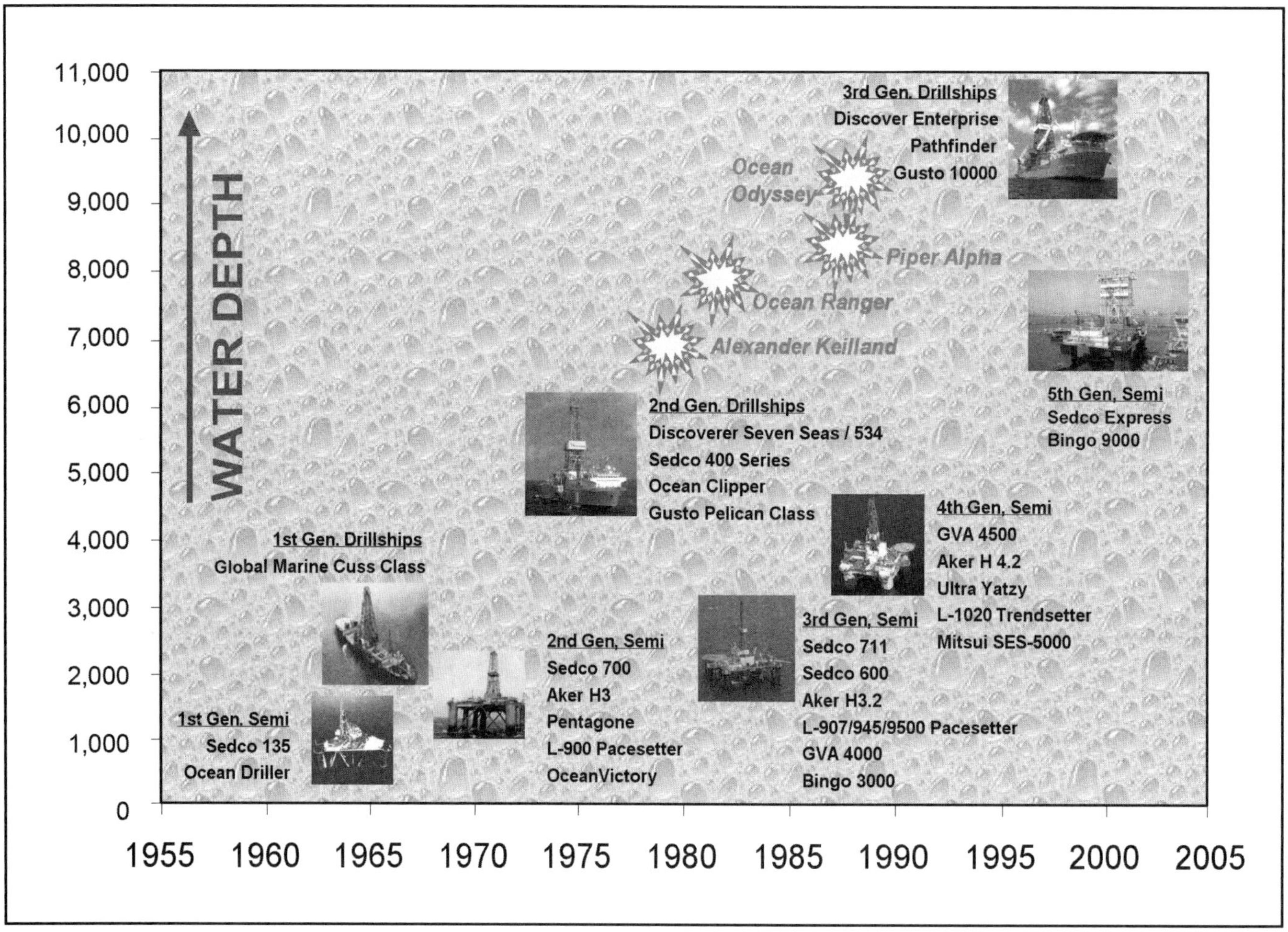

Figure 11.1 Floating drilling rigs generation time line

Time Line of Technical Capability

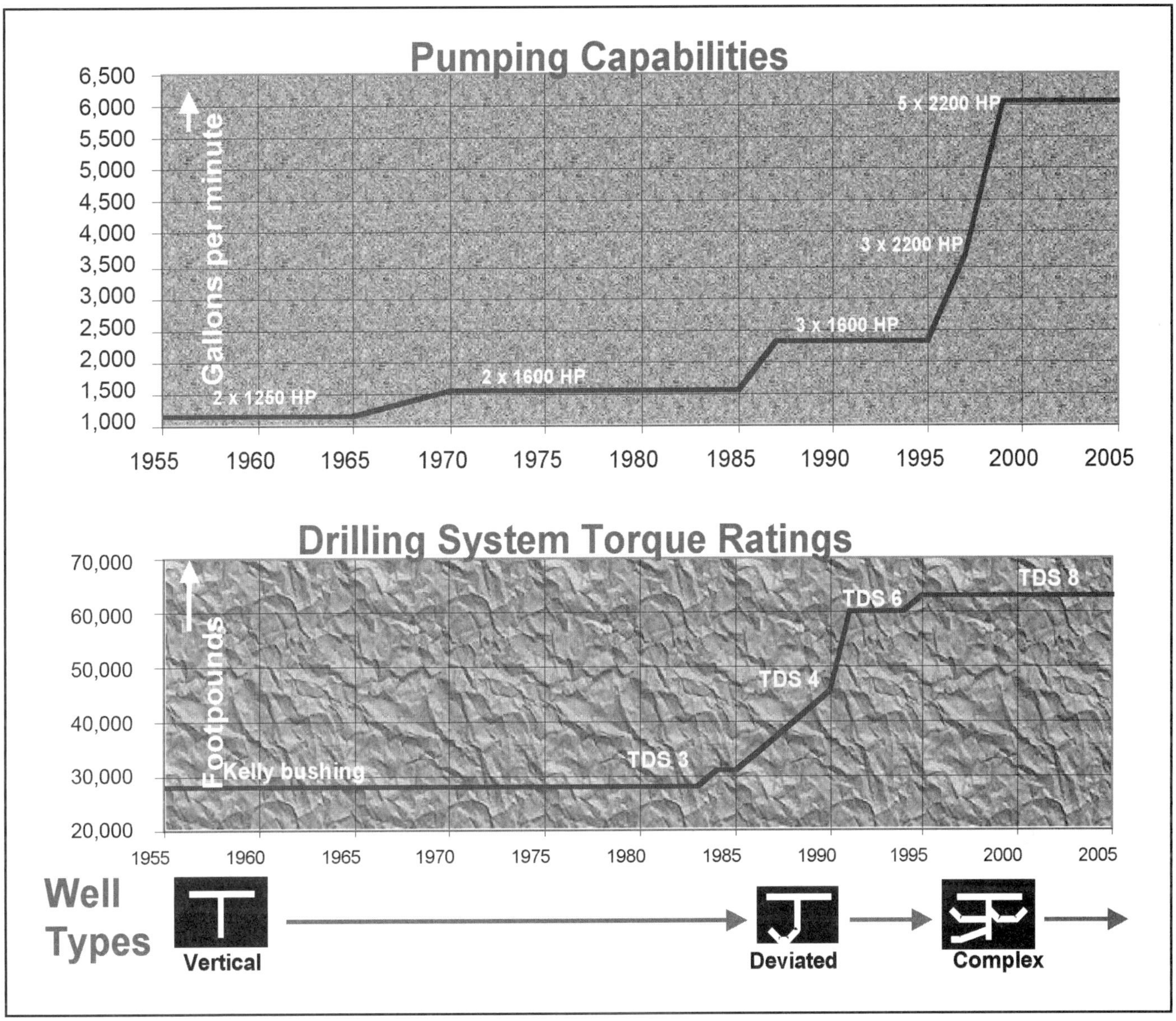

Figure 11.2 Time line of technical capability

Comparison of Ocean Victory Class Rig Specifications

◆ PROSPECTOR

Name:	Ocean Prospector
Rig Type:	Semisubmersible
Rig Design:	Ocean Victory
Year Built:	1971
Yard Built:	Mitsubishi, Hiroshima, Japan
Class:	ABS
Registry:	Panama
Water Depth–ft:	1,500
Drilling Depth–ft:	25,000
Quarters:	77 + 3 bed hospital
Dimensions:	338' x 263' x 126'
Helideck:	82' x 79' for S-61
Drilling Draft:	70'
Variable Deckload:	Operating: 2,250 LT/Transit : 1,341 LT
Operating Displacement:	23,722 LT
Bulk Mud & Cement:	9,320 cubic ft
Liquid Mud:	1,447 bbl
Fuel Oil:	7,746 bbl
Drill Water:	13,520 bbl
Potable Water:	355 bbl
Sack Storage:	4,000 sacks
Drawworks:	Continental Emsco C-3-II w/1⅜" drill line
Derrick:	Lee C. Moore 40' x 40' x 152', 1,000 kips, Cantilever mast
Top Drive:	N/A
Pipe-Handling System:	N/A
Rotary:	National D-495, 49½"
Mud Pumps:	(2) Continental Emsco FB-1300
Main Engines:	(4) Fairbanks Morse 38-D-8⅛
Annular BOP:	(1) Shaffer 18¾" 5 M
Ram BOP:	(2) Cameron "U" (double) 18¾" 10 M
Riser:	Regan 21" FD-8
Riser Tensioning:	640 kips
Solids Control:	(3) Brandt LCM-2D, Cascade system
Cranes:	(2) Seaking SK-3500 w/130' boom
Mooring System:	(8) 2¾" x 5,200' chains, 15 ton Moorfast anchors
Dynamic Positioning:	N/A

ENDEAVOR

Name:	Ocean Endeavor
Rig Type:	Semisubmersible
Rig Design:	Ocean Victory
Year Built:	1975
Yard Built:	Transfield, Fremantle, Australia
Class:	ABS
Registry:	Panama
Water Depth–ft:	2,000
Drilling Depth–ft:	25,000
Quarters:	82 + 3 bed hospital
Dimensions:	323' × 292' × 128'
Helideck:	83' × 83' for S-61
Drilling Draft:	70'
Variable Deckload:	Operating 2,250 LT/Transit 1,456 LT
Operating Displacement:	23,127 LT
Bulk Mud & Cement:	9,600 cubic ft
Liquid Mud:	1,830 bbl
Fuel Oil:	6,972 bbl
Drill Water:	10,984 bbl
Potable Water:	620 bbl
Sack Storage:	4,000 sacks
Drawworks:	Continental Emsco C-3-II w/1⅜" drill line
Derrick:	Lee C. Moore 40' x 40' x 180', 1,400 kips static hook load capacity, Cantilever mast
Top Drive:	Varco TDS-4S w/PH-85 pipe handler
Pipe-Handling System:	BJ Type V lower racking arm, Varco Model 2000 Iron Roughneck
Rotary:	Continental Emsco T-495, 49½"
Mud Pumps:	(2) Continental Emsco FB-1300
Main Engines:	(4) EMD 16-645-E8
Annular BOP:	(2) Hydril GL 18¾" 5 M
Ram BOP:	(2) Shaffer SL (double) 18¾" 10 M
Riser:	Vetco 21" MR6-B
Riser Tensioning:	640 kips
Solids Control:	(3) Brandt triple tandem, (3) Derrick Flo-Line Cleaners, Cascade system
Cranes:	(2) Favco Mod. 80,000 w/120' booms
Mooring System:	(8) 3" x 5,300' ORQ+20 chains, 15 ton Moorfast anchors
Dynamic Positioning:	N/A

VICTORY

Name:	Ocean Victory
Rig Type:	Semisubmersible
Rig Design:	Enhanced Ocean Victory
Year Built:	1972
Yard Built:	Avondale Shipyard, New Orleans, LA
Class:	AMS AI (M) CSDU
Registry:	Panama
Water Depth–ft:	5,000
Drilling Depth–ft:	25,000
Quarters:	104 + 2 bed hospital
Dimensions:	336' x 290' x 128'
Helideck:	83' x 83' for S-61
Drilling Draft:	74.5'
Variable Deckload:	Operating 5,000 LT/Transit 3,500 LT
Operating Displacement:	32,838 LT
Bulk Mud & Cement:	13,800 cubic ft
Liquid Mud:	3,094 bbl
Fuel Oil:	6,664 bbl
Drill Water:	13,569 bbl
Potable Water:	812 bbl
Sack Storage:	4,000 sacks
Drawworks:	Continental Emsco C-3-II w/1⅝" drill line
Derrick:	Dreco 40' x 40' x 180', 1,400 kips static hook load, Cantilever mast
Top Drive:	TDS-4S w/PH85 pipe handler
Pipe-Handling System:	N/A
Rotary:	National D-495, 49½"
Mud Pumps:	(3) Continental Emsco FB-1600
Main Engines:	(5) Caterpillar D-3516
Annular BOP:	(2) Shaffer 18¾" 10 M
Ram BOP:	(2) Shaffer SLX (double) 18¾" 15 M
Riser:	Shaffer 21" FT
Riser Tensioning:	1,280 kips
Solids Control:	(5) Derrick Cascade shakers
Cranes:	(3) SeaTrax 6032 w/140' booms
Mooring System:	(8) 3¼" x 4,200' ORQ+20 chains, (8) 3½" x 8,800' mooring wires, Bruce 10 MT MK-4 anchors
Dynamic Positioning:	N/A

New Builds

ADVANCES IN DEEPWATER DRILLING TECHNOLOGY

Reengineering Deepwater Drilling

Enterprise-class drillships

Express-class semisubmersibles

Enterprise-Class: A New Industry Standard

- Industry's first dual activity rig
- Can work in water to 10,000 ft
- Design efficiency gains 15%–40%

Vessel Overview

Length	835 ft
Breadth	125 ft
Depth	62 ft
Draft	42 ft
Displacement	100,000 mt
VDL capacity	20,000 mt

Comparison to DS: Enterprise-Class Enhanced/Semi: Sonat Enhanced GVA 4500

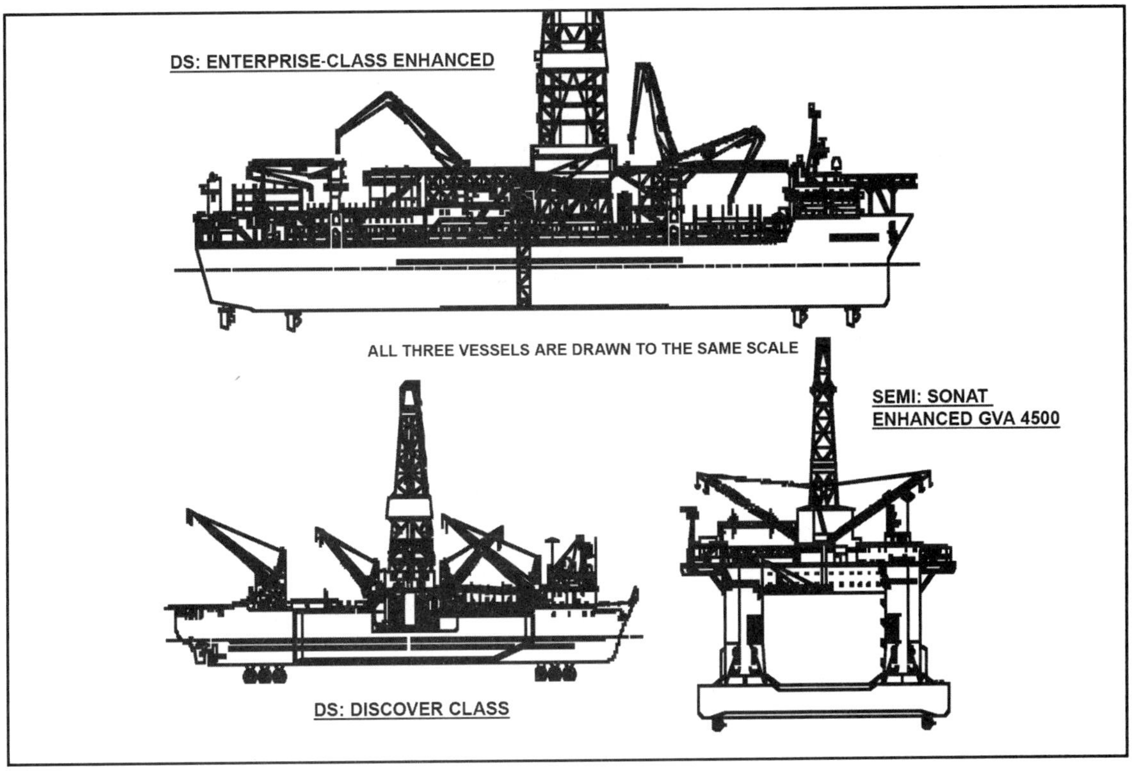

Comparison to Astrodome

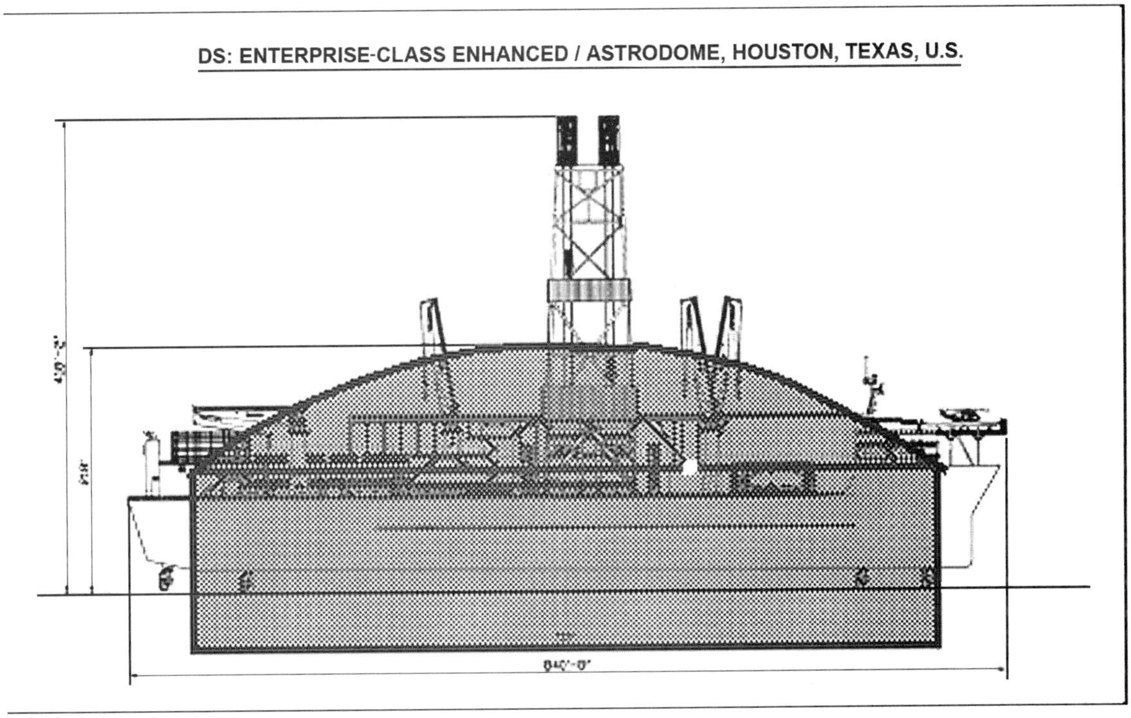

Deck Space

- Deck space for 20,000' of marine riser
- Plus two well's worth of consumables
- Automated tubular handling and craneage (4 × 60 mt knuckle-boom)

Dual Activity

Two full-service drilling stations housed underneath a single derrick.

Derrick 2 × 2,000,000 lb capacity
135 ft stands
80 ft × 80 ft × 226 ft

Drawworks 2 × 5,000 hp

Mud pumps 4 × 2,000 hp, 7,500 psi

Motion Compensators 2 × 500 ton (crown mounted)

Top drive 2 × 1,150 hp

>15,000 bbl fluids handling system

Derrick Tubular Capacity

- 450 stds of $6\frac{5}{8}$" dp (>60,000')
- Plus 20 stds of HWDP
- Plus 46 stds of drill collars or BHA components
- And 84 stds of $9\frac{5}{8}$" casing (>10,000')

DS: Enterprise-Class Enhanced—Automated Pipe-Handling Systems

DS: Enterprise-Class Enhanced—Drill Floor

Direct-Acting Tensioners

Driller and Assistant Driller Workstations

Work Sequence

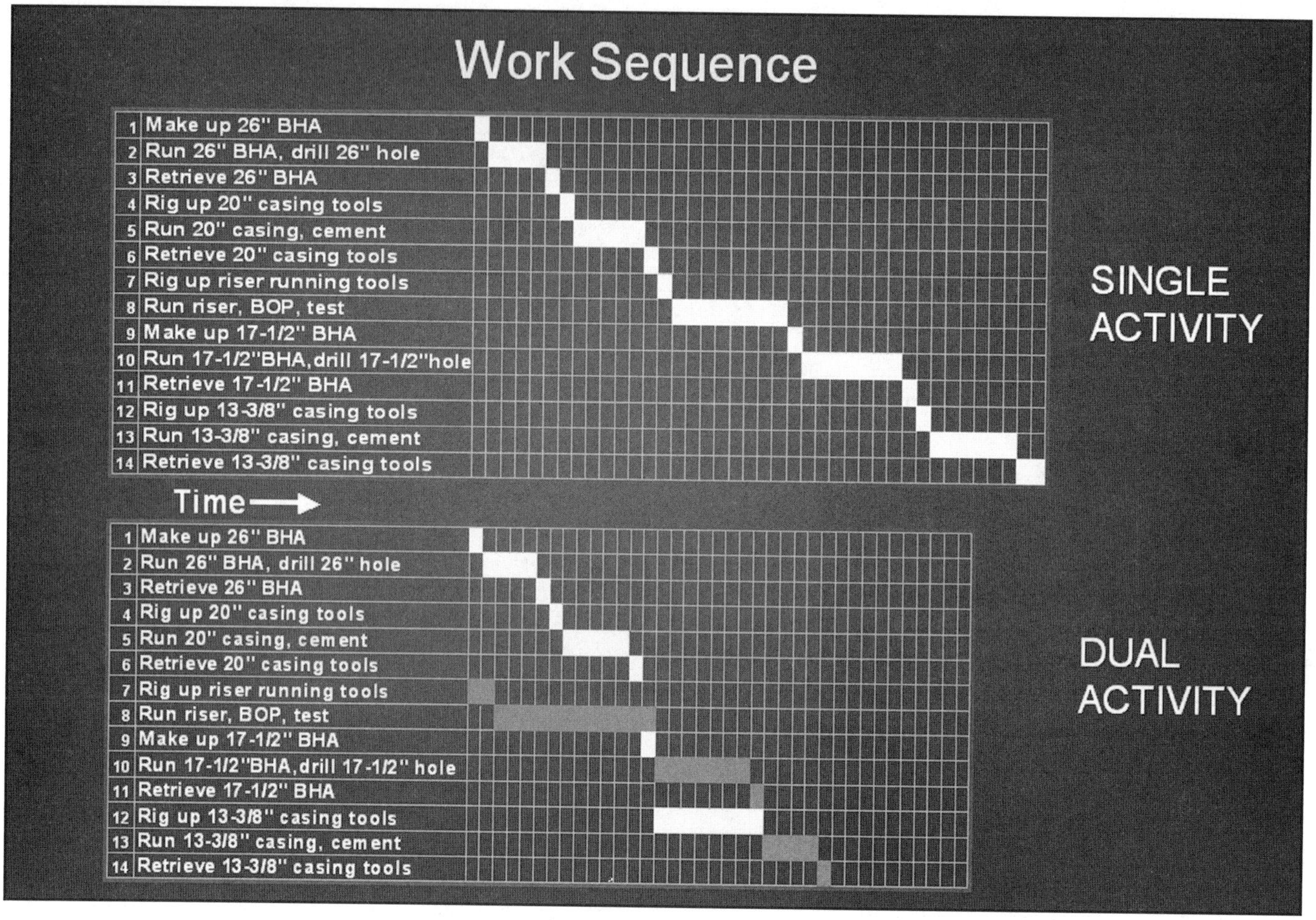

Semi: SF Express: Another Offshore Milestone

- Industry's first tri-act derrick
- Enhanced communications among service providers
- Can work in up to 8,500 ft of water
- Design efficiency gain 25%

Tri-Act Derrick

Derrick Tri-act 2,550 mt capacity
135 ft stands
190 ft high

Drawworks Hitec/Dreco AHC
6,800 hp

Mud pumps 3 × 2,200 hp,
7,500 psi

Motion Compensators N/A (crown mounted)

Hull Overview

Length	349 ft
Breadth	226 ft
Height	111 ft
Draft	65 ft
VDL capacity	6,000 mt

A NEW ERA OF DEEPWATER DRILLING

CHAPTER 12
Drilling Problems

INTRODUCTION

Drilling problems cover nonroutine events such as:

- Well control
- Stuck pipe
- Fishing
- Lost circulation
- Hole stability
- Hydrates
- Mud contamination
- Hole cleaning
- Formation damage

Well control and stuck pipe will not be covered in this manual. Some of the above subjects have been covered in earlier sections of this manual and will not be discussed further.

The key to dealing with drilling problems is to be aware of what is likely to occur and to have contingency plans and equipment in place to effectively deal with them.

FISHING

There is a multitude of fishing tools available to cover a whole range of scenarios.

However, the single most important rule is to always have sufficient fishing equipment available on the rig to make a first attempt at fishing any tool that is run in the hole. To accomplish this, it is important to have a detailed drawing of all tools, including wireline logging tools that show outside and inside dimensions.

Of course, if logistics is an issue then additional equipment can be held on site.

A typical fishing equipment list includes

- Overshots
- Sufficient grapples (spiral and/or basket) to cover all sizes plus over and undersize
- Overshot lip guides and extensions
- Fishing jars and accelerators
- Bumper subs
- Taper taps
- Safety joints
- Reverse circulating junk baskets
- Mills

Always ensure that the dimensions of any fishing tools are recorded prior to being run in the hole. DO NOT RELY ON GENERIC SCHEMATICS FOR MEASUREMENTS. This applies also to replacement tools (crossovers, etc.).

Always ensure that all relevant personnel are aware of how particular tools operate.

Always ensure that tools have been redressed correctly prior to being run in the hole. If possible, perform a function check.

Always ensure that fishing tools are included on preventative maintenance routines. All elastomers (e.g., overshot packers) have a finite shelf life. Ensure that they are stored correctly and replaced regularly.

LOST CIRCULATION

Lost circulation can occur at any time. Prevention measures should be established during well planning and lost circulation treatments should be specified in the drilling fluids program.

Preventative Practices

- *Minimize annular loading*—increases in ECD due to excessive drilled cuttings is a common cause of lost circulation, especially in top hole sections. This can be minimized by controlling ROP.
- *Maintain good drilling fluid properties*—maintain gel strengths, yield point, and viscosities at levels that will effectively clean the hole. Maintain low solids levels with efficient use of solids control equipment.
- *Minimize swab and surge pressures*—break circulation regularly on trips in the hole. Bring the pumps up slowly after connections. Rotate the pipe before turning on the pumps. Reduce tripping speeds if losses occur.
- *Keep ECD to a minimum*—reduce annulus restrictions. Keep hydraulics to the minimum level required to clean the hole. Consider controlling ROP.

Remedial Action

Losses should be dealt with as soon as they occur.

The hole should be kept full at all times, even if this means topping up the annulus with seawater.

In some wells it may be desirable to have a pit ready mixed with an LCM pill or to run LCM in the drilling fluid system. Alternatively, especially in the reservoir section, LCM such as calcium carbonate can be run as a weighting agent.

LCM pill formulations are generally designed around the severity of losses, using smaller concentrations for seepage losses and higher concentrations for total losses. Usually a combination of granular, flake, and fibrous material in different grades is mixed and pumped to cure the losses. Typical formulations and concentrations are shown below.

LCM Type	Seepage 1–10 bbl/hr	Partial 10–50 bbl/hr	Severe 50–500 bbl/hr
Granular	5–10 ppb	10–15 ppb	20–25 ppb
Flake	5–10 ppb	5–10 ppb	10–15 ppb
Fibrous	0–5 ppb	10–15 ppb	15–20 ppb

Granular LCM—Nut plug, calcium carbonate, graphite
Flake LCM—Cellophane, mica
Fibrous LCM—Cellulose, wood fiber, pulverized formica

Care must be taken not to plug the drill string (nozzles or downhole tools such as MWD).

If total losses occur or the severe loss pill does not work, then consideration should be given to using barites or diesel oil, and bentonite (gunk squeeze) or diesel oil, bentonite and cement, or a sodium silicate and cement pill.

HOLE STABILITY

Hole instability generally occurs in shales. Shales have low permeability and are partially composed of clay minerals. The main causes of shale instability are:

- Naturally fractured shales
- Incorrect mud weight
- Pressure invasion
- Fluid invasion and hydration stress
- Drill string vibration

Naturally Fractured Shales

With naturally fractured formations the rock is mechanically incompetent before drilling takes place. This is a problem that has to be lived with and problems should be minimized by establishing procedures that optimize hole cleaning, reduce pressure fluctuations, and minimize drill string vibration. Back reaming may be required on trips.

Incorrect Mud Weight

There is a stress increase in the rock around the borehole as the well is drilled. This stress increase is counteracted, in most cases, by the hydrostatic pressure of the mud column.

If the rock stress is greater than the mud hydrostatic pressure, i.e., the mud weight is too low, then the shale can fail and fragment. Once failure has occurred, the fragmented shale can easily fall into the well.

If the mud hydrostatic pressure is too high, then the rock can break down and losses can occur.

The stress distribution around the borehole is dependent on the inclination and azimuth. This tends to be worse in deviated wells, which is why higher mud weights are often required to drill deviated wells.

Borehole stability charts can be generated by computer models to show the minimum required mud weight to drill at certain inclinations and azimuths.

Pressure Invasion

In permeable formations, a filter cake is generated on the borehole wall that acts as an impermeable membrane.

Shale has extremely low permeability and no filter cake is formed. The mud hydrostatic pressure is therefore directly in contact with the formation and will equalize with the pore pressure around the borehole. With time, the hydrostatic pressure will gradually reach into the formation. This is known as pore pressure invasion.

As the pore pressure invasion occurs, the stress levels in the rock increase. If the stress levels reach a point whereby they are greater than the mud hydrostatic pressure, the rock will fail and fragment.

Swabbing can often cause the rock to fail, as this lowers the mud hydrostatic pressure to a point where the rock stress is greater, initiating failure and fragmentation.

Pressure invasion can be reduced by using a nonwater-based mud.

Fluid Invasion and Hydration Stress

All shales have the potential to hydrate, swell, and disintegrate when in contact with water. The degree of hydration is determined by the charges present on the clay platelets.

When pore pressure invasion occurs at the borehole wall, hydration can result in the failure of the shale, resulting in soft cuttings and clay balls.

Hydration can be reduced by using salts (KCl) and/or polyglycerols or nonwater-based muds.

HYDRATES

Hydrates are solid in nature, have a tendency to adhere to metal surfaces, and have a lattice-type structure in which gas molecules are completely enclosed within the crystal structure of frozen water.

Hydrate formation is a function of pressure, temperature, gas composition, and the aqueous phase composition.

Natural gas molecules ranging from methane to isobutane, hydrogen sulfide, and carbon dioxide can form hydrates.

As water depths increase, so the likelihood of hydrates occurring increases due to higher pressures and lower seabed temperatures.

The risk of hydrate formation should be minimized by using sufficiently inhibited mud systems and appropriate well-control practices. In addition, contingency plans should be established that outline the course of action to be taken if hydrates form.

Inhibition

Inhibition of the mud system can be achieved using either thermodynamic or kinetic inhibitors.

Thermodynamic inhibitors lower the activity level of the aqueous phase of the mud, suppressing the temperature required for hydrate stability at a given pressure. Typical thermodynamic inhibitors are salts, alcohols, and glycols as shown below.

Salts	Alcohols & Glycols
Sodium Chloride	Methanol
Potassium Chloride	Ethanol
Calcium Chloride	Glycerol
Sodium Formate	Ethylene Glycol
Potassium Formate	Propylene Glycol
Sodium Bromide	Polyalkylene Glycol
Calcium Bromide	
Zinc Bromide	

Mud systems containing as much as 20–26% by weight of sodium chloride with polymers have been used in water depths greater than 7,500 ft to prevent hydrate formation.

The use of kinetic inhibitors is still in its infancy. Essentially they slow down the rate of hydrate formation. Much is still to be learned about how they work.

Well Control

Hydrates are most likely to form while circulating out a gas influx through the choke and/or kill line. The increased velocity and expansion cooling in the lines, or shut-in periods with the gas influx close to seabed, offer ideal conditions for hydrates to form.

Obviously the best means of preventing hydrates forming in this instance is to avoid a gas influx. However, contingency plans need to be in place should hydrates occur during a well-control event. This could include running coiled tubing or a second string of pipe inside the riser and circulating hot mud or heat generating chemicals at the top of the BOP.

Well Testing

Always include a chemical injection sub, preferably below seabed, and inject methanol or ethanol to suppress the formation of hydrates.

A ported slick joint is required to allow the chemical injection line to pass below the BOP. A suitable quantity of check valves should be installed above and below the ported slick joint.

CHAPTER 13
Advances in Technology

HORIZONTAL DRILLING

Horizontal drilling has become commonplace during the last decade and now covers the range of well types noted below. In general, horizontal wells are drilled in development areas where the formations and pressures are known. However, there is an extra time element required to plan and design a horizontal well—it will probably take twice as long to plan, design, and order the equipment items, and take approximately 50% extra time to drill. This is due to the additional cost of specialized equipment, safety constraints, and time taken to achieve the build along the horizontal leg. Also, the longer the horizontal section that needs to be drilled, the lower the build rate. The majority of horizontal wells are drilled using medium radius builds.

Additional factors need to be considered when drilling a horizontal well, especially the need for primary well control where there is a greater requirement to maintain constant bottomhole pressure during a well kill. Hole cleaning is also more difficult due to the presence of 'dune buildup' across the build sections, and can also increase the chances of swabbing. Horizontal drilling has become the norm in many areas where the requirement is to maximize production. The current distance records are hampered only by the equipment limits of the respective drilling rigs, especially the hoisting capacity and maximum flow rates and pump capacity.

Definition

Horizontal drilling is the process of directing a drill bit to follow a horizontal path, approximately 90 degrees from vertical.

Purposes

Maximize production
Enhance secondary production
Enhance ultimate recovery
Reduce the number of wells required to develop a field

Main Types

Short radius (1°–4°/1 ft) shallow wells, can go from vertical—horizontal in 50 ft
Medium (8°–20°/100 ft) fractured reservoirs, need 300 ft to achieve build
Long radius (2°–8°/100 ft) offshore, inaccessible reservoirs, need 1,500 ft
Ultra short radius (almost no build)

Applications

Tight reservoirs (permeability < 1 md)
Fractured reservoirs
Economically inaccessible reservoirs
Heavy oil reservoirs
Channel sand and reef core reservoirs
Reservoirs with water/gas coning problems
Stratified thin reservoirs

Constraints

Cost
Well spacing and lease restrictions
Reservoir characteristics
Production methods
Amount of reach
Rig constraints–lifting capability, pumps, etc.
Availability of equipment–survey, coring, LWD tools, etc.
Kickoff depth constraint
Horizontal displacement constraints

MULTILATERAL WELL DRILLING

Rather than attempting to move hydrocarbons into vertical wellbores that may not be well positioned, the industry is now resorting to the use of horizontal, multilateral, and multibranch wells that move the wellbore closer to the hydrocarbons in place.

Multilateral well systems allow multiple producing wellbores to be drilled radially from a single section of a 'parent' wellbore. A major difference between this method and conventional sidetracking is that both the parent wellbore and the lateral extensions produce hydrocarbons.

Because only a single vertical wellbore is required, multilateral well designs require less drilling time, often have fewer equipment and material requirements, and increase hydrocarbon production.

Multilateral drilling and completion methods have been practiced since the mid-1940s. The first applications were developed for mining, where multiple bores were drilled from the parent shaft. These short, directional displacement bores were achieved with bent subs and the conventional rotary drilling technology of the time. Several patents were issued covering multilateral or multibore tools and methods for use in mining (Gilbert, Rhem), but the technology was not initially used in the oilfield.

For years, hydraulic fracturing (although not really a competitor of modern multilateral drilling) provided a real large exposure between the well and the reservoir. However, with the significant advancements in horizontal drilling technology in the mid-1980s and its evolution into multilateral drilling in the mid-1990s, the performance of a vertical well with a hydraulic fracture can now be readily surpassed by a properly oriented horizontal or multilateral well in a really anisotropic reservoir. Furthermore, horizontal wells provide better results in reservoirs with large gas caps or water aquifers.

The first extensive modern application of multilateral drilling was in the Austin Chalk formations in Texas during the late 1980s. High initial production rates and high decline rates required increased reservoir face exposure for the achievement of maximum production in the shortest possible time.

Definition

The drilling of multiple drain holes from a single vertical, horizontal, or inclined wellbore to improve hydrocarbon recovery and minimize the cost of dollar per barrel.

Purposes and Drivers

Cost
Slot consideration
Increase reserves

Increase overall recovery
Reduce the overall cost per barrel recovered
Reduce the number of wells required for the reservoir
Optimize secondary recovery
Extend the life of the well

Constraints

Complexity—six different types
Limited well applications at present—18,000' MD
Cementing and junction stability
Whipstock orientation
Ability to reenter both multilaterals
DRA required—including drilling, completion, and reservoir
Mechanical type
Budgetary contingency
Select good candidates—better on multiwell programs
Lead times—average time 6 months

SLIM-HOLE AND COILED-TUBING DRILLING

Slim-hole drilling is reemerging technology expected to make step-improvements in the efficiency and economy of well construction. Wells are normally classified as slim holes if the production interval is intentionally drilled with a bit diameter less than 4.75 in. (Hough, 1995). McCann *et al* (1993) suggested a 'narrow gap' criterion (a drill string-to-hole diameter ratio greater than 0.8) to distinguish slim hole from reduced-bore and conventional wells. Either definition is acceptable.

The primary slim-hole drilling techniques include

- continuous boring,
- conventional rotary drilling,
- downhole-motor drilling, and
- coiled-tubing drilling.

They share several mud related traits, including small circulating and pit volumes, small pumps, and high frictional pressure gradients (inside the drill string and/or annulus). The key mud-related differences among the methods are the rotary speed of the drill string and the annular gap, both of which can vary widely.

Continuous coring, adapted from the hard-rock mining industry, requires the use of solids-free drilling fluids. High speed rotation (up to 1,000 rpm) creates extraordinary centrifugal forces that can cause mud solids to cake on the inside walls of the drill rod (drill string) and interfere with wireline recovery of the core barrel. Innovative mud formulations based on formate brines (primarily potassium formate, up to 1.6 sg) have provided the best results (Downs, 1993). High density brines made with formate salts are nonhazardous and compatible with both conventional oilfield polymers and formation waters containing sulfates and carbonates. Continuous coring operations also typically use very narrow annular gaps, some less than 0.25 in. Annular pressure losses, which can uncharacteristically exceed drill-rod pressure losses, become critical for lost circulation and well-control concerns. Pipe rotation also increases annular pressure losses (McCann *et al*, 1993). The resulting high overbalance may contribute to formation damage.

Rotary speeds for conventionally drilled slim holes (Sagot and Dapuis, 1994) are considerably lower (usually less than 350 rpm), minimal for downhole motor drilling, and nonexistent for coiled-tubing drilling. Ultra-low solids content is not required, although solids (especially fines) should be minimized and carefully controlled for maximum performance.

Different types of drilling fluids have been used successfully for slim-hole wells. However, biopolymer/brine fluids have distinct advantages. Brines provide solids-free density, inhibition and improved temperature stability. Biopolymers such as xanthan and welan gum are compatible with brines and provide excellent viscosity, suspension, and drag reduction in turbulent flow (at low solids content). Starch or starch derivatives can be added to control fluid loss and acid-soluble, sized calcium carbonate can provide bridging, if required. Goodrich *et al* (1996) reported successful use of solids-free xanthan/brine drilling fluids for a coiled tubing project in Prudhoe Bay, Alaska. Claims include increased horizontal reach, improved hole cleaning, better formation stability, lower pump pressures, and reduced pipe sticking tendencies. Fluid loss is controlled by penetration of a highly viscous filtrate into the formation rather than a conventional filter cake.

Slim-Hole Drilling

Definition

Slim-hole drilling is defined as holes drilled with bit sizes < 7" diameter while a slim-hole completion is one with the production casing size less than 4 inches diameter.

Applications

Exploratory wells
Deepening existing wells
Horizontal and multilateral wells
Wells with low production

Benefits

Lower cost option (smaller equipment size, less materials, less man power, less space)
Less environmental impact
Less risk from a cost viewpoint

Limitations

Equipment availability and loading limits (also rig equipment)
Well control
Fishing tools and ability to do remedial work
Limited to shorter wells

Coiled-Tubing Drilling

Coiled tubing consists of:

- Coiled tubing
- Tubing injector head
- Coiled-tubing reel
- Wellhead blowout preventer stack
- Hydraulic power-drive unit
- Control console
- Downhole mud motors and MWD tools

Constraints

Coiled tubing size (1–3 in. in diameter)
Shallow wells and less than 8½" hole size
Reentry hole sizes are limited to < 2⅜"

Multilateral holes < 3¾"
Limited hydraulics for annular hole cleaning
Buckling limits
Fatigue life limits

Advantages

Safety
Efficiency
Small rigs—much less expensive
Fewer personnel
Less impact on the environment
Lower overall drilling cost

UNDERBALANCED DRILLING

For the drilling of fractured chalk, limestone, or depleted reservoirs, a very low solids drill-in fluid is preferred because the fluid that invades can be removed by production. High solids loading in fluids can cause mud that enters the fracture system to 'gel up'. Gelled mud may be impossible to remove. Some fractured limestones or chalks also have a low matrix permeability so that fluid loss can occur from the fractures. These formations are sometimes drilled underbalanced with solids-free brines. Underbalanced drilling permits oil to be produced as the well is being drilled, and the oil is separated from the solids-free drill-in fluid at the surface.

Definition

Underbalanced drilling is a process where the wellbore drilling fluid pressure gradient is less than the formation pressure gradient which is being drilled.

Requirements

Special rotating BOP stack or coiled tubing for drilling
Surface separating equipment (four-phase separation)
Drilling fluid whose weight can be reduced by injecting nitrogen gas or air
Preferably a closed flow circulating system

Applications

Pressure depleted zones
Weak and permeable formations
Formations susceptible to high production damage

Benefits

Reduced formation damage
Faster rate of penetration (ROP)
Can eliminate stuck pipe
Can eliminate mud losses in highly
permeable formations
Improved formation evaluation

MWD, LWD, AND GEOSTEERING

No other technology used in well construction has evolved more rapidly than measurement-while-drilling (MWD), logging-while-drilling (LWD), and geosteering (fig.13.1). Early in the history of the oilfield, drillers and geologists often debated environmental and mechanical conditions at the drill bit. It was not until advances in electronic components, materials science, and battery technology made it technically feasible to make measurements at the bit and transmit them back to the surface, that the questions posed by pioneering drillers and geologists began to be answered.

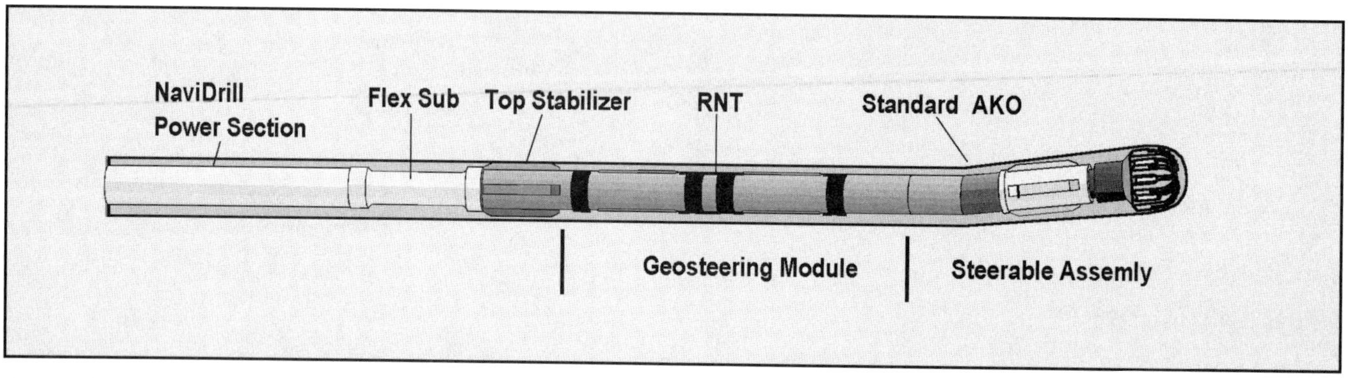

Figure 13.1 Geosteering

The first measurements to be introduced commercially were directional, and almost all the applications took place in offshore, directionally drilled wells. It was easy to demonstrate the savings in rig time that could be achieved by substituting measurements taken while drilling and transmitted over the technology of the day. Single shots (downhole orientation taken by an instrument that measures azimuth or inclination at only one point) often took many hours of rig time since they were run to bottom on slick line, then retrieved. As long as MWD achieved certain minimum reliability targets, it was less costly than single shots, and it gained popularity accordingly. Achieving those reliability targets in the harsh downhole environment is one of the dual challenges of MWD and LWD technology. The other challenge is to provide wireline quality measurements.

In the early 1980s, simple qualitative measurements of formation parameters were introduced, often based on methods proven by early wireline technology. Geologists and drilling staff used short, normal resisitivity and natural gamma ray measurements to select coring points and casing points. However, limitations in these measurements restricted them from replacing wireline for quantitative formation evaluation.

Late in the 1980s, the first rigorously quantitative measurements of formation parameters were made. Initially, the measurements were stored in tool memory, but soon the resistivity, neutron porosity, and gamma density measurements were transmitted to the surface in real time. In parallel with qualitative measurements and telemetry, widespread use of MWD systems (combined with the development of steerable mud motors) made horizontal drilling more feasible and, therefore, more common.

Soon, planning and steering horizontal wells on the basis of a geological model became inadequate. Even with known lithology of offset wells and well defined seismic data, the geology of a directional well often varied significantly over the horizontal interval that steering geometrically (by using directional measurements) was quickly observed to be inaccurate and ineffective. In response to these poor results from geometric steering, the first instrumented motors were designed and deployed in the early 1990s. Recent developments in MWD and LWD technology include sensors that measure the formation acoustic velocity and provide electrical images of dipping formations.

Definition

Capability to drill, log, and navigate in long and medium-radius wells, to stay in the pay zone, reduce hole exposure, and improve deviation control.

Applications

Long deviated and ERD wells
Wells with thin or disjointed reservoir sections
Designer profile wells—i.e., horizontal and multilateral wells
Wells with small, faulted areas of production

Benefits

Can drill, log, and navigate in long- and medium-radius wells
Can change hole trajectory while rotating—full 3-D control
Now have FE and MWD tools available
Can delineate reservoirs in much shorter time scale
Higher value wellbores—able to stay in the pay zone
Can reduce hole exposure and improve deviation control
Reduced number of wells

In horizontal wells, when using FEMWD tools, need to consider

- Bending limits of the tools
- Position of sensors from the bit
- Depth of investigation of the sensors
- Environmental restrictions—drilling fluid, etc.
- Real time vs. memory data

CORING

Detailed information from target formations is essential of both primary and secondary recovery programs. Core samples can yield this critical subsurface information. With quality cores, oil companies can more fully understand formation characteristics and more efficiently achieve production objectives.

High quality cores provide the most accurate lithology, porosity, and permeability information for building the geologic model of the reservoir. Such models are important tools, for example, in evaluating horizontal and vertical permeability.

Core samples can provide the petrophysicist and the reservoir engineer with accurate saturation, wettability, and electrical properties of the formation. When secondary displacement is the objective, core sample data are essential.

Core quality is the key. The sample must be obtained without altering its native (or in situ) properties. Informed application of specialized tools and techniques can produce quality core samples.

Coring technology advances over the past few years include:

- Coring while drilling (BHI coredrill)
- Low invasion coring
- Gel coring
- Antiwhirl technology
- High-pressure coring

CHAPTER 14
Subsea Systems

INTRODUCTION

The exploration for oil and gas offshore began in the late 1800s, and in 1896 an offshore well was drilled off the coast of California. In 1938, the discovery of the Creole field 2 km (1.24 ft) from the coast of Louisiana in the Gulf of Mexico heralded the beginning of the move into open, unprotected waters. In this instance, a 20 m (65 ft) by 90 m (295 ft) drilling platform was secured to a foundation of timber piles set in 4 m (13.12 ft) of water. Typically, these pioneering offshore wells utilized piers to create a platform above the prospect which thus enabled them to drill vertical wells into the target.

As the search for oil and gas reserves has continued to intensify, so exploration has moved into increasingly deeper waters. The first subsea well was completed by Shell in 1960 and came on stream in January 1961. This marked both the successful conclusion of many years of R&D and the beginning of a new era in subsea production. Nowadays, subsea completions are a commonplace option and it is the mode of production that is changing. Originally, such wells would have been tied back directly to a platform but now alternatives exist and can be ranked for any particular field depending on cost and water depth.

CURRENT SUBSEA DEVELOPMENTS

Current subsea development options include:

- Tension leg platforms without storage (TLP)
- Floating production vessels without storage (FPV)
- Floating storage units (FSU)
- Floating production storage and offloading vessel (FPSO)
- SPAR buoys for storage or production and storage
- Deep draft semisubmersibles with storage and offloading (DDSS)

FPSO

The idea of FPSOs has been around for many years and the concept has been utilized since the 70s when conversions from existing tankers was the norm. In the late 80s, the Petrojarl heralded the first of the turret systems and was marketed as a testing and early production system. Since then, various turret designs have appeared and include those on the Gryphon, Uisge Gorm, Captain, Anasuria, and Foinavon.

Definitions

Floating—the body is in equilibrium when floating. This excludes TLPs which use buoyancy to maintain equilibrium. The unit must have a displacement and buoyancy compatible with its payload requirement, a form compatible with its station-keeping requirement, and be able to provide a safe, stable platform as a working environment.

Production—the vessel could contain primary and secondary processing equipment to treat live well fluids; e.g., oil/water separation. These are field specific and can range from a single stage separation to a full blown separation, compression, and injection system with its associated power requirements.

Storage—able to store significant quantities of oil until it can be removed by shuttle tanker. This could be due to the lack of an effective export option in the vicinity other than a shuttle tanker or to the poor quality of the crude which would incur a high pipeline tariff. Note that lack of sufficient storage could be detrimental in the long term if production has to be halted because of a logjam in the export route (planned shutdown excepted).

Offloading—contains a means by which oil can be transferred from storage to either a shuttle tanker or alternative export source. Direct unloading is permissible only if there are no weather implications; i.e., the FPSO can weathervane. If that is the case, then a remote-loading buoy may be required. Such remote buoys would include:

- Surface loading buoys (CALM systems)

- Loading towers
- Subsurface buoys (UKOLS, STL systems)
- Trailing hoses (surface or submerged)

The evolution of the FPSO has resulted in there now being three generic types recognized:

1. Monohull ship-shaped structures
2. Semisubmersible structures
3. Spar structures

Monohull Ship-Shape

These are by far the most common and evolved from the modified tanker parentage. Their size is dictated by the storage capacity required and the sea-keeping capabilities. Consequently, the available deck space for processing and utility equipment is usually more than adequate. For operating in benign areas, a box shape could provide an acceptable solution while in more severe environments, a longer, thinner vessel would be required.

Because of the large lateral excursions coupled with the need to be able to weathervane, these units can only be used with subsea completions, wet trees, and are connected to the wells with a flexible riser system. Drilling and intervention activities are not practical from such systems.

Semisubmersible Structures

These encompass all conventional and deep draft designs. The former includes all design types which are capable of being moored with a catenary system in water depths in excess of 70 m (230 ft). Production to these semisubmersible designs requires dynamic risers due to the deck motions being too great to allow surface wellheads. There are various flexible configurations available for connecting the seabed flow lines to the production facilities on the rig. The number of risers is not seen to be a severe constraint for semisubmersibles, the deck load capacity being the more critical factor. Drilling and intervention activities can be conducted from the vessel if located directly above the wellheads.

SPAR System

The SPAR is essentially a vertically moored hull of around 200 m (656 ft) in height and up to 40 m (131 ft) in diameter. Mooring is usually provided by a catenary mooring system that restricts vessel movements and permits the use of rigid risers and dry trees. The risers are supported either by hydraulic tensioners or buoyancy modules. As the SPAR can carry solid risers, then it can also be equipped with drilling and intervention equipment. While the design has already been used extensively for storage purposes (e.g., Brent Spar), it is now being considered as able to provide a production option as well; e.g., Neptune Field for Oryx (Kerr McGee).

Tension-Leg Platform

This is essentially a vertically-tethered platform. The platform is floated over the wellhead area and tethered at each corner by tubular members that are attached to piles. When ballasted up, the buoyancy in the vessel is sufficient to place the tubulars in tension. This design effectively eliminates vessel movement and enables the use of rigid risers and dry trees. The TLP, once moored, effectively behaves as a platform. It differs only in that, for drilling and intervention purposes, the rigid production riser and surface tree are replaced with a drilling riser and subsea BOP and the work conducted as if on a semisubmersible.

SUBSEA TIE-BACK METHODS

Introduction

Subsea production systems can be connected to the host facilities using conventional steel pipelines and in-field flow lines. For the steel pipelines, there are three methods of laying them:

- Conventional S-lay
- J-lay
- Laying from a reel ship

Conventional S-lay is the oldest method of installing a subsea pipeline offshore. Lengths are welded together horizontally and launched over the stern of a ship over a long stinger while the ship 'crawls' forward on anchors or uses thrusters (DP). Used typically in shallow to medium waters, it thus avoids subjecting the pipeline to excessive stresses or having to use unacceptably heavy (thick) pipe. This method is quick, but it does have limitations dictated by water depth and line pipe size. It is also labor intensive; i.e., costly.

J-lay involves welding pipe in a vertical orientation thus eliminating the need for a stinger. The advantages of this are that pipe can be laid in deeper waters and a greater selection of vessels can be used. The downside is that it is a slower process than S-lay due to the vertical welding position.

Laying from a reel ship necessitates pipe to be prefabricated in a linear onshore facility and then wound onto a reel. This can then be loaded onto a specially adapted vessel. Any pipe deformation while being spooled onto the reel is removed during the laying process by passing the pipe through a straightening machine. Any welding is conducted beforehand on shore, thus making this process a more cost effective one. The laying of the pipe is much faster as a result of this but it is restricted to pipe diameters of 16" or less. This, however, covers most subsea developments.

Regardless of the laying method, all pipelines must be able to withstand the stresses encountered during laying and the thermal expansion forces during production. The latter is dealt with by provision of expansion spools in the form of expansion loops and burial of the pipeline to restrain against upheaval buckling, provide thermal insulation, and protect against fishing activities.

Normally, a pipeline is installed by laying from one target area to another. The pipeline ends are normally terminated in the form of pipeline end modules (PLEMs) which consist of some form of skid (with or without isolation valves) supporting the end of the pipeline and its termination above the mud line. The actual pipeline mouth should be equipped with either a flange or clamp connector to assist in connecting to whatever facility—tree, manifold, or platform interface. A survey then determines the actual shape of the tie-in spool required to connect the pipeline end with the 'host facility'. This can then be fabricated onshore and fitted with minimal offshore modifications. Commonly, these connections used to be made up with the assistance of divers but are more commonly now being engineered to allow hookup with ROV assistance, vital as the industry moves into the deeper water areas.

Flexible Flow Lines/Pipelines

Flexible pipelines are basically thermoplastic pipes with helically-wound steel strips incorporated to increase strength. Advantages over conventional line pipe:

- Not subjected to the same laying stresses since flexible
- Tie-in tolerances at the ends are greater since the need for tie-in spools is eliminated.
- Thermal stresses are accommodated by the flexible nature of the pipe.

However, there are temperature limitations (112°C / 233.6°F) and size limitations 16" at present.

Rigid Risers

On fixed platforms, they are clamped to the legs and connected to import/export flow lines on the seabed. A variation of this is the J-tube which is a rigid riser incorporating a 90° bend on the seabed. A smaller diameter flow line can be pulled through this to the platform top sides.

They also occur on TLPs and SPARs where heave and large excursions have been eliminated.

Flexible Risers

Flexible risers are used for connecting flow lines to semisubmersible production systems, FPSOs, or other floating systems such as loading buoys which are subjected to significant wave and weather-induced motions. They are similar to flexible pipelines in construction but the steel armouring is specifically designed to withstand the dynamic and fatigue stresses induced by the vessel motions. They are normally attached to a riser base at the seabed which is used both as an anchor and as a support to mount the connectors.

Buoyancy aids are normally installed at some mid-water depth which creates a riser with an 'S' shape. These provide sufficient slack in the riser to allow for the most extreme vessel motions and a means of controlling the slack.

Catenary Risers

Steel catenary risers provide an alternative to the flexible riser option. These are steel risers which hang vertically from a floating vessel and take up a natural 'J' form at the seabed. As the vessel heaves, the riser lifts off and touches down on the seabed continuously varying the contact point. The pipeline has a thicker wall and additional reinforcement throughout the touch down zone to resist any abrasion.

Pipeline Bundles

These comprise the flow lines, umbilical cables, etc., bundled inside an external carrier which can also be filled with an insulating material. Once fabricated and tested (in the UK, sites at Tain and Wick), the bundles are launched across the beach and towed out to location. Currently, the maximum length practical for towing purposes is deemed to be about 4.97 miles (8 kms). Longer lengths of final bundled lines necessitate linking series of individual bundled lines once on location.

Connection Methods

The simplest form of connecting pipeline or spool ends onto manifolds is using diver intervention. However, the current trend is now to move to remote-activated systems such as the pull-in systems utilizing a winch to draw the pipeline/spool end towards the manifold, and then making up the two mating hubs with an ROV (remotely operated vehicle) activated clamp. A variation would be to tie the rope end onto the structure and an ROV onto the pipeline end and allow the ROV to winch the pipeline end into position.

As an alternative to pull-in methods, the tie-in can be made using a vertical connection system. Thus, an upward facing termination hub is fitted to the structure while the pipeline/spool end has a corresponding mating hub. The pipeline/spool end is 'dropped' onto the upward facing hub and the connection secured with an ROV installed clamp.

SUBSEA TEMPLATE AND MANIFOLD OPTIONS

Conventional Template

Conventional subsea templates comprise a seabed-mounted structure secured by suitable means dependent on the soil conditions. The functions of the structure include acting as a drilling template, providing support for the Christmas trees, enabling commingling of production streams, providing an injection manifold along with all the connection systems, controls, and chemical injection equipment. The template may be designed to protect equipment from trawling activities and dragging anchors depending on the level of fishing activity. It will certainly provide some dropped objects, probably restricted to lighter items such as drill pipe sections. This will probably be in the form of hinged panels that can be raised by either ROV or remotely operated tools (ROT) to give access for intervention work.

The well bay isolation valves in a production or injection manifold are remotely-operated hydraulically-actuated valves with ROV override as with the Christmas tree valves. A test header may also be included on the manifold to enable individual wells to be isolated, and the production stream conveyed back to a test separator on the host facility via a specific test flow line. A cheaper testing option would involve the use of a multiphase meter sited on the manifold and suitable pipe work to enable the flow from individual wells to be routed through it, thus eliminating the need for a costly test flow line. Well testing is a necessary, regular activity during production and is used to aid reservoir management and maximize recovery.

When two or more wells are producing into the same manifold, the wellhead pressures must be balanced to prevent cross-flow occurring. To control individual wellhead pressures, chokes are installed on the flow path from each well and are capable of adjusting the pressure in a series of finite steps. The temperatures and pressures of the individual flows are constantly monitored by the production control system, both upstream and downstream of the chokes, which can either be mounted on the trees or on the manifold.

The largest conventional template installed is the Saga Snore template in Norway which has a 20 well capacity and weighs 222 tonnes.

Expandable Moon Pool-Installed Templates

The original template and manifold designs have been modified recently to provide equipment that can be installed through the moon pool on a semisubmersible rig thus eliminating the cost of an additional heavy lift vessel. The most successful design to date is the HOST template system (hinge over subsea template) manufactured by FMC Kongsberg. This comprises a central rectangular section on which a number of 'wing' sections (usually four) are hinged. The wings are hinged vertically during deployment through the moon pool. After securing the structure onto its preinstalled foundation, the wings are allowed to fall outwards into a horizontally hinged position, each one forming a well bay guide. The wells are then drilled and Christmas trees run and installed in the normal manner. A manifold module can then be run and installed on the central bay and the Christmas tree flow lines hooked up, horizontally for the HOST system or vertically for similar concepts.

Clusters and Satellites

As an alternative to the template system (conventional or HOST type where wells, trees, and manifolds are all contained within a single structure), a well cluster can be utilized. This consists of a central manifold containing the well bay control valves, the umbilical termination unit, and control and chemical distribution systems. It may also contain the choke valves and Christmas tree control modules. The actual wells are removed from the manifold, the distance being anything up to several kilometres which permits smaller, lighter structures to be used, and simplifies the tree/manifold hook up; i.e., greater tolerance due to greater distance apart. This type of well configuration also has more flexibility since wells may be sited at the optimum locations rather than simply in a single drilling center. The disadvantages of clusters are that they require more subsea connections, there are in-field jumpers and umbilicals to lay, and more rig moves may be required.

For both templates and clusters, it is usual to incorporate dual production headers joined by a pigging loop which will permit 'round trip' pigging; i.e., launching a pig from the host facility down one pipeline and returning up the other.

Note: A pig is a device propelled through a pipeline by fluid pressure to carry out various inspection and maintenance tasks in the pipeline. These include corrosion monitoring and scraping scale/wax deposits off the pipeline.

Intervention

Both template and cluster systems require careful consideration regarding intervention and repair/maintenance activities during the life of the field. Although reliability of components has increased, failure resulting in the loss of production, however temporary, cannot be tolerated. Consequently, a great many of the more vulnerable components have inherent redundancies built in. Other components are designed to be easily retrievable—for instance, control modules and multiphase meters can be retrieved and replaced by diver, ROV, or ROT.

Other components liable to failure, such as choke valves and well bay control valves, are available with inserts which can be removed and replaced by special tooling without recovering the whole valve body.

All remotely operated valves should be fitted with ROV overrides so that, in the event of failure of the remote control system, production can be maintained by ROV intervention. As a last resort, the complete manifold system should be designed such that it can be retrieved without disturbing the trees should component failure warrant it.

All of these intervention activities and their access requirements need to be given full consideration during the design of templates and manifolds, and they should also be proven during SIT (system integration testing) onshore.

Deepwater Options

The basic options described above are all applicable to deepwater developments. The chief differences lie in the installation and in the operation/intervention methods and techniques. The diver intensive options become unacceptable due to risk and technical difficulty.

Guideline and Guidelineless Systems

In water depths up to around 500–600 m (1,640–1,968 ft), the accepted practice is to run subsea equipment on tensioned guidelines acting as a course orientation tool by being attached to guideposts on the seabed and to the rig at their upper end. The equipment being run is skidded into the moon pool and the four guide wires are latched into the guide funnels and frame members. The equipment can then be lowered to a depth just above the seabed structure on slack guide wires with the rig off station to minimize any dropped object risk. The guide wires can then be tensioned as the rig is moved over the structure and the equipment can be landed out on the guideposts.

In water depths beyond 500–600 m (1,640–1,968 ft), the guideline system becomes impractical due to the excessive lengths and increased risk of entanglement. Instead, equipment is run fitted with large diameter entry funnels and matching reentry cones, either funnel-up or funnel-down, irrespective of whether the rig is moored using anchors or dynamic positioning. Both parts to be mated are equipped with transponders and final positioning is undertaken by matching the signals. Course orientation may be affected initially using ROV assistance.

Template Design

The template must fulfill the following functions:

- Provide a solid foundation for the equipment to be installed and be capable of withstanding all imposed loads:
  - Static loads
  - Dynamic installation loads
  - Drilling loads static and dynamic
  - Dropped objects loads
  - Snagging loads normally 60 T—the breaking load of a normal trawl.
- Provide protection for the installed equipment from impact due to dropped or dragged objects.
- Provide access to permit all routine operational, intervention, repair, and maintenance activities by whatever means—diver, ROV, ROT.
- It must be as lightweight and compact as possible to permit installation by the smallest possible vessel if not the drilling rig itself.

SUBSEA WELLHEAD SYSTEMS

Introduction

The subsea wellhead provides a seabed location for suspending and sealing the wellbore casing strings. Originally, they were designed around a two-stack system; i.e., able to work with 21¼" 2,000 psi and 13⅝" 10,000 psi BOP stacks at different stages in a well. Most modern subsea wellheads are now designed around the use of a single 18¾" BOP stack rated at up to 15,000 psi.

These wellheads should be able to withstand the loads imposed by the BOPs and their associated risers. They should also enable a drilling guide base to be replaced by a production guide base thus converting the well to a producer.

Initially, the basic wellhead component serves to guide equipment to the correct seabed location via guide wires. Then, after drilling the top hole, the 30" housing on top of the cemented 30" conductor pipe provides a structural foundation. The 18¾" high-pressure housing run on top of the 20" casing allows drilling to continue through higher pressure regimes with full mud circulation and pressure control. Subsequent casing strings are run through the BOP and landed off in the wellhead, one on top of the other, as the casing sizes diminish. Interfaces exist between the BOP hydraulic connector and 18¾" housing and between the casing hangers and the casing tubulars. The correct profile must be machined on the outside of the housing to suit the BOP connector and the correct thread cut in the casing hangers to suit the casing. A compatible connector and extension must also be welded to the 20" and 30" housings.

Design Issues

The major issue is the actual casing scheme to be run. However, no matter what scheme is ultimately chosen, the basic wellhead parts—hanger and pack-off interface—will be of standard design. At an early stage, consideration should be given to suspending the total casing string weight in the wellhead housing along with the simultaneous application of full-bore pressure end load. At 15,000 psi, there can be a problem with the limited shoulder available at the bottom of the housing to take this load. In some cases, a hardened insert is used to distribute the load into a greater area to reduce the bearing stress. In other cases, load rings are used to transmit the load from each hanger into grooves in the wellhead housing thus reducing the bottom shoulder load.

Subsea wellhead equipment and tools are designed according to API 17D specifications. While basic dimensions, performance requirements, and pressure ratings are determined by API standards, the design and operation of running tools is at the manufacturer's discretion. There is, therefore, a wide variety of different tool designs to accomplish the same task.

Standard subsea wellhead system pressure ratings are 2,000, 5,000, 10,000, or 15,000 psi. Most wellheads are now rated at 15,000 psi although they may be used much less than this. All pressure containing and controlling parts are designed to meet all material requirements of NACE Standard MR-01-75 (sulfide stress cracking resistant metallic materials for oilfield equipment). The wellhead housing, casing hanger bodies, and casing hanger pack-off are classed as pressure containing equipment.

Wellhead Components

Temporary Guide Base (TGB)

The TGB provides a guide template for drilling conductor hole and stabbing the conductor housing. It compensates for misalignment due to irregular seabed conditions and also provides a support for the drilling guide base. It also provides an anchor point for the guidelines.

It is 'temporary' inasmuch as it serves to provide initial guidance until the drilling guide base goes down. After that, it merely serves to anchor the guide wires. It is also 'permanent' in that it is not recovered after the well has been abandoned.

Design loads include ballast, guideline tension, and the weight of the 30" conductor. Any soil reaction load should also be catered for. Typically, the TGB should be capable of supporting a minimum static load of 175,000 lb on the central cone while the TGB is supported at 90° on four locations. These must be a minimum at a minimum radial distance of 62" from the center.

Drilling Guide Base (DGB)

This provides entry to the well prior to the BOP being installed and, along with the guide posts, gives guidance for running the BOP or Christmas tree. It provides structural support, final alignment for the wellhead system, and a seat and lockdown for the conductor housing. Earlier guide bases used to be permanently attached to the 30" housing. Current designs allow the DGB to be recovered from the wellhead and replaced with a production guide base if required.

Design loads include the weight of the conductor housing and pipe, flow-line pull-in, connection or installation loads, and environmental loads. Typically, the DGB must be capable of supporting a minimum static load of 175,000 lb at its center while being supported at its outer edges.

Guide posts must be designed and manufactured out of 8⅝" OD pipe and have a minimum length of 8 ft (26 m). They must be on a 6 ft radius from the center of the DGB and be replaceable in the field. Most are now designed to allow ROVs to perform this task, and use with a TGB means that the posts will have to be slotted to take the guidelines. The guide base itself is usually constructed from 'I' beam or box section and suitably painted.

Conductor Housing

The conductor housing is attached to the top of the conductor pipe (usually 30" OD) either mechanically or butt welded. The outer profile is machined to accept a drilling or retrievable guide base. The lower part of the housing contains ports to allow cement to escape from within the 30"/20" annulus. Internally, the housing is machined to accept both a 30" housing running tool and the 18¾" high-pressure housing.

Design loads include wellhead loads, riser forces, DGB loads, environmental loads, pressure loads, and thermal loads. The housing should be capable of supporting a minimum static load of 175,000 lb and have a minimum working pressure rating of 1,000 psi.

In general, the conductor housing has a 30" nominal OD with typical OD of 37.38" and ID of 26.2". The 30" nominal OD is usually standard due to rotary table constraints. When selecting the ID of the housing, bit gauge diameter for the following section plus ⅛" clearance should be allowed.

18¾" Wellhead Housing

The wellhead housing seats within the conductor housing and supports subsea equipment (BOP, tree, riser) and the subsequent casing strings and tubing hanger. The BOP latches on to the top of the housing using a standard wellhead connector; e.g., H4. The wellhead housing should also be able to accept a tubing hanger and either a conventional or horizontal tree.

The central outer section of the wellhead housing features a lock-down mechanism that secures it to the conductor housing. Above this, there is an upward facing profile with a special gasket pocket to create a secure pressure seal with the BOP.

The subsea wellhead housing is treated as a piece of pressure-controlling equipment and the design loads considered include riser forces, BOP loads, subsea tree loads, pressure, radial loads, thermal loads, environmental loads, flow-line loads, suspended casing loads, conductor housing reactions, tubing hanger reactions, and hydraulic connector clamping forces.

Standard wellheads are rated at between 2.5 and 3.0 million ft•lb in bending and as much as 7 million lb in tension although these limits can be exceeded by manufacturers. Bending loads are governed by the wall thickness of the high-pressure housing at the connection to the BOP and how the high- or low-pressure wellheads interact together. A standard wellhead system will have a high-pressure housing with an OD of about 27". To increase the bending load capacity, the high-pressure housing OD can be increased in size (OD up to 30") while the high/low pressure wellhead housing interface is strengthened.

For deepwater application, the bending rating has been increased to as much as 7 million ft•lb. The high- or low-pressure housings can also be rigidly locked together to minimize any damage from vortex induced vibration derived from high currents impacting on the riser.

The bending strength of wellhead connectors is dependent on the axial load and internal pressure. Most standard wellhead connectors are limited to 3–4,000,000 ft•lb although some suppliers can supply connectors for higher bending loads with capacities up to 7 million ft•lb bending load.

The pressure-tight seal between the housing and BOP is critical. A proprietary gasket is used and the seal surface on the housing is usually inlaid with Inconel 625 corrosion resistant alloy.

There are no penetrations within the body of the wellhead housing. The base is butt-welded onto the surface casing and both the 30" and 20" casings provide structural support. The minimum bore of the wellhead housing is dependent on the nominal system designation and BOP configuration as illustrated in the following table.

Casing Hanger

The subsea hanger is installed on top of each casing string and supports the string in the wellhead housing. It is normally run through the riser and BOP and then landed off in the wellhead. It has provision for a pack-off assembly and should be able to support loads generated by BOP test pressures above the hanger and subsequent casing strings. A lock-down mechanism can be used to restrict movement due to thermal expansion or annulus pressure. An external flow-by area allows returns to flow past the hanger during cementing operations and is designed to minimize pressure drop should cuttings be returned. The hanger is usually supplied with a pup joint above to reduce the risk of damage during handling. The working assembly comprising casing hanger, pack off, running tool, and cement plug launcher is normally preprepared and racked in the derrick prior to running casing.

Nominal System Designation (in.-psi)	BOP Stack Configuration	Housing Working Pressure (psi)	Minimum Vertical Bore (in.)
18¾–10 M	Single	10,000	17.56
18¾ –15 M	Single	15,000	17.56
16¾ –5 M	Single	5,000	15.12
16¾ –10 M	Single	10,000	15.12
20¾–21¼–2 M	Dual	2,000	18.59
13⅝ –10 M		10,000	12.31
21¼ –5 M	Dual	5,000	18.59
13⅝ –15 M		15,000	12.31
18¾ –10 M	Dual	10,000	17.56
13⅝–15 M		15,000	12.31

Casings hangers are categorized as either intermediate or production casing hangers and are treated as pressure controlling equipment. Design loads to be considered include suspended weight, overpull, internal and external pressure, and thermal, torsion, radial, and impact loads.

The profile and design of a casing hanger is determined by the load and pressure ratings. These in turn are a function of the material used and wall section of the hanger as well as the equipment in which the hanger is seated. The following requirements need to be examined in casing hanger design and it is rated according to these requirements:

- *Hanging capacity*—The hanging capacity of casing from the hanger. A box thread is usually used for the casing/hanger connection.
- *Pressure capacity*—The rating of the thread capacity at the bottom of the hanger may limit the pressure capacity of the hanger.
- *BOP test pressure*—This is the maximum pressure that can be applied to the upper portion of the hanger body and casing hanger pack-off assembly.
- *Support capacity*—This is the rated weight that the casing hanger(s) are capable of transferring to the wellhead housing or previous hanger(s) plus the internal pressures.
- *Flow-by area*—This is the minimum cumulative cross-sectional area between a landed hanger and the wellhead housing bore. This is important during cementing operations when mud and cement return flow through the cavity. The maximum particle diameter which can pass through this circulation path is also important.

Casing Hanger Pack-off Assemblies

The casing hanger pack-off provides pressure isolation between the casing hanger and the wellhead via a metal-to-metal seal. It can be run with the casing hanger or separately and can be actuated by torque, weight, and/or hydraulic pressure. Generally, for a certain wellhead size, the pack-off assemblies for use after each successive casing string are identical. Should a pack-off assembly fail to seal and it cannot be removed/replaced, a contingency emergency pack-off can be run. This uses sealing surfaces adjacent to the standard pack-off seal area within the wellhead housing and depends on elastomer seals.

The subsea casing hanger pack-off must be treated as pressure controlling equipment and the rated working pressure must be equal to that of the hanger. Design loads to be considered include setting loads, thermal loads, pressure loads, and releasing/retrieval loads.

Bore Protectors and Wear Bushings

A bore protector protects the casing hanger pack-off sealing surfaces inside the wellhead housing before the casing hangers are installed. After the casing hanger is run, a corresponding size wear bushing is installed to protect the remaining annular sealing surface and the previously installed casing hanger pack-off assemblies and casing hangers. They are not considered to be pressure-retaining devices although wear bushings may be designed to accommodate BOP pressure test loading.

Design loads for bore protectors and wear bushings must include BOP test pressure loads and radial loads. BOP test loads can impose considerable compressive axial loads on wear bushings especially via tapered shoulders. The pack-off should also have some means of retaining it in the locked down position.

Corrosion Cap

The corrosion cap serves to protect the subsea wellhead from contamination by debris, marine growth, or corrosion. These are normally nonpressure retaining and lock onto the external profile of the wellhead or conductor housing. If a pressure retaining cap is used, a means of sensing and relieving trapped pressure prior to releasing the cap must be provided. There is usually a means of injecting corrosion inhibitor beneath the cap to provide long-term protection of the wellhead sealing surfaces since the cap is normally installed for temporary abandonment purposes.

There are no critical design issues with the corrosion cap. It must fulfill its functional requirements.

Running, Retrieving, and Testing Tools

These tools are used to assist with various subsea tasks such as running, retrieving, and testing purposes, guidance, casing, and housing suspension, annulus sealing, and providing protection. Example selection of tools would include:

- Temporary guide base running tool (simple 'J' tool)
- 30" housing running tool (cam actuated, right-hand rotation to release)
- 18¾" wellhead housing running tool (cam actuated)
- Casing hanger running tool/pack-off installation tool (single trip run hanger, set pack-off)
- Wear bushing running/retrieving tool
- BOP test tool
- Emergency pack-off recovery and installation tool

The running tools are torqued to the drill pipe running string. If being run in open water as opposed to the riser, more robust connections are required (6⅝" reg); those to be run in the riser can utilize 4½" IF connections.

Design loads should include suspended weight, bending loads, pressure, torsion loads, radial, overpull, and environmental loads. The tools should have adequate necks for the tongues to grip, and also

- It is possible that the load capacity of the tool string may be less than its running string connection strength.
- Through bores should be adequate for circulation and the passage of tools such as cement darts.
- Flow area past the tools should be sufficient to minimize any swab/surge while tripping and should permit the tools to drain while pulling out.
- Air vent ports need to be included in the open water 30" and 18¾" housing running tools to allow air to be expelled when the casing is run into the sea.

TUBING SUSPENSION EQUIPMENT

Tubing Hanger

The dual-bore tubing hanger is used to suspend and seal the tubing into the wellhead such that a Christmas tree can be installed later. The tubing hanger must also be able to communicate the production and annulus bores and the various hydraulic and electrical downhole functions. To do so, it must be correctly orientated within the wellhead to ensure that the connections mate with those in the tree.

The hanger is concentric but it is oriented by a helix within the tree spool to ensure that all the connections line up with those in the tree correctly. Integrity monitoring of the electrical and hydraulic lines is conducted while running in to detect any damage before the hanger is landed.

The primary component is the main body. This contains both production and annulus bores, bores for safety valve control lines, and for downhole electrical functions. Two sealing options are used. Either a nose seal can be fitted to the bottom end of the tubing hanger to seal on the production casing or a pack-off can be provided to seal between the hanger and the wellhead.

The horizontal tree hanger does not seat in the wellhead. The tree is installed and then the tubing hanger is set in the tree body. The advantages of this option are that the completion and hanger can be pulled later without disturbing the tree and flow line and larger tubing sizes can be accommodated.

Here, the primary component is also the main body of the hanger which contains a main through bore and side outlet. A wireline plug profile is incorporated into the vertical bore above the side outlet for a metal-metal sealing wireline plug. It also carries upper and lower pack-off seals and an orientation sleeve attached to its lower end. There will also be seal mechanisms for safety valve control lines and electrical connections for pressure/temperature gauges.

Mechanical interface problems can arise if the wellhead and tubing hanger are supplied by different companies. In this case, it is essential that detailed information on the wellhead is given to the completion equipment supplier. Other interfaces to beware of are wireline plug profiles in the production and annulus bores, the production and annulus box threads, and the electrical penetrator.

Design issues center around the production casing and tubing size. For a dual-bore hanger, the production tubing offset must be such that the tubing will still fit in the production casing. The coupling diameters must also be considered. The pressure rating of the pack-off seal and tubing hanger landing shoulder should exceed that of the production and annulus bores. They should also be able to contain any pressure leaks should the SCSSSV control line leak into the tree/wellhead cavity during production. API 17D provides guidance on pressure ratings for this eventuality. Third-party wireline plug details should also be carefully considered during hanger design.

Tubing Hanger Running and Orientation Tool (THROT)

The THROT is used to run, orientate, and lock the tubing hanger and completion string into the wellhead through the drilling riser and BOP. It is normally run on the same workover riser joints as used to run the tree. Downhole electrical and hydraulic functions are routed through the THROT to connect into an umbilical, thus allowing continuous monitoring while running/setting the hanger. Recent developments include use of a subsea test tree within the THROT and the use of a mono-bore riser.

The THROT for a horizontal tree tubing hanger is simpler because it does not have an orientation helix and can be run on a dedicated mono-bore riser or completion tubing.

The lower end of the THROT interfaces with the latch profile on the tubing hanger. Its upper end interfaces with a standard workover riser joint and a small umbilical. A helix on the THROT engages with a hydraulic pin on the BOP to orientate the tubing hanger in relation to the wellhead. However, prior to this, the THROT must first be aligned at surface. The horizontal tree THROT requires no alignment prior to being run.

Regarding THROT design issues, it is important to know which RAMS or annulars will be used to provide a seal when pressure testing the pack-off seal. If the rig is unknown, then sufficient range should be provided on the sealing spool to accommodate most eventualities. Similarly, any spare choke and kill side outlets must be determined in order to position the helix.

The tensile capacity of the THROT should be able to support the expected string weight. The THROT external dimensions need to be considered regarding any possible restrictions; e.g., BOP flex joint. Similarly, the internal dimensions of the THROT should be compared with any wireline tools likely to be run.

The parts of the THROT should be properly keyed to ensure rotational rigidity between the key mating with the tubing hanger and the slot in the orientation sleeve.

Horizontal Tree Landing String

Often referred to as the completion/workover riser (C/WO riser) it is a dedicated modularized tubular member connecting the subsea equipment to surface. There are two types—dual-bore and mono-bore. The latter is normally used in conjunction with a horizontal Christmas tree, otherwise known as a spool tree or side valve tree. A key safety feature of the landing string is the ability to unlatch from the tubing hanger yet still provide the two-barrier statutory requirement between the production tubing and the surrounding environment. A subsea completion test tree is often incorporated above the tubing hanger running tool for this requirement. The test tree typically comprises ball valves in both the production and annulus bores and a connector referred to as the emergency release latch. The time allowed for the complete disconnection of the landing string depends on whether it is deployed from an anchored vessel or a dynamically positioned (DP) vessel. The latter case must be able to release quickly to accommodate a drive-off from location due to bad weather or thruster failure.

Functions

The principal function of the landing string is to run the completion tubing into the hole through the BOP and riser.

The tubing hanger is landed off within the tree and locked in place using the THROT. In this mode, the riser is run in conjunction with the intervention equipment (subsea test tree, retainer valve, lubricator valve) to provide a tubular conduit through which the initial well test can be carried out. This system is primarily used on fixed or anchored installations. With a DP vessel, there is the need for a subsea control module (SCM) to increase the response times of the conventional direct hydraulically-controlled version.

Once the well test is complete, the tubing hanger bore protector can be pulled on wireline through the completion/workover riser and a plug set into the mating profile in the tubing hanger. The dedicated hydraulic umbilical providing control functions to the THROT is clamped to the riser.

The landing string can also be used for intervention work on a subsea horizontal tree. It is possible to reinstall the BOP and marine riser and then reestablish the landing string latched on to the tubing hanger. However, in many cases, it would be considered a more cost-effective solution to run the completion/workover riser in open water in conjunction with an emergency disconnect package (EDP) and a lower riser package (LRP). This would require only one trip. In this mode, the C/WO riser would provide a pressure conduit through which wireline plugs could be pulled and set, coil-tubing techniques employed, and the well could be produced to surface. A larger dedicated hydraulic umbilical would also be required to provide the necessary control functions for the EDP, LRP, and Christmas tree.

Design

For a specific riser pressure rating and bore size, consideration should be given to the structural strength, fatigue life, and operating limitations. These are governed by vessel type, water depth, sea states, current velocities and directions, vessel characteristics, and vessel offset. The tubing hanger mode is the less onerous scenario since the riser is protected by the marine riser in this instance.

Principal Operations and Interfaces

The bore through the riser normally matches the nominal bore through the subsea horizontal tree and must allow passage of wireline plugs and bore protectors from within the tubing hanger. Each riser joint is made up of a production tubular with welded box and pin connectors on the ends. All the axial and bending loads are taken by the production tubular as is the pressure end load. A standard riser joint is 45 feet and the string is supported by a riser spider at the rotary table during running/retrieval operations. The spider is hydraulically operated and closes on a landing shoulder provided just below the box connector of each joint. A separate shoulder on the underside of the box connector can be supported by a standard set of elevators and allows the riser string to be raised or lowered through the spider. The overall length of the riser can be adjusted using pup joints to allow the surface equipment to be conveniently positioned relative to the rig floor.

Riser Spider

This supports the landing string with tubing attached at the rotary table. It also centralizes the string while being run/ retrieved through the rotary table bushings which also helps to prevent any damage occurring to the umbilical normally clamped to the riser.

Surface Joint

Referred to as the landing joint or cased wear joint, this is required in both modes and is situated directly below the surface tree. It is essentially a standard length joint with a 16" shroud running almost the entire length. Once the LRP or tubing hanger is landed, the surface joint slides up and down through the rotary to accommodate rig or riser movements, normally 15 ft plus or minus from the nominal position. The shroud centralizes the riser and provides protection for the umbilical which is contained within a slot running the length of the shroud.

Stress Joint

This provides the interface between the EDP and the standard riser when in 'Christmas tree' mode. This joint is normally tapered, thickest at its lower end, and able to provide extra bending capacity at the base of the riser.

Tension Joint

This is a derivative of the standard joint and is positioned directly below the surface joint where it provides a method of applying additional tension to the riser within the moon pool area using the marine riser tensioning system. It has a tension ring situated halfway along its length equipped with four padeyes for attaching the lines. Thus, the riser load is shared between the derrick and the marine riser tensioning system which offers a greater operating envelope.

Lubricator Valve

This valve is positioned just below the slash zone and its function is to provide pressure containment of the landing string production bore when wireline or coil tubing activities are being deployed through the surface tree swab valve. It comprises a ball valve capable of shearing either wireline or coil.

Standard Joints and Pipe Utility Piece (PUP) Joints

The standard joint is 45' long and comprises a main tubular which provides pressure integrity and tensile capacity within a given depth of water. The end connections must enable the passage of the control umbilical. The joints are run pin up. Double shoulders are provided at the upper end of each joint, one to locate a single joint elevator and one to allow landing in the spider. The seal faces may require exotic alloy inlays to prevent corrosion from acid gases (carbon dioxide and hydrogen sulfide).

Subsea Accumulator Module (SAM) and Control Module (SCM)

When a C/WO riser is being run in water depths greater than 500 m, the SCM and SAM are usually incorporated. They enable an emergency disconnect from either the tree or locked-down tubing hanger in less than sixty seconds. This is effected using electrical signals from a surface control module fed via an armored umbilical to the SCM. Here, a solenoid valve on the SAM is opened allowing hydraulic pressure to function the necessary valves. The distance from the hydraulic accumulators to the valves is about 90' (27 m) which promotes a far more rapid response than if hydraulic pressure had to be communicated from the surface down to the valves.

Retainer Valve

The retainer valve is located at the lower end of the C/WO riser and its primary function is to contain fluid within the riser during an emergency disconnect. Hence, it is used in tubing hanger mode only. It should also be able to bleed off any trapped pressure between the upper valve of the subsea test tree and retained ball valve.

It comprises a ball valve which is able to retain the internal pressure of the production riser, seal off against hydrostatic pressure which might be in the region of 5,000 psi at 2,500 m (8,202 ft) and should be able to shear wireline and coil tubing. It should also provide a conduit for hydraulic and electrical control lines to both the subsea test tree and tubing hanger running tool. It also provides a smooth outer diameter for the elastomeric seal of the BOP annular to close around. It should be rated to the same bore and pressure as the tubing hanger within the subsea horizontal tree. The lower connection of the retainer is attached to a shear sub which will enable the BOP to shear through the production tubular and retain well pressure should the latch fail at the top of the test tree. Thus, on several counts, the elevation of this section of the pipe is critical.

Subsea Test Tree

The subsea test tree is the last piece of equipment between the C/WO riser and the tubing hanger running tool and is only used in the tubing hanger mode within the marine riser. Its primary function is to provide the statutory two barriers between the production fluid and marine environment in the event of an emergency disconnect. It comprises an upper connector to the shear sub (referred to as a latch that can be hydraulically actuated) and two ball valves, the lower to shear coil/wireline and seals, and the upper to provide a second seal. These must be sized and pressure rated to suit the tubing hanger. The valve closure sequence is normally lower valve, upper valve, retainer valve, and then latch release.

Emergency Recovery Tools

These tools are primarily used to recover the tubing hanger in the event of failure of the THROT. They are normally purely mechanical devices with simple operating mechanisms.

The emergency pack-off recovery tool is run on drill pipe to recover the tubing hanger pack-off, thus releasing the tubing hanger lock-down ring. It engages the threads on the inside of the pack-off allowing this to be retrieved with a straight pull. The tubing hanger emergency recovery tool is run and landed off inside the tubing hanger. Rotation to the left forces the lock ring out and this allows the tubing hanger to be recovered with a straight overpull.

Design capacities should be clearly stated and include a significant safety margin since it is likely that these tools will experience heavy usage through jarring and overpulls.

These tools should be able to engage with the tubing hanger under any circumstances; e.g., with wireline plugs installed.

Tubing Hanger Handling and Test Tool

This enables the tubing hanger to be handled safely and pressure tested. For a dual-bore tubing hanger, the test tool comprises both a production stinger and an annulus stinger each equipped with elastomer seals.

The tool should be able to interface with the tubing hanger and seal at the required pressure. It should also be able to lock with the tubing hanger using the latch profile on the tubing hanger.

Space-Out Measurement Tool/Lead Impression Tool

These tools help to determine the height of the production casing hanger relative to the lock-down profile in the bore of the wellhead. The lead impression tool provides a basic measurement while the space-out tool gives a more accurate figure.

These tools are run on drill pipe and activated by closing the rams or annular and then pressuring up on top of the tool.

There are no design issues with these tools. The space-out tool may require calibrating due to its greater degree of accuracy.

BOP Orientation Pin

The purpose of the orientation pin is to engage with the (THROT) helix and orientate the dual-bore tubing hanger within the wellhead. It is mounted to a spare choke or kill line side outlet on the BOP stack and can be hydraulically extended into the BOP bore at the appropriate time. No pin is required with a horizontal tree.

The pin is simply a spring-retract hydraulic piston and must be rated to the operating pressure of the BOP since it could be exposed to wellbore fluids.

Design issues should consider the consequences of a stem packing leakage when the stack is exposed to test or wellbore pressure. There may be a requirement for ROV override of the pin in the event of hydraulic or spring failure.

The main concern with the hydraulic orientation pin is that, if it fails to retract, it may be impossible to pull the THROT or any other downhole tool through the BOP to the surface. This could have serious consequences for well control.

SUBSEA CHRISTMAS TREE SYSTEM

Control Systems

The control system on a subsea well allows surface control of the tree valves, downhole safety valves, chokes, chemical control, and metering valves. The system can also be used to monitor process variables such as pressure, temperature, flow rate, and valve and choke position. There are five basic methods of implementing control:

Direct Hydraulic

A dedicated hydraulic hose for each subsea function runs from the control panel via an umbilical to the valve actuator. The control panel valve, when opened, allows hydraulic fluid from a supply header to flow into the actuator. As the pressure on the actuator piston increases, the spring compresses and the valve opens. Closing the control panel valve allows the pressure to bleed off which in turn allows the subsea valve to close.

Topside requirements	–	Hydraulic power unit
	–	Manual control valves
Subsea equipment	–	Umbilical
	–	Umbilical termination unit
Advantages	–	Low cost, reliable, easy access for maintenance
Disadvantages	–	Slow response time over greater distances, busy umbilical, absence of any subsea monitoring facilities

Piloted Hydraulic

Similar to the direct hydraulic control with the addition of a subsea control module to improve response time. The subsea control module contains a number of pilot valves (one per function) and a subsidiary hydraulic reservoir charged by the surface supply header.

To open the subsea valve, the surface valve is operated. This communicates hydraulic pressure to the subsea control panel-mounted pilot valve which opens allowing the hydraulic fluid from the subsidiary subsea hydraulic cylinder to flow into the valve actuator and thus open the valve. The valve is closed by moving the surface control panel-mounted valve to the vent position; this bleeds off the hydraulic pressure throughout the line.

Topside requirements	–	Hydraulic power unit
	–	Manual control valves
Subsea equipment	–	Umbilical
	–	Umbilical termination unit
	–	Hydraulic jumper hose
	–	Subsea control module

Advantages	–	Simplistic design
	–	Low cost, high reliability
	–	Easy access to critical components
	–	Quicker response time due to the reduced volume needed to pump
Disadvantages	–	Response time still slow with long umbilicals
	–	The umbilical comprises many hydraulic hoses, two per function
	–	There is no subsea measurement/monitoring system

Sequenced Hydraulic

To simplify the previous configuration, it is possible for multiple pilot valves to be controlled by a single surface pressure regulator. In this instance, the different pilot valves are actuated at different pressures.

Topside requirements	–	Hydraulic power unit
	–	Manual control valves
Subsea equipment	–	Umbilical
	–	Umbilical termination unit
	–	Hydraulic jumper hose
	–	Subsea control module
Advantages	–	Low cost, high reliability
	–	Easy access to critical components
	–	Reduced number of hoses in umbilical
Disadvantages	–	Response time still slow with long umbilicals
	–	The sequence of valve operation is fixed unless subsea intervention
	–	There is no subsea measurement/monitoring system

Electrohydraulic

The pilot valves are fitted with solenoids to enable them to be electrically operated from the surface. To open a subsea valve, the appropriate surface switch is closed which energizes the solenoid valve and opens the pilot valve. Hydraulic fluid can then flow from the subsea accumulator into the valve actuator. The valve is closed by deenergizing the solenoid (power off) which allows hydraulic fluid from the subsea valve to drain back into the accumulator.

Topside requirements	–	Hydraulic power unit
	–	Manual switches
	–	Electrical power unit
Subsea equipment	–	Umbilical
	–	Umbilical termination unit
	–	Hydraulic jumper hose
	–	Subsea control module
Advantages	–	Can operate over greater distances with reasonable response time
	–	Use of electrics allows valve position feedback
	–	Reduced number of hoses in umbilical
Disadvantages	–	Reliability more critical due to use of subsea electrics
	–	Higher cost—electrical controls and more complex valves/umbilical

Multiplexed Electrohydraulic

This system revolves around the subsea electronics module (SEM) situated in the subsea control module (SCM) and the master control station (MCS) at surface.

Subsea control module (SCM)

This contains a microprocessor which monitors and stores inputs (temps, pressure) and acts on requests from the MCS for data or valve commands.

Master Control Station (MCS)

This is essentially a computer equipped with a modem to communicate with the SCM. It provides the following functions:

1. Communicates with the SCM
2. Provides an operator interface with the system
3. Displays process equipment
4. Means of controlling subsea equipment
5. Performs automatic functions; e.g., emergency shutdowns
6. Historical data storage

Topside requirements	– Hydraulic power unit (HPU) – Master control station (MCS) – Electrical power unit (EPU)
Subsea equipment	– Umbilical – Umbilical termination unit – Hydraulic jumper hose – Subsea control module (SCM)
Advantage	– Can operate over greater distances with good response time – Use of microprocessor allows measurement of variables – Few hydraulic hoses plus electrical wiring in umbilical – A single communication line operates many valves
Disadvantages	– Reliability more critical due to use of subsea electronics – Higher cost—electronics, computer, and software

Larger Systems

The above describes a single tree configuration. For a subsea development using multiple trees, it is normal to control multiple wells, each with its own SCM with unique address, from the same surface MCS. A subsea distribution unit (SDU) delivers the individual hydraulic and electrical connecting lines from the umbilical to each SCM.

Valves

There are several types of valves in use—needle valves for sealing off hydraulic control fluid and gate valves for sealing off well fluids.

The following features support gate valve use for sealing well fluids:

1. Bidirectional (seals against flow in both directions)
2. Full bore access
3. Metal sealing between gate and seat, seat and body, and body and bonnet
4. Sealing between stem and body either elastomeric or thermoplastic
5. Pressure locking of gate to seat will not occur for single slab gate

Needle valves are normally manually operated by either diver or ROV intervention. Gate valves, however, can either be manually or hydraulically operated. Although essentially similar internally, manual and hydraulic gate valves differ only in the stem and stem/bonnet seals. A hydraulically-actuated valve relies on the actuator to provide a linear force to open/close the valve. A manual valve requires a diver/ROV to rotate the valve stem to open/close the gate. A nonrising valve stem is preferable since the sealing part of the stem is protected and not exposed to seawater.

A subsea Christmas tree normally carries the following gate valves:

Production Master Valve (PMV)

This is normally open during production. Used to shut in the well during an emergency or when the tree cap is removed. Hydraulically operated for more rapid response, the valve operates in a fail-safe close mode. Some operators may elect to have a second manually operated master valve below the PMV as a backup.

Production Swab Valve (PSV)

Used only on conventional trees. Isolates well flow to the tree cap during production (valve closed) and allows vertical access during workover operations (valve open). The PSV can only be opened during workover operations and can either be hydraulically operated via the workover control system (fail-safe close) or manually operated by ROV.

Production Wing Valve (PWV)

Used to control the flow of well fluids. Normally hydraulically operated (fail-safe close).

Annulus Master Valve (AMV)

Normally closed during production operations. Will be open if gas lift is required. Used to isolate the annulus if well fluids are discovered in the annulus due to packer or tubing failure. Acts as backup to the annulus wing valve.

Annulus Swab Valve (ASV)

Function and operation as for production swab valve.

Annulus Wing Valve (AWV)

Normally closed during production. If gas lift is required, it is used to control the fluid flow from the gas-lift line into the annulus. Hydraulically operated (fail-safe close).

Crossover Valve

Normally closed during production. Enables communication between the production and annulus bores during workover riser/flow-line flushing or pressure testing.

Chemical Injection Valve (CIV)

Normally situated between the production master and production wing valves. Isolates the chemical injection line from the surface. The valve is normally between 0.5" and 0.75" in size and is normally hydraulically operated (fail-safe close) for rapid response.

In addition to the above, there are usually manual needle valves (ROV operated) for isolating the tree connector test line and downhole safety valve (DHSV) hydraulic control line.

Actuators

Most of the valves on a subsea tree need to be remotely operated and this is accomplished using actuators. These are normally hydraulically driven because:

1. Power source (hydraulic pump) is readily available
2. Reliable
3. Fast response time
4. Hydraulic pumps can generate high pressure thus enabling a compact design

Actuators are designed as 'fail-safe closed' for subsea tree applications to ensure that, in the event of loss of hydraulic pressure, the well will be shut in. A backup facility in the form of a diver/ROV override is also normally provided and incorporated in the ROV panel usually located at the edge of the tree frame for ease of access.

The valve size is usually dictated by the size of the bore required. The size of the valve stem should be sufficient such that the force generated by the line pressure on the sealing diameter of the stem will close the valve if hydraulic pressure is lost. For shallow water application, 3,000 psi is the most common design operating pressure. For deep water or big valves (5" plus), an operating pressure of 5,000–10,000 psi may be used to keep the size of the actuator to a minimum. According to API 17D requirements, the actuator should be able to operate under the most demanding conditions likely without exceeding 90% of the design operating pressure.

All subsea actuators require some form of compensation system to counterbalance the static head of the hydraulic operating pressure from surface. This is achieved through the use of a bladder-type hydraulic accumulator on the back side (spring side) of the actuator piston. To comply with API 17D, the back-pressure must remain at a minimum of 100 psi with zero psi line pressure at the maximum rated operating depth.

Water-based hydraulic fluid is the most common type. Use of more exotic fluids will need to consider elastomer compatibility and any density difference over seawater in connection with piston sizing especially in deep water applications.

DIVING METHODS VS ROV USE

The development of subsea experience and technology has enabled a wider range of diving activities to take place at increasingly greater depths. In air diving, surface decompression procedures are widely used to maximize bottom time. However, depth is the limiting factor in air diving as the partial pressure oxygen rises with the depth increase and becomes toxic within the blood stream, and will cause the diver to lose consciousness. Therefore, the diver's tasks are limited to maximum depth of approximately 17 m (55 ft). Consequently, tasks that are carried out at greater depths require saturation diving and mixed gases such as Trimix and Heliox. This is now a common and commercial method of carrying out subsea operations at extended depths.

The current depth limit set in the UK is 198 m (650 ft). However, the maximum depth that a commercial diver can operate is calculated on an economically viable basis rather than the effects on their physiology. With increased depths, there comes increased safety risk and the diver's ability to carry out his tasks. Moreover, the decompression issues surrounded in extended depth make operation physically and economically unjustifiable. Therefore, the industry drive is to make diving operations safer, less risky, and commercially viable.

The advantages and disadvantages of the various methods are summarized in the table below:

Intervention Method	Advantages	Disadvantages
ROV	Depth (1,000 m / 3,280 ft) Cost Less downtime—mechanical, weather Adaptable Good future Safer	Lack of sensory feedback Lack of spatial awareness Tooling expensive Affected by swell Cameras remote from site Cannot interrogate Totally reliant on video
Manned Submersible	Depth (1,000 m / 3,280 ft) On-site observer Diver for specific operations Suited to on-off applications Operate in currents	Battery life Lack of access Unsuitable for mid-water work Deployment costly Manipulators restricted
Atmospheric Diving Suits	No decompression Depth (700 m / 2,296 ft) Some dexterity Suited for on-off applications	Unsuitable for inspection Pilot fatigue Lack of access Totally dependent on visibility Limited sensory feedback
Diving	Spatial awareness Dextrous Interrogate Adaptable Trainable	Depth constraint (300 m / 984 ft) Human risk Expensive Sea state restrictions Umbilical drag Performance variances

CHAPTER 15
Completion Equipment

INTRODUCTION

Completion design is the process of converting a drilled wellbore into a safe and efficient production or injection system.

Prior to starting any design the following information is required:

- *Reservoir parameters*
    - Porosity, permeability, homogeneity, thickness, angle, water/gas/oil pressure profiles
- *Rock characteristics*
    - Rock strength, formation damage potential
- *Production constraints*
    - Fluids handling, injection pressures
- Fluid characteristics
    - Density, composition, gas-oil ratio, toxicity, pour point, scaling tendency, wax, asphaltene, CO_2, contaminants
- *Well appraisal data*
    - Rates, pressure, temperatures, samples
- Facilities information
    - Control line pump pressures, flow-line sizes, sampling/testing/monitoring, safety constraints
- *Drilling data*
    - Well profile, casing program (and constraints), safety valve depth constraints
- *Field economics*
    - Time frame and importance of fluids, life of field, trade off between capital expense and operating expense, tax implications

Some of the above information might not be readily available or can be reached by discussion with other members of the project team. If a specific tubing size is required to meet a flow rate, then it needs to fit inside the production casing, so discussion is needed with the drilling engineer.

The information is used to determine what type of completion is run, the tubing size, material specification, and the additional completion equipment used.

Tubing design (similar to casing design) is undertaken. The design needs to accommodate collapse, burst, and tensile load cases for the complete life of the well.

Generally speaking, the simpler the completion the greater its reliability.

COMPLETION TYPES/CLASSIFICATION

There are a number of ways of classifying completions. However, the main types are as shown below.

Interface between wellbore and reservoir	–	Open hole
	–	Cased and uncemented
	–	Cased and cemented
Production method	–	Flow naturally
	–	Require artificial lift
Stage of completion	–	Initial
	–	Recompletion
	–	Workover

Interface Between Wellbore and Reservoir

Open-Hole Completions

In an *open-hole* or *barefoot* completion, the production casing is set in the caprock above or just into the top of the pay zone, while the bottom of the hole is left uncased. Often, the final drilling of the pay zone is carried out with special nondamaging drilling fluids or an underbalanced mud column. This form of well completion dates back to the days of cable-tool drilling, but is rarely used today.

Advantages	Disadvantages
Exposure of entire pay zone to the wellbore	Inability to control excessive gas-oil and/or water-oil ratios (except in the case of bottom water)
No perforating expense	Need to set casing before drilling or logging the pay
Less critical need for precise log interpretation	Difficulty of controlling the well during completion operations
Reduced drawdown because of the large inflow area	Unsuitability for producing layered formations consisting of separate reservoirs with incompatible fluid properties
Slightly reduced casing cost	Inability to selectively stimulate separate zones within the completion interval
Ease of deepening the well	Need for frequent cleanouts if the producing sands are not completely competent or if the shoulder of the caprock between the shoe and top of the pay is not stable
Relative ease of converting the well to a liner completion	
No risk of formation damage resulting from cementing casing	

Cased and Uncemented

To overcome the problems of collapsing sands plugging the production system, the early oil producers placed slotted pipe or screens across the open-hole section as a downhole sand filter.

The simplest and oldest liner completion method involves running slotted pipe into the open hole. The slots are cut small enough that the produced sand bridges off on the opening rather than passing through. This method is still used in some areas today, but because it entails many of the same disadvantages inherent in open-hole completions (i.e., lack of control), its use is not widespread.

For very fine sands, wire-wrapped screens or sintered bronze are used in place of machine-cut slots. This technique is a reasonably effective sand-control method in uniform coarse sands with little or no fine particles (e.g., in California).

Sometimes this is the only sand-control system that can be used because of pressure loss and placement considerations (e.g., in unconsolidated heavy oil sands). In general, however, the uncemented liner completion is no longer recommended because

- sand movement into the wellbore tends to cause permeability impairment by the intermixing of sand sizes, and of sand and shale particles.
- fine formation sands tend to plug the slots or the screen.
- at high rates, the screen often erodes as formation sand moves into the wellbore.
- poor support of the formation can cause shale layers to collapse and plug the slots or screen.
- formation failure can cause the liner itself to collapse.

To overcome these problems, operators have resorted to more effective sand-control methods, such as gravel packing, in which the annulus between the screen and the open hole is filled with coarse graded sand, or the use of prepacked screens. In some cases, even where sand control is planned, it may be best to employ a cased and perforated completion with an external gravel pack—this configuration has become the norm for light oil and gas developments because of the flexibility it provides.

Cemented and Perforated Completions

By far the most common type of completion today involves cementing the production casing (or liner) through the pay zone, and subsequently providing communication with the formation by perforating holes through the casing and cement. Ideally, perforations should penetrate any damaged zone around the original wellbore and create a clean conduit within the undamaged formation. If the well is cased and unperforated during the early stages of the completion operation, well control is easier and completion costs may be reduced.

Using various depth control techniques, the required sections of pay are perforated and opened to flow, thereby avoiding undesired fluids (gas, water), weak zones that might produce sand, and unproductive sections or shale barriers.

This selectivity, which is completely dependent on a good cement job and adequate perforating, also allows a single wellbore to produce several separate reservoirs without their being in communication. This is done by setting isolating packers within an unperforated section of the pipe. Selective perforation can also be used to control the flow from, or stimulation of, various parts of the pay. By shutting off or partially plugging selected perforations, injected fluids (water, stimulation fluids, or cement) can be diverted into less permeable zones.

Cementing casing at TD rather than completing the well open hole can reduce the likelihood of well-control problems. Moreover, the decision to set production casing can be deferred until the open-hole logs of the prospective pay zone have been evaluated, substantially reducing the dry-hole costs if the hole is dry.

In summary, the advantages of cased and perforated completions include:

- safer operations
- more informed selection of the zones to be completed
- reduced sensitivity to drilling damage
- facilitation of selective stimulation
- possibility of multizone completions
- reduced dry-hole costs
- easier planning of completion operations

This type of completion is generally used unless there is a specific reason to prefer an open-hole or uncemented liner completion. Even where sand control is planned, perforated completions with internal gravel packs have become the norm for light oil and gas developments because of the flexibility provided.

Production Method

Flow Naturally

Naturally flowing is when a reservoir can flow to surface, at an economic rate, due to the reservoir pressure without any external assistance.

Tubingless Completions

Tubingless completions are a low-cost completion type but rarely used, as the production casing is exposed to reservoir pressure and corrosion. In addition, additional safety devices cannot be installed.

Tubing Completions

By far the most common method of completing a well is to run single or multiple tubing strings into the well, depending on the number of zones to be completed.

Artificial Lift

When a well will not flow naturally to surface at an economic rate, then artificial lift techniques can be used to boost production. Common methods of artificial lift for offshore operations are:

- Gas lift
- Electric submersible pumps

Gas Lift

Gas lift is used to lighten the liquid hydrostatic head, reduce the fluid viscosity, reduce pipe friction, and supply additional energy to help lift the contents of the tubing out of the well.

Require gas-lift mandrels to be installed at selected depths along the tubing string to optimize the flow rate.

Gas-lift valves add potential leak paths to the tubing string and gas-lift wells often require additional wireline intervention to change out gas-lift valves to ensure that optimum well performance is maintained.

Electric Submersible Pumps (ESP)

ESPs are used to move large liquid volumes with low gas-to-liquid ratios from water wells, high water cut oilwells, and high deliverability oil wells.

As they are powered by electricity, a power cable must be run down the outside of the tubing and through the wellhead.

Reliability of this equipment can vary significantly due to a multitude of reasons.

COMPLETION EQUIPMENT

This section is designed to explain the major components that are found in completions. It is not suggested that every completion has all of these components.

The reason for including components will vary considerably. The following reasons should be considered:

- Primary pressure control (such as a tree)
- Regulating flow (from the reservoir or further up the tubing)
- Emergency control of flow (e.g., a safety valve)
- Injection of chemicals
- Data acquisition
- Pressure or temperature control (e.g., insulation)
- Artificial lift
- Workover capabilities

Christmas Tree

The Christmas tree is the pressure control system located at the wellhead. The tree consists of a series of valves that provide the interface between the reservoir, completion, and through to the production facilities. There are many other aspects to the purpose of the Christmas tree such as:

- To provide a pressure-tight barrier between the reservoir and surface.
- A method that allows controlled production or injection.
- To kill the well prior to workover operations or maintenance.
- A system that permits the deployment of intervention work strings.

The type of Christmas tree to be used has a large impact on the completion costs, the completion design, and the intervention capabilities. Essentially there are two main types of tree—these are surface trees and subsea trees.

Surface Trees

- Conventional flanged trees are used on most land and low-to-moderate rate offshore wells.
- Solid block trees are sometimes used offshore, especially for high-pressure or high-rate wells under critical service conditions since they reduce the number of turbulence raisers and potential leak points. Under highly turbulent conditions, a Y configuration is available for the side outlets.
- Horizontal spool trees. In this configuration, the tree is installed before the BOPs. The casing and tubing are then installed through the tree. All the tree valves are located on the side of the vertical bore and are, therefore, 'horizontal'.

Subsea Trees

- Conventional dual-bore trees. These have two vertical bores for access to the production tubing and the annulus. They are designated by the size of the bores; e.g., 4 × 2, 5 × 2, or 7 × 2. As the access to the annulus is through the tree, any special considerations such as gas lift must be addressed.
- Horizontal spool trees. As in the surface horizontal trees, the BOP is run above the tree.
- In-line trees. These are a new development where the tree valves are housed within the wellhead and are part of the completion.

Tree Choice

Horizontal trees have the key advantage that the tree does not have to be pulled in order to pull the tubing. This has the potential to save considerable time and money when performing interventions such as electrical submersible pump (ESP) replacements. In addition, as all the valves are away from the vertical bore, the installation and landing off of additional strings inside the tubing is easier. This opens up more opportunities for retrofit gas lift, coiled tubing deployed pumps, and velocity strings.

A second master valve is usually required for sour and critical wells and where valve servicing is difficult (e.g., on subsea and satellite wells). This may also be specified in local regulations. Some companies specify a second manual master valve below any valve that is part of the emergency shutdown (ESD) system to facilitate servicing because of the risk of wear from spurious shutdowns.

Two side outlets are usually specified for sour high-pressure and critical wells to allow permanent installation of a kill line. This may also be specified in the local regulations, facility safety philosophy, or certification requirements. It is also convenient for valve equalization and well cleanup.

Where frequent through-tubing workover or wireline operations are expected, a swab valve is usually specified. Alternatively, the swab valve can be temporarily installed onto, or in place of, the tree cap during well entry operations.

The production wing valve, choke, and flow-line valve arrangement must consider how the well is to be brought on and shut down, how off-take will be controlled, and how the main ESD valves will be equalized. If a motorized adjustable choke is used, this should normally be of the full-closing type that can withstand the jet erosion during opening.

The pressure losses across the Christmas tree, choke, and flow-line connection must be considered in the well deliverability analysis, and can often be significantly reduced at the maximum well capacity by avoiding turbulence and sudden direction changes.

The location of chemical injection points and the rate of fluid supply should be addressed, particularly where chemical injection is required into the tubing to protect the downhole safety valve (DHSV) and tree from gas hydrates or to equalize the ESD valves before opening.

If the tree valves are automated, it is highly desirable to have a staged ESD sequence in which the wing- or flow-line valve closes first to stop the flow before the SCSSV and master valve start to close. In cold climate gas wells, it is common for the lowest of the master valve to be further delayed on low level alarms to ensure that some methanol is dumped onto the SCSSV before the well is fully isolated.

Wellhead

The wellhead transfers the casing and completion loads to the ground via the surface casing and provides a seal system and valves to control access to the tubing and annulus. It is made up of one or more casinghead spools, the tubing head spool, the hanger, and the Christmas tree.

Wellhead specifications are laid out in API Specification 6A, which was extensively revised in October 1989.

It will primarily be the responsibility of the casing design to specify the requirements for the wellhead. The completion will, however, impact the wellhead selection in a number of ways:

- Loads will be transferred from the tubing to the wellhead through the tubing hanger. This is not normally a problem unless the wellhead is on a tension leg platform (TLP).
- With a surface wellhead, there may be the requirement for injection into the annulus through the wellhead (gas lift, jet pumps, inhibitors, or injection water). The metallurgy and size of the port will need to be considered for pressure drops, erosion, and corrosion aspects as well as pressure and temperature limits. Additional valves (actuated or manual) may be required for integrity assurance or control.
- Certain parts of the wellhead will be exposed to annulus fluids (i.e., the production casing hanger). This may impact their metallurgy.
- The monitoring of annulus pressures may be required. This is relatively easy on an accessible surface wellhead. For a subsea wellhead, this may require special nonintrusive sensors.

Tubing Hanger

The tubing hanger's function is to transfer the weight of the tubing to the wellhead and to contain the casing or tubing annulus fluid.

There are five types of tubing hanger system in common use:

- Mandrel (doughnut) compression hangers (metal-to-metal or elastomeric type)
- Ram-type tension hangers
- Slip and seal assemblies
- Direct suspension from the tree (e.g., horizontal trees)
- Submudline tubing hangers or tubing hanger packers

The main problem with hanger selection occurs where the tubing is to be landed in tension (e.g., in some gas wells), when this is applied mechanically rather than hydraulically.

The number of vertical bores required through the tubing hanger for flow or supply conduits, control lines, chemical injection lines, and cables should be specified. These can be sealed with a stab seal or an annular seal-ring seal on an extended neck hanger. On subsea wells, a vertical bore is also required for the annulus access, and proper orientation of the hanger with respect to the guide base must be addressed.

The method by which the main bores will be plugged at surface during removal of the BOPs or Christmas tree should be considered. There are two main options:

- Use of a plug profile in the hanger. This can be for either a plug or check valve run on rods or, more commonly today, a conventional nipple profile. It is particularly useful to have a tubing hanger running tool and landing string that is slightly larger than the tubing being run. This allows the use of a nipple profile that does not impact the rest of the completion.
- Use a wireline profile located below the hanger. This arrangement is most useful in subsea wells if, when running plugs into the nipple, the running tool would sit inside the tree and, if stuck, may prevent closure of valves. Note: by having the nipple below the hanger, it is harder to access if plugs are stuck or debris falls on top of plugs. On a subsea well annulus bore, the use of a nipple below the hanger is recommended as, in this case, a stuck plug can, if required, be punched above.

Subsurface Safety Valve (SSSV)

Subsurface safety valves (SSSVs) are installed below the wellhead to prevent uncontrolled flow in an emergency situation. In the worst case, when the wellhead has suffered severe damage, the SSSV may be the only means of preventing a blowout. The correct design, application, installation, and operation of this equipment is fundamental to the safety of the well. SSSVs should be considered for all wells capable of natural flow. In many locations, government regulations require the use of SSSVs. In areas like offshore UK, Norway, and USA, government regulations dictate that the valve must comply with API Specification 14A.

Definitions

SSSVs can either be surface-controlled or subsurface-controlled. Subsurface-controlled valves are controlled by well pressure, by the flow itself, or as a result of a pressure differential caused by the flow. This type of valve's dependency on well conditions as a means of control makes them inherently less reliable than surface-controlled valves, and their application is therefore limited. Surface-controlled subsurface safety valves (SCSSV) are normally closed, and they are usually held open by an external pressure applied from surface. Some SCSSVs are controlled by electric, electromagnetic, or acoustic signals. However, by far the most common form of control is hydraulic pressure applied from surface via a control line. When the hydraulic pressure is lost, the valve is closed by means of a spring acting on the closure mechanism. In order to close the valve, this spring must overcome the hydrostatic pressure in the control line. Each SCSSV therefore has a maximum safe setting depth. Regulations in most offshore locations require the use of SCSSVs.

Closure Mechanisms

SCSSVs have three main closure mechanisms:

- Flapper
- Ball
- Poppet

Flapper and ball closure mechanisms are the most commonly used. Poppet mechanisms are sometimes used in equalization devices (see below) and annular systems. As the name suggests, annular systems are used to isolate the annulus; e.g., in concentric gas-lift systems or subsea wells.

The flapper type of mechanism is now strongly preferred to the ball mechanism as a result of:

- Greater reliability
- Simpler design
- Less prone to seal damage
- In the event of a failure, the valve can be pumped through at sufficiently high rates to kill the well.

To open the valve, pressure is applied via the control line, compresses the closing spring, and moves the flow tube down onto the flapper. As the flow tube continues to move down, the flapper rotates about its hinge into the flapper housing. When control line pressure is removed, the spring forces the flow tube up, allowing the spring on the flapper to bring the flapper into the closed position. In the closed position, the flapper is held closed by the differential pressure across the valve.

The ball valve is operated in a similar manner, except in this case the hydraulic pressure acts on a piston, which rotates the ball by means of a mechanical linkage.

The poppet mechanism simply relies on a conically-shaped poppet sealing against a metal seat.

Certification

All subsurface safety valves require certification:

- All critical components are traceable to mill heat reports.
- Design qualified through function test.
- A product manufactured, inspected, and functionally tested to API 14A.

API 14A has three standard classifications of service:

Class 1 *Standard Service*. Suitable for service in oil and gas wells where it is not exposed to sand production or stress corrosion cracking.

Class 2 *Sandy Service*. In addition to Class 1, the valve is also suitable for use in oil or gas wells where solids, e.g., sand, could be expected to cause valve malfunction or failure.

Class 3 *Stress Corrosion Cracking Service*. In addition to Classes 1 and 2, the valve is suitable for use in oil or gas wells where corrosive agents could cause stress corrosion cracking. Within Class 3 there are two subdivisions, 3S for sulfide stress cracking and 3C for chloride stress cracking.

Types Of SCSSV

In addition to the type of closure mechanism, SCSSVs can be further subdivided into four main categories:

- Wireline or tubing-retrievable.
- Nonequalizing or self-equalizing.
- Concentric or rod piston.
- Single control line or dual balanced lines.

A valve may have any particular combination of these features; e.g., tubing-retrievable, self-equalizing concentric piston with a single control line. The selected configuration will be governed by well conditions, the completion design, and previous experience.

Selecting a self-equalizing valve provides operational flexibility in that no external source of pressure is required to pressurize above the valve prior to opening the valve. However, the self-equalizing feature introduces an additional potential failure mechanism and must therefore have an impact on reliability. When to use self-equalization will depend on the operating environment and whether such a feature will provide a significant advantage in operating the field; e.g., self-equalizing valves are a virtual necessity on unmanned satellite platforms in the North Sea.

The hydraulic power for SCSSVs is delivered by means of a single concentric piston or one or more rod pistons mounted radially around the valve. Limits on the available control line pressure dictate a maximum setting depth for a valve. If a valve is to be set deeper than about 800 ft (244 m), e.g., subsea or below permafrost, then the large spring force and resulting high opening pressure required by a concentric valve make this valve inappropriate, and a rod piston valve should be utilized.

Tubing-Retrievable or Wireline-Retrievable

The statistics on reliability indicate that tubing-retrievable downhole safety valves are more reliable than wireline-retrievable valves. Mean time to failure is approximately 6 years for wireline valves and 15 years for tubing-retrievable valves. Tubing-retrievable valves are also full bore and allow for easier access. Tubing-retrievable valves can usually be converted into wireline-retrievable valves through wireline intervention. Tubing-retrievable surface-controlled subsurface safety valves are therefore the most reliable and common valves. Their position is dictated by various considerations:

1. The valve OD often impacts casing designs. For example, 7" valves require 10½" casing. 5½" valves may also not fit inside 9⅝" casing if large cables or lines are required to go around the valve or if a pressure rating of 10,000 psi or more is required. 4½" valves may not fit inside 7" casing.

2. They should be below platform piles or the probable crater depth in the event of a blowout.
3. If they are of a self-equalizing design, they should ideally be positioned below the hydrate formation depth (determined from the geothermal gradients).
4. If adjacent drilling is going on with simultaneous production, the downhole safety valve can be used during the drilling of the top hole section. It is closed to mitigate the consequences of a collision. This avoids the need to run plugs into and out of the well.
5. Local regulations.

Most designs prefer to be located above or below kickoff points.

The setting depth of the valve must be such that in the event of a control line leak, the valve fail safes in the closed position. This is a function of the valve design and the annulus fluid. It is also vital that the control system is capable of opening the valve with the highest expected tubing pressure at the valve. There have been instances of control systems that have been designed with inadequate pressures.

Tubing-Retrievable Surface-Controlled Subsurface Safety Valve (TRSCSSV)

Hydraulic pressure acting on the concentric rod piston arrangement, which is connected to the flow tube, operates the subsurface TRSCSSV. As the rod piston moves downward, the flow tube in turn pushes through the flapper valve and thereby provides flow through the valve to surface. When hydraulic pressure is bled off the control line, the power spring pushes the flow tube and rod piston upward and allows the flapper valve to close. In order to close the valve, this spring must overcome the hydrostatic pressure in the control line. Each SCSSV therefore has a maximum safe setting depth.

TRSCSSVs incorporate a system whereby a wireline-set safety valve can be installed. This operation would only be carried out should the safety valve fail to hold differential pressure or fail to function. There are typically wireline runs associated with the installation of the insert safety valve. Generally, these are as follows:

1. Run in hole and lock out the flapper valve.
2. Run in hole with communication tool and function communication system to open position. This provides communication through to the tubing via the safety valve control line and control chamber.
3. Run in hole with and set the insert safety valve. The insert safety valve straddles the communication system thus providing hydraulic integrity necessary to the operation and functionality of the insert safety valve.

Wireline-Retrievable

As with the tubing-retrievable valve, the wireline-retrievable valve is operated from a surface control panel and via a control line that connects into the safety valve nipple. When assembled to an appropriate lock system, the safety valve is installed and retrieved utilizing wireline and landed in a safety valve nipple.

Concentric or Rod Piston

The hydraulic power for TRSCSSVs is delivered by means of a single concentric piston or one or more rod pistons mounted radially around the valve. The newer generation of valves tends to be of the rod piston variety and are particularly appropriate for hostile environments as the smaller diameter rod piston allows the use of metal-to-metal sealing throughout the valve. Rod piston valves do, however, have the disadvantage of having less opening and closing force and as such are more prone to sticking when scale, solids, or asphaltenes are present. In addition to the factors above, past experience of the valve's performance in similar conditions should be used to decide both the types of valve and the particular model.

Surface Controlled

The hydraulic system for an SCSSSV comprises surface-control panel, control line and accessories, and control-line fluid. Viscosity, density, and cleanliness are important factors in selecting a control-line fluid. The density obviously affects the hydrostatic head on the valve, and the viscosity has an impact on the bleed-down time. Decreasing viscosity with temperature is a critical factor in arctic service and North Sea subsea wells, and special liquids need to be used to take account of this. Mineral oil, known as HLP 32, is used as the standard control-line fluid on North Sea platforms. On subsea wells, a 5:1 glycol erifon mixture diluted to a total of 35% glycol and fresh water-based HW 540 is typically used.

Velocity Controlled

The velocity safety valves are normally open-type valves that are closed by well flow in excess of the valve's preset closure flow rate. If loss of normal well control occurs at the surface, any increase in flow velocity causes the valve to close, thus shutting in the well downhole. This type of valves' dependency on well conditions as a means of control makes them inherently less reliable than surface-controlled valves, and their application is therefore limited. The valve will reopen once pressures across the closed pressure are equalized. The velocity safety valves are normally equipped with an equalizing subassembly which permits self-equalizing prior to retrieving the valve, and are compatible with a variety of wireline locks. Adjustment of the valve closure mechanism is accomplished by changing the orifice size and internal spring rate. This requires specific well information and is accomplished through the use of a computer program designed for this purpose.

Injection-Controlled

The injection-controlled valves are normally closed valves that are designed to automatically shut in the tubing in the event injection is stopped or flow reverses. Injection pressure opens the valve by creating a pressure drop across the orifice that allows the spring guide to push past the flapper to the open position. When injection ceases, the spring guide allows the flapper to close.

Annular Safety System (ASV)

The annulus safety system is based around a retrievable packer designed to anchor the tubing string in the casing at a point below the wellhead. The pack-off tubing hanger (POTH) seals the area between the casing ID and the tubing OD and also provides seal bores suitable for the production of an isolated flow path for gas injection. The tubing anchor is designed primarily to allow retrieval of the upper completion without the need for rotation. The annulus safety valve provides a method of controlling the annulus fluid flow and will close automatically should control fluid pressure be lost.

The complete system will serve as a redundant wellhead capable of carrying the weight of the tubing string safely in the event of catastrophic failure.

The slip-and-packing element system used in the system hanger is designed to distribute the loads from both the sealed-in pressure and the full weight of the supported tubing string safely into the casing. The system is designed to transmit these loads without damaging the casing, even if the casing is unsupported.

The hanger is retrieved on drill pipe using a specially designed retrieving tool.

Primary applications include gas-lift installations and wells where pressurization of the upper casing annulus as a result of a downhole leak is not permissible. Secondary applications include tension leg platforms where the system can be used to guarantee that minimum tubing string weight is suspended from the platform.

Nonequalizing or Self-Equalizing

Nonequalizing valves are superior in sealing reliability to self-equalizing valves, this due to a simpler design and reduced leak paths. They, however, require the tubing pressure to be equalized above the safety valve in order for it to be opened. Selecting a self-equalizing valve provides operational flexibility in that no external source of pressure is required to equalize the pressure above the valve prior to opening the valve. However, the self-equalizing feature introduces an additional potential failure mechanism and must therefore have an impact on reliability. When to use self-equalization will depend on the operating environment and whether such a feature will provide a significant advantage in operating the field; e.g., self-equalizing valves are a virtual necessity on unmanned satellite platforms in the North Sea.

The equalizing feature will also determine the depth of the valve. If equalization is required, it is better to have the valve shallow in order to reduce the amount of liquids that have to be pumped. Likewise, wax or hydrates may force a self-equalizing valve to be set deeper.

Single Control Line or Dual Balanced Lines

To overcome some of the problems of limited setting depth, a dual control line balanced valve can be proposed. In a balanced valve, a second control line is run to the valve and filled with the same fluid as the main control line. The fluid in the second line balances the hydrostatic pressure on the piston regardless of setting depth. In theory, this should give the valve unlimited setting depth. However, the time required to displace the fluid to surface limits the valve's response time and hence setting depth. Although these valves have a deep setting capability, they are not recommended for a number of reasons. In particular, they are prone to failing open. If gas migrates into the balance line and reduces the hydrostatic pressure, the original hydrostatic pressure in the control line is no longer balanced and can be sufficient to open the valve without any applied surface pressure. This will cause the valve to fail open.

A valve may have any particular combination of these features; e.g., tubing-retrievable, self-equalizing, concentric piston with a single control line. Well conditions, the completion design, and previous experience will govern the selected configuration.

Blast Joints, Flow Couplings, and Pup Joints

Flow Couplings

Flow couplings are short sections of thick-walled pipe. They are manufactured from bar stock (as are most completion accessories). The flow coupling is normally manufactured with a wall thickness equal to that of the tubing ID through to coupling OD. Flow couplings should be considered above and below any major ID change; i.e., crossovers, tubing-retrievable safety valves, wireline nipples, etc. The purpose of this is to withstand any erosion caused by turbulent flow through differing tubing IDs and therefore promotes enhanced completion reliability of the production life of the well.

Accelerated erosion is induced by landing nipples and by any item that causes an abrupt change in flow area. Erosion is accelerated due to abrupt entry to a restriction and an abrupt exit. Therefore, a flow coupling below a landing nipple is as important as one above. The most critical area is at the point of entry below the subsurface control device. A good practice is to have a lower flow coupling length that will cover the subsurface flow control below the nipple. Above the flow coupling or a safety valve, an adequate length is twenty times the ID, although a minimum of 36" is recommended.

Blast Joints

Flow couplings are designed to withstand internal erosion caused by turbulent flow. Blast joints differ by withstanding erosion externally, and are normally positioned either side of a sliding sleeve situated at perforated production zones where the jetting action of the fluid can erode the outside of the tubing. A blast joint is a joint of tubing with enhanced wall thickness, and is usually manufactured from a heat treated alloy. 415H alloy has been shown by tests to be a good material. Tungsten carbide or stellite can be used as a coating.

Pup Joints

Pup joints are short tubing joints that give flexibility in attaining a desired tubing length; e.g., when spacing out the completion. This is important particularly while landing off the completion. They are also utilized above and below completion accessories as part of a completion module assembly and transported to the well site. This allows for easy, safe handling and controlled makeup torque below and above the assembly. It is recommended that all completion modules have a 5' or greater pup joint at each end.

When using pup joints for space-outs, a limited range of sizes is required. As tubing is made up in length ranges, it is easy to put aside extremes of the range during the tubing layout. In this manner, only 3 or 4 pup joints are required (say 5', 10', and 20').

Landing Nipples

Landing nipples, also known as wireline or seating nipples, provide a location point for various flow-control devices in the production string. Typically, landing nipples are short tubular sections with an internally machined profile. This profile usually consists of a landing and locking profile to locate and hold the wireline lock, and a polished packing bore or sealing section.

There are two main categories of landing nipples, nonselective and selective:

- *Nonselective nipples*, or what are commonly called no-go nipples, rely on the nipple having a smaller ID (no-go) than the lock. This reduction in ID can either be at the top (top no-go) or bottom (bottom no-go) of the nipple and is used to locate the lock. Once the lock is located on the no-go, it is then in the correct position to allow the locking dogs to be jarred into the locked position.
- *Selective nipples* utilize a different method of locating the wireline lock and do not rely on a reduction in ID. There are two basic methods, one where the nipples have different selective profiles or, the alternative, where the profiles are all the same and the selectivity is achieved by the running and setting operation. With different selective profiles in the nipples, the locking dogs on the lock must match the appropriate nipple. Each lock can therefore be run through a series of nipples until it reaches the nipple with the matching profile. Using the running and setting operation to achieve selectivity allows all nipples to be accessed with one lock, whereas the selective nipples obviously require a different lock for each nipple.

Selective nipples have the advantage of being able to maintain the same ID throughout the completion, whereas each no-go nipple requires a step-down in ID. This can be important in smaller completions.

However, this advantage does have an associated drawback in that all the selective nipples have the same packing bore. This means that if a plug is to be located in the bottom nipple, the packing stacks have to be jarred through all the packing bores in the upper nipples. This exposes the packing stacks to a high risk of damage before it reaches the appropriate nipple and, for this reason, selective nipples are not recommended, particularly in large tubing sizes.

The location and size of each wireline nipple should be carefully considered in the planning stages of the completion to allow maximum versatility in the positioning of various flow-control accessories.

Wireline nipples may be used for the following operations:

- Land blanking plugs as a barrier or to test the production tubing.
- Land blanking plugs to provide a means of setting hydraulically-actuated completion tools.
- Land velocity type safety valves—subsurface-controlled safety valves (SSCSV).
- Land chokes to reduce surface flowing pressures or have pressure drops downhole to prevent surface freezing in gas production.
- Land valve catchers to aid recovery of accidentally dropped gas-lift valves during change outs.
- When installed above blast joints with a polished nipple below, separation sleeves may be installed to repair any damaged or eroded blast joints.
- Land instrument hangers with devices such as flow meters, thermometers, and pressure gauges.

Wireline Locks

A basic wireline lock consists of a fishing neck, expander mandrel, locking dogs or keys, retainer sleeve for the dogs, springs for the dogs, a packing mandrel, and a packing stack. Although locks differ in their configuration, the basic function of the components remains the same.

The fishing neck can either be internal or external and is used to attach the lock and associated equipment to the running or pulling tool. The expander mandrel when collapsed forces the dogs out into the profile in the landing nipple. Various methods are used to hold the mandrel in this collapsed position, thus locking the lock in the nipple. The Camco DB lock uses a collet mechanism, whereas the Baker Sur-Set uses a 'C'-ring. The running procedure for the locks depends on the selective mechanism. No-go locks and locks using selective profiles are run on wireline, and the spring-loaded keys automatically locate the locks in their respective nipples. The locking mandrel is then jarred (driven) down, locking the keys securely in place. Where the lock itself is selective, the running procedure is as follows:

- Lock and running tool are run through the nipple.
- The assembly is then pulled back through the nipple and this activates the locking keys.

- The assembly is again lowered into the nipple, and the keys automatically locate in the nipple profile.
- The lock mandrel is then driven down, locking the keys into the profile.

With the lock set, the running tool is recovered by upward shearing, releasing it from the lock mandrel. This procedure applies to all types of lock.

The recovery is similar:

- The pulling tool is run and located in the fishing neck of the lock.
- Upward jarring is used to pull the lock mandrel up and allowing the keys to be released. Continued upward jarring frees the packing stack from the seal bore.
- The lock can then be recovered from the nipple.

Controlled ID Joint

Controlled ID joints are essentially an evolution of flow couplings that provide a location and anchoring point for various flow-control devices in the production string. Unlike landing nipples, controlled ID joints do not have an internally machined profile. The flow-control devices that are designed to locate and seal in controlled ID joints are equipped with a bidirectional slip system similar to most retrievable packers.

Sliding Sleeve

The sliding sleeve is a device that allows communication between the tubing and the annulus for well-kill operations, circulation for tubing or annulus, and selective zone production. The sliding sleeve is essentially full-opening with an inner sleeve that can be opened or closed by means of a wireline shifting tool. The main applications for a sliding sleeve include providing circulation to the annulus pre- or post-workover, multizone completions, and annulus communications for power fluid; e.g., jet pump installation. Sliding sleeves used to have a poor reputation for reliability in some areas, particularly where the sleeve remained unused for long periods. However, recent modifications and utilizing nonelastomeric seal technology has enhanced their efficiency. Sliding sleeves by design, however, create a potential leak path in the completion string and for this reason sliding sleeves are not generally recommended. With no sliding sleeve in the completion, a tubing punch is used to achieve communication prior to a workover.

As standard, the sliding sleeve is normally equipped with a nipple profile in the upper sub of the sleeve, with a polished seal bore above and below the ports, to accept a variety of locks, packing devices, and accessories.

Sliding sleeves may be used to establish communication between tubing and annulus for such operations as:

- Displacing the tubing or annulus fluid after the Christmas tree is installed.
- Selective testing, treating, and production of individual zones in a multizone selective well.
- Using tubing to 'kick off' the annulus to tubing in a dual string completion.
- Producing more than one zone through a single tubing string.
- Killing a well by circulation.
- Gas lifting.
- Landing a blanking plug in the nipple profile to shut in the well or when testing tubing.
- Landing commingling chokes in nipple profile.
- Circulating inhibitors for corrosion control.

Mandrels

Side-Pocket Mandrels (SPMs)

The side-pocket mandrel (SPM), originally designed for gas lift, can also be used as an alternative circulating device. The SPM uses valves that can be set or retrieved on wireline, using a kickover tool, which positions the device in the side pocket. When no communication is required, a dummy valve is located in the SPM. Working valves are usually activated by annular pressure and are used for gas lift, circulation, and chemical injection. SPMs have both merits and drawbacks as a communication device. Unlike the sliding sleeve, the flow-control device can be removed without pulling the tubing. The mandrel also allows unrestricted flow through a full bore. However, the SPM has a large OD and requires a relatively large casing. Another disadvantage of the SPM is that debris can accumulate in the side pocket, making setting and retrieving the valve difficult.

Gauge Mandrel

Surface Readout

The downhole permanent gauge is a pressure monitoring system which facilitates real time surface data acquisition of downhole pressure fluctuations and reservoir depletion. These systems are commonly installed in subsea developments where subsurface monitoring of the bottomhole conditions are an essential part of production management. Cost issues are also a deciding factor to the implementation of these systems; e.g., costs associated with wireline run memory gauges where a jackup or semisubmersible rig would have to be positioned over the well to support wireline operations. There are also other issues attributed to an intervention style data acquisition such as personnel and equipment costs, equipment failure, miss-runs, safety, and lost production during these operations. In recent years, enhancements made to downhole permanent gauge mandrels have confirmed to provide improved long-term reliability and data accuracy.

Reservoir pressure monitoring is useful, for example, for pressure maintenance information for preventing gas breakout in the reservoir.

- Bottomhole flowing pressure information. This can be useful so that the well can be controlled to prevent asphaltene deposition in the reservoir, for example, or prevent gas breakout in the near wellbore.
- Well productivity information. Monitoring the well productivity will give early indications of many productivity problems. This enables preventative action or early remedial action to be scheduled. Examples include fines production or scale buildup.
- Natural flow well performance. Knowing the bottomhole pressure allows the tubing performance curves to be accurately correlated. It also allows possible identification of increases in frictional pressure (scaling or corrosion of tubing) or lifting problems (liquid loading, for example).
- Artificial lift troubleshooting. Bottomhole flowing pressure (and other measurements perhaps) allows accurate estimates of the gas injection depth during gas-lift or free gas production for the suction of ESPs.
- Production allocation and well testing. If downhole flow rates can be obtained, then the requirement for regular well tests reduces. This is especially useful for subsea wells tied back with a common flow line.

The position of the gauge is therefore important.

Nonsurface Readout Gauge

There are a variety of intervention systems for gauges:

- Traditional tail-pipe nipple and perforated joint in tail pipe. This allows a memory gauge to be positioned below the packer. This system is tried and tested. The major problem is that the gauge creates a low flow spot in the well. This allows debris to build up on top of the gauge, making it hard or impossible to remove or change out the gauge.
- Fluted gauge hanger in nipple. This system uses any nipple to hang off a gauge. The problem is that the flow rate may be restricted either by the gauge or the lock in the nipple.
- Retrievable gauges in a mandrel. Conventional gas-lift mandrels can be used to house gauges. These gauges can either be internal or external sensing. They can be preinstalled in the completion and can be retrieved or replaced using conventional kickover tools. This technique is particularly suitable for use during stimulations, when gas-lift valves often have to be pulled to allow annulus pressure to be increased. The mandrels can also be constructed without a flow path to the annulus, thus avoiding a potential leak path.
- Interrogatable gauges. This system also uses a gauge mandrel, but the gauges do not have to be pulled for the data to be extracted. An inductive coupling allows extraction of the data using slick-line techniques. This avoids multiple wireline runs purely to extract data and allows data to be extracted if the gauges are stuck in the mandrel.

- Deployment of fiber optic sensors through a control line. The sensor highway system uses a conventional control line and fluid drag to deploy sensors (pressure, temperature, strain, noise, magnetic fields, etc.) down a control line. The control line loops downhole and then back to surface.

Chemical Injection Mandrel

A chemical injection mandrel is a method of providing continuous downhole chemical injection to treat wells for the prevention of downhole corrosion, scaling, asphaltene formation, etc. The chemical injection mandrel is installed in the completion string with a dedicated control line connected. When using injection mandrels, it may be possible to minimize the use of high chrome and exotic materials in completions.

The main problem with injection mandrels is the blockage of the injection line. For this reason, as large a bore as possible is recommended and ⅜" or ½" is preferred. The standard control line size of ¼" is very easily blocked.

Expansion Devices and Anchoring Methods

Polished Bore Receptacle (PBR)

The polished bore receptacle (PBR) is designed for use in those applications where extreme tubing movement needs to be accounted for. Where large tubing movements are encountered, correct selection of seal arrangements and elastomers is critical to maintaining integrity. Leaks in PBR assemblies in water injection wells have led to moves to design water injection completions without moving seals. The retrievable PBR can be latched into a retainer production packer or liner hanger by means of an anchor tubing seal assembly. The PBR seal assembly is held in place inside the honed bore during run in by a shear ring. When combined with hydraulic set packers, the entire completion can be run in and set on the production tubing string in one trip. If desired, the well can be flanged up and the tubing string displaced prior to setting the packer. The entire assembly above the packer is retrievable in two trips. The seal assembly is retrieved when tubing string is recovered. The PBR housing is easily recovered in a second trip using the specifically designed retrieving tool.

The PBR has advantages over expansion joints in that the continuous tubing conduit is used offering distinct advantages for running intervention tools. The inner mandrel is normally equipped with a guide to assist plugging devices, etc., back into the tubing conduit during intervention operations.

Expansion Joint Seal Assembly

The expansion joint is a telescoping tool that compensates for tubing movement. It is mainly intended as a simple space-out device for landing the surface tubing hanger. This is particularly useful when there is another fixed point in the completion string relatively close to the surface; e.g., an annulus safety system.

These seal assemblies perform the same function as a PBR and are essentially an inverted PBR with the seals located in the overshot and sealing on a polished tube. Because of the problems in sealing on the outside of a curved polished surface, especially with long seal movement; i.e., extra long tubing seal receptacles (ELTSR), this type of assembly is more difficult to manufacture than a PBR. ELTSRs have tended to be run in conjunction with permanent packers and therefore do not usually have as large an ID as the PBR.

Anchor Seal Nipple

The anchor seal nipple is an anchoring and sealing device that connects the retrievable tubing string to the upper bore of the retainer production packer. The latch component of the seal nipple provides positive engagement with the packer, and the seal unit maintains the pressure integrity of the connection. The anchor seal nipple can be supplied with appropriate seal units.

The anchor seal nipple to production packer system is required to meet the following:

- To provide a sealing safety barrier at the bottom of the tubing as near the productive zone as practicable. This is required to protect the production casing from the corrosive elements of the reservoir products and to protect

the production casing from any high pressures experienced during operations, such as well kill or stimulation.
- To facilitate well workover of damaged tubing without exposing the production zone to damaging fluids. This is achieved by means of placement of a wireline nipple profile in the production packer tail-pipe assembly.
- To provide a tubing anchor point to minimize tubing movement or allow the attachment of a tubing expansion device.
- To assist in well-killing operations by providing a positive safety barrier near the reservoir, which will result in the requirement for lower specific gravity or kill-weight brines.
- Pressure integrity assurance to the liner top.
- Maximized ID.

At the top of the packer is a square thread together with a seal bore. In most applications, this packer is run with an anchor (fixed) seal, which is located in the packer. Although the packer is permanent, the tubing can be removed by picking up approximately 5,000–10,000 lb and rotating approximately 14–15 times to the right.

Shear Release Anchor

The shear release anchor is a modified version of the conventional anchor and is equipped with a shear ring or shear pins. The anchor design to be retrieved by a straight pull through the tubing. Particular attention to injection and stimulation stress analysis is required as the tubing contraction could exceed the force required to shear out the anchor. Picking up approximately 5,000–10,000 lb and rotating approximately 14–15 times to the right can also disengage the anchor.

Hydraulic Release Anchor

The hydraulic release anchor tubing seal assembly is a tool ideally suited to be run between numerous hydraulically set packers run in tandem where downward body movement is necessary to complete pack-off. Once the release mechanism has functioned, the tool is then free to stroke within the packer upper seal bore due to tubing movement caused by pressure or temperature changes in the well.

The release mechanism is actuated hydraulically with no tubing manipulation required.

To suit multizone completion designs, tandem where packer downward body movement is necessary.

No rotation required for retrieving. The anchor retrieves fully when the packer is retrieved.

Production Packer

The production packer is a mechanical device design to provide a seal area between the casing ID and the tubing OD. The packer is also equipped with a slip system that ensures that it is firmly anchored to the casing/liner.

Depending on the well, packers are used for one or more of the following reasons:

- To provide a sealing safety barrier at the bottom of the tubing as near to the productive zone as practicable. This is required to protect the production casing from the corrosive elements of the reservoir products and to protect the production casing from any high pressures experienced during operations, such as well kill or stimulation.
- To facilitate well workover of damaged tubing without exposing the production zone to damaging fluids. This is achieved by means of placement of a retrievable plugging device in the production packer tail-pipe assembly.
- To provide a tubing anchor point to minimize tubing movement.
- To assist in well-killing operations by providing a positive safety barrier near the reservoir, which will result in the requirement for lower specific gravity and kill-weight brines.
- To improve vertical flow conditions and prevent erratic flow and heading cycles.
- To separate pay zones in the same wellbore in a multiple production string arrangement.
- To pack off perforations rather than squeezing cement (bridge plugs).

- To facilitate gas lift or hydraulic power fluid off the formation.
- To install a casing pump.
- To minimize heat losses by use of empty annulus or thermal insulator.
- Pressure integrity assurance to the liner top.
- To isolate casing leaks.
- To facilitate temporary well service operations (e.g., stimulations, squeezes) or well testing with drill stem test (DST).

Packer Components

There are certain basic components that are common to all production packers.

The seal assembly consists of the elastomer packing element together with backup rings. This provides isolation between the annulus and the underside of the packer by packing off against the casing wall. One problem with packing elements is seal extrusion, which is overcome by the backup rings. An alternative technique is to use a combination packing element, with a softer inside element providing the seal, and harder elements on the outside backing up the soft seal and preventing extrusion. The differential pressure a packer will hold is dependent on the stress induced in the element.

This type of packer is set by means of hydraulic pressure in the tubing. The pressure acts on the piston, forcing the piston along the lock ring and pushing the bottom slip over the cone. The cones are basically metal wedges which force the toothed slips out into the casing wall. In this case, the slips on either side of the packing element are opposed (hold in opposite directions) and full circle. As the packer is set, these circumferential slips split into segments as they grip the casing wall. Once the bottom slip contacts the casing wall, the load on the cone builds and the pin locating the cones shears, allowing the element to compress. As the element compresses, the pin in the upper cone shears, forcing the upper slip out into the casing.

Movement of the packing element while under differential pressure will usually result in a packer failure. To avoid this, the lock ring, or ratchet, locks the stress in the element and prevents the slips from disengaging.

At the top of the packer is a square thread together with a seal bore. In most applications, the packer is run with an anchor (fixed) seal which locates in the packer. Although the packer is permanent, the tubing can be removed by picking up approximately 5,000–10,000 lb and rotating approximately 15 times to the right. In the unlikely event that the packer needs to be removed, this can only be done by milling.

Permanent Versus Retrievable

One-trip permanent completions have gained widespread acceptance within the oil industry because of cost saving associated with reducing the time taken to install the completion in the well and bring the well into production. One-trip single string packer completions are relatively straightforward and consequently have become a well proven technique. In fact, it is probably fair to say that most wells in the North Sea are equipped with hydraulic set permanent packers.

Retrievable packers do have their place however and, provided the well conditions are not too severe and not too many pressure reversals are encountered over the life of the well, the retrievable packer may perform satisfactorily.

Permanent Packer

Permanent packers because of their simple rugged construction are also inherently stronger and will generally have a larger through bore in any given casing size than the equivalent retrievable packer.

Advantages

- The packing element system is more resistant to 'swab-off' during completion installation.
- Mechanical strength—once set in the casing, the permanent packer is stronger and is more resistant to high loading in tension or compression.
- Full-cycle slips distribute hydraulic and mechanical loading and minimize casing damage.

- Generally have a larger ID through the packer.
- Normally have a higher differential pressure capability than retrievable packers.
- Few if any O-rings are required.

Disadvantages

- Can only be removed from the wellbore by milling.
- Not reusable once removed from the wellbore.

Retrievable Packer

Retrievable packer element design and construction is not so well boxed in as the permanent packer. Hydraulic and mechanical forces imparted to the element can cause some extrusion and movement of the rubber creating a potential leak path during pressure reversals.

Advantages

- Can be removed from the wellbore intact without milling.
- Certain types of retrievable packer can be retrieved with the completion string.
- Once removed, may be reusable after redress (depending on the severity of the well conditions; i.e., how much corrosion pitting, etc.).

Disadvantages

- If the packer cannot be removed by normal means, milling can be long and problematic.
- Corrosion of the retrieving mechanism may impair retrievability.
- May not be so easy to provide compatibility with well conditions as certain components may require high-strength materials.
- If corrosion resistant materials are used, galling may occur and impair retrievability.

In a nutshell, the benefits of a permanent packer over a retrievable packer can be best summed up as 'strength' and 'durability'.

However, a new generation of retrievable packer has been developed which can withstand greater loads than most permanent systems. The packer is essentially an adaptation of the annular safety system which is designed to withstand loads above 500,000 lb in tension and the same in compression.

Advantages

- If preset, it can be removed from the wellbore intact without milling.
- In workover mode, can be removed from the wellbore intact without milling.
- Once removed, may be reusable after redress (depending on the severity of the well conditions; i.e., how much corrosion pitting, etc.).
- Eliminates the requirement for an expansion device as the packer can withstand the tubing loads.
- Tubing is connected directly to the top of the packer and eliminates any potential elastomeric leak path.
- Corrosion of the retrieving mechanism is protected by a puncture retrieval system.

Disadvantages

- If the packer cannot be removed by normal means, milling can be a long and problematic operation.

Retrieving and Milling

There are generally four methods of packer retrieval. This depends on the individual packer design.

Permanent packers are the most difficult, time-consuming, and costly to retrieve. In order to retrieve a permanent packer, the anchor has to be backed out of the packer and the tubing removed from the well. A milling tool is then made to the drill string and run into the well. If the milled packer is to be retrieved from the well, then the milling tool has to be engaged and the catch sleeve unjayed. An overpull will confirm the milling tool has engaged the packer. The rotary table and mud pumps can now be started and the weight slowly set down on the packer. As the milling tool cuts down over the outside of the packer, it will eventually drop onto the catch sleeve of the milling tool. Once this happens, the rotary and the mud pumps are stopped. The packer can now be pulled from the well.

Retrievable packers (depending on their design) can be retrieved in one of two ways:

1. Using retrieving tools.
2. Tubing manipulation.

When retrieving tools are utilized, the packer is retrieved in a similar fashion to the permanent packer. The anchor and tubing have to be removed first, and a separate trip with a work string is required. In this case, the retrieving tool takes the place of the milling tool. Once the retrieving tool has engaged the packer, a straight pull upward is all that is required to retrieve the packer.

Some retrievable packers can be retrieved by tubing manipulation. This can take the form of a straight pull or tubing rotation.

Mechanical-Set Packer

Mechanical-set packers are frequently deployed in multizone completions, well tests, and when utilized as a sump packer for water shutoff. When completing a well with a mechanical-set packer, the packer is required to be installed prior to the tubing. To install the packer requires a separate run with a hydraulic setting tool. This is normally performed in one of two methods:

1. On an electric line setting tool. This system is generally recommended for setting packers vertically on a low-angle well as deployment of the packer is relatively quick, efficient, and cost effective.
2. On a drill pipe setting tool. These setting assemblies are particularly useful for setting packers in high-angle, deviated wells.

Regardless of what method is adopted, the hydraulic force applied to the setting sleeve from the setting tool initiates the setting action from the top of the packer downward. This force is transmitted to the upper slips through to the packing element and on to the lower slips. As a result, the slips and packing element expand outward which, in turn, anchor and pack off against the casing. The pack-off force is governed by a shear stud or ring contained within the setting tool. The shear stud or ring shears as the pack-off force exceeds their predetermined values allowing the setting to disengage from the packer and be retrieved.

Hydraulic-Set Packer

The hydraulic-set packer is a development of the mechanically-set packer. The advantage with this system is that it can be run and set in a single trip with the completion tubing string. If required, the completion can be deployed and the tubing hanger landed and tested prior to setting the packer. The setting system is actuated by installing a plugging device located below the packer and pressuring up the tubing above; this allows pressure to enter the setting port and actuated setting piston. The setting piston transmits the force through the lower slips packing element to the upper slips. As with the mechanical-set system, the slips and packing element expand outward which in turn anchor and pack off against the casing. The setting piston cross-section area and hydraulic pressure govern the pack-off force.

Dual Bore Packer

Dual bore packers are commonly utilized in multistring completions and, by design, can allow independent production of each zone through separate tubing strings, and can permit gas or water/gas injection in one zone while producing from the other.

Dual bore packers are installed and set by installing a plugging device located below the packer and pressuring up the tubing above. This allows pressure to enter the setting port and actuate the setting piston.

Retrieval is normally achieved by a straight pull through the tubing. Particular attention to injection and stimulation stress analysis is required as the tubing contraction could exceed the force required to retrieve the packer.

Mill-Out Extension (MOE)

Most production packers are run and installed with a mill-out extension mounted on the bottom guide immediately below the packer. The MOE will have a larger ID than the packer and production tubing below. This will facilitate the catch sleeve on the milling tool.

Seal-Bore Extension (SBE)

A seal-bore extension is used to provide an additional sealing bore when a long seal assembly is run to accommodate tubing movement. The seal-bore extension has the same ID as the packer ID. Packers that have continuous seal-bore extensions and no mill-out extension can be milled and retrieved utilizing a specially designed milling tool and spear. Alternatively, they can be milled over and pushed to bottom.

Seal-bore extensions are normally supplied in standard lengths and can be joined together with concentric couplings to achieve any length required.

Remote-Actuated Tools

Remote-actuated tools, sometimes referred to as 'smart' tools, are items of equipment that do not require any intervention runs, such as wireline, coil tubing, or snubbing, to operate them. These tools are relatively new and are designed and ideally suited to highly deviated or horizontal wells where intervention would be extremely costly and difficult or even impossible. The following operations can be achieved with remote-actuated tools:

- Pressure test tubing prior to setting hydraulic set packer.
- Setting hydraulic set packer.
- Provide a barrier from the reservoir to surface.
- Provide a barrier from the tubing to reservoir (loss circulation device).
- Provide a means of attaining annulus to tubing communication.

Remotely-actuated tools can be operated in a variety of different ways, the most common being hydraulic. However, this method is not always ideal and can be restricted by completion component design limitations. Consequently, by utilizing a combination of tools which are operated in different ways, the completion objectives can still be accomplished.

Schlumberger Liner Top Isolation Device (LTIV)

The Schlumberger LTIV is a device located at the liner top that isolates the reservoir from the tubing and can hold pressure in either direction. It has a ball sealing system that, when in the closed position, provides protection to a preperforated reservoir or sand screens and allows the installation of the upper completion.

The LTIV is a pressure cycle-operated tool that utilizes a nitrogen chamber and an indexing mechanism for its operation. A series of pressure cycles applied in the tubing open the ball valve and gain full-bore access to the reservoir. The ball valve can also be opened using a contingency intervention tool should the system fail. The LTIV can also be set up on surface to open at a predetermined number of cycles.

Ocre Tools

Mutifunction cycle tool—full-bore isolation valve (MFCT-FBIV)

Ocre tools (MFCT-FBIV) are designed to install the completion self filling and, when desired, isolate the tubing from the annulus and set a hydraulic-set packer. Tubing pressure integrity is maintained by a flapper valve which holds pressure from above and below and, therefore, can be classed as a barrier. When open it provides full bore access below and above the completion. The Ocre tools are also manipulated by pressure cycles and utilize a ratcheting system and can also be set up to operate at a predetermined number of cycles. The flapper valve can also be opened using contingency intervention tools should the system fail.

Hydrostatic Tools

Hydrostatic tools utilize a hydrostatic chamber and a rupture disc system to actuate. The reservoir has to be isolated by casing in order for this type of tool to operate. A hydrostatically operated tool does not require a plugging device located in the tubing below and is equipped with an atmospheric chamber that is actuated simply by pressuring up the whole wellbore, unlike conventional hydraulically operated tools that require a pressure differential from tubing to annulus.

Close attention must be given to the pressure rating of the rupture disc as this is affected by hydrostatic pressure. If this is overlooked, then it could cause the tools to operate prematurely.

Baker Oil Tools Edge System

The EDGE system uses a pulse frequency-based communication technique to actuate and manipulate downhole completion tools equipped with onboard electronics. The downhole tools (which may include permanent or retrievable packers, sliding sleeves, etc.), are programmed at surface to recognize one of 16 discrete actuation commands from a portable terminal unit. Once in position, the tools are actuated using a sophisticated, computerized Supervisory Control and Data Acquisition (SCADA) system to precisely control the actuation frequency imparted onto the tubing fluid. The SCADA system also allows the operator to gauge the quality of pulses delivered, thus ensuring that communication has been established. As many as 16 devices can be actuated independently of one another in a single wellbore; and, since the system is independent of absolute pressure in the tubing, there is no need to install a tubing plug to set a packer.

By using systems equipped for electronically enhanced remote actuation, tubing tests can be performed at the full rating of the tubing. The advantage with this system is that there are no limiting pressure cycles involved.

Halliburton Mirage Disappearing Plug

The mirage disappearing plug can withstand pressure from above and below and can be considered as a barrier. This device is also a pressure cycle-operated tool, and the plug itself is made from a salt-based material, which is soluble in fluid. After a predetermined number of pressure cycles, the plug disintegrates allowing full-bore access above and below.

PES Anvil Plug

The PES anvil plug has a solid metal plate that can withstand pressure from above and below and is considered to be a barrier. When on depth, the application of 4,000 psi tubing pressure initiates the opening mechanism. Subsequent low-pressure cycles are required to complete the initiating procedure. When the last pressure cycle is bled off, a sleeve powered by hydrostatic pressure pierces through the metal plate and folds it back into a recess in the body of the plug in turn providing full bore access.

Tail Pipe

The tail pipe is the section of tubing below the packer. It can consist of liner seals, nipples, entry guides, or remote-actuated tools (for setting the packer). It can be only a few feet or several thousand feet (in the case of a packer set at wireline accessible angles and a liner set horizontal).

Perforated Joint

The perforated joint is located in the tail-pipe section of completion below the production packer and above either a landing nipple or controlled ID joint. Its purpose is to provide an alternate flow path in cases where wireline deployed memory gauges and flowmeters are required for monitoring bottomhole conditions in wells where production management is essential.

However, there is an inherent disadvantage with this monitoring system whereby, under production condition, the pressure drop across the perforated joint can cause debris fallout which in turn is deposited on top of the wireline lock. This has caused several serious instances of stuck tools, fish, and troublesome workovers.

Wireline Entry Guide (WEG)

The half or full mule shoe wireline entry guide (WEG) is situated on the bottom of the tail-pipe section. The purpose of the WEG is to provide a means of guiding the completion tail-pipe section through the liner top and into the liner section. It is also utilized as a means of guiding intervention tool strings such as wireline and coil tubing, back into the tail pipe without hanging up. Considerable care is required in the design to ensure that both internal and external bevels are appropriate for the access required. A test installation and use of typical intervention tool strings are also recommended.

Self-aligning mule shoes are also available which should require no rotation to ensure that they are engaged.

TYPICAL COMPLETION PROGRAM

A typical completion program will have the following steps.

Assume that the well has been cased and cemented and that perforation will take place after installing the completion.

1. Displace drilling mud to completion fluid (including any cleanup pills deemed necessary)
2. Pressure test production casing
3. Run gauge ring/junk basket to packer setting depth
4. Run tubing, including packer, safety valve, and any other completion equipment
5. Space out as required
6. Land tubing hanger
7. Displace tubing to underbalance fluid
8. Set packer
9. Pressure test tubing
10. Pressure test annulus
11. Test safety valve
12. Install barriers (wireline plugs) as required by barrier philosophy
13. Nipple down BOP
14. Install and test Christmas tree
15. Hook up to production facilities
16. Recover plugs
17. Perforate well
18. Offload and produce well

TYPICAL WORKOVER PROGRAM

1. Bullhead kill well
2. Set plug in tail pipe
3. Establish tubing to annulus communication above packer (tubing punch or sliding sleeve, etc.)
4. Circulate tubing and annulus to kill-weight fluid
5. Install barriers (wireline plugs) as required by barrier philosophy
6. Remove Christmas tree
7. Nipple up BOP
8. Recover barriers
9. Sever tubing above packer if required
10. Recover tubing
11. Recover packer
12. Perform remedial work as required
13. Recomplete well as per typical program above. If the well is live at this stage, then the packer setting operation and displacing to the tubing to underbalanced fluid will require additional barriers to be in place.

CHAPTER 16
Technical Limit Drilling

INTRODUCTION

Technical limit drilling is a performance improvement process that advocates the pursuit of sound engineering and proper planning to both the onshore planning and offshore execution phases of well construction.

It is nothing new and is certainly not rocket science. It was first called technical limit by Woodside, operating on the northwest shelf of Australia in the early 1990s. This was based on the improvements that Unocal, Thailand achieved in the late 1980s and early 1990s. Since then Shell has adopted a similar principle and called it drilling the limit (DTL) and Amerada Hess called it to the limit (T^2L).

The technical limit philosophy is based around two questions relating to performance.

- Where are we now?
- What is possible?

Where Are We Now?

This question can be answered by reviewing past performance or historical data.

Essentially it involves breaking the well down into discrete phases and identifying the conventional lost time or downtime that has occurred. This data can then be reviewed and steps taken to prevent this downtime from occurring again.

Examples of this include:

- Determining the root cause of hole instability on directional wells.
- Running vibration subs to eliminate downhole vibration that was responsible for multiple twistoffs.
- Utilizing a hydraulic swivel packing instead of a conventional swivel packing to eliminate rig downtime.

What Is Possible?

Once the current level of performance has been established, the question then becomes one of where could our level of performance go or what is possible?

This is a two-stage process that first challenges existing practices (that's the way we've always done it, is no longer an acceptable answer) and secondly starts asking, what if?

Challenging existing practice focuses on what is known as invisible lost time (ILT). ILT is time that is not classed as downtime, but is time that is nevertheless not productive or is inefficient.

Examples of this include:

- Slow ROP
- Excessive connection time caused by outdated practices

Asking what if focuses on enhancements to equipment or identifies new technology that will improve the overall time taken to perform a specific operation.

Examples of this include:

- Rotary steerable tools
- Multilaterals
- Dual activity derricks

Perfect Performance

The objective of the previous two steps is to come up with the perfect well or technical limit as in Figure 16.1.

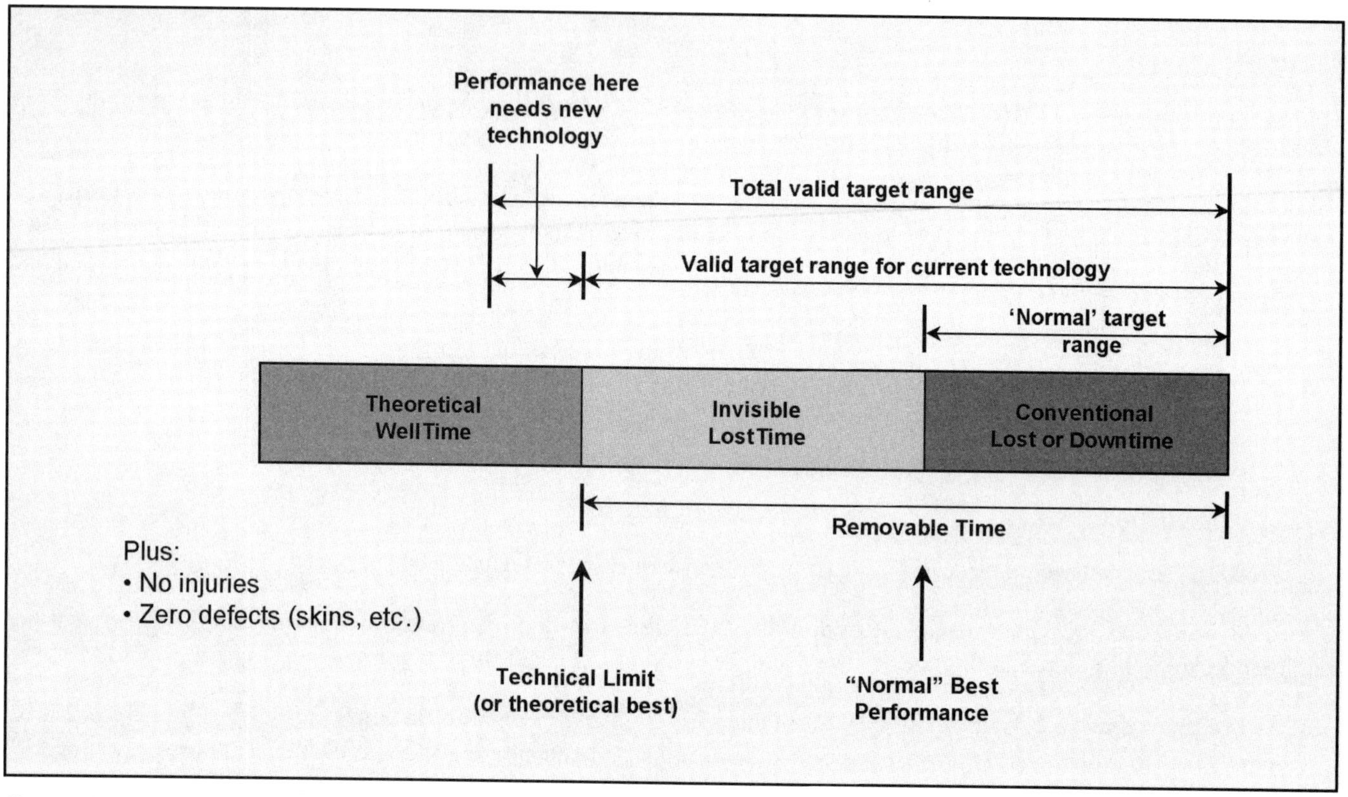

Figure 16.1 Technical limit or perfect well

BASE ASSUMPTIONS

To achieve technical limit, the following assumptions are made:

- No activity is rushed; safety is not compromised.
- All people are competent and perfectly informed as to their tasks.
- All tools, materials, and people are available precisely when needed.
- All tools (including the rig) are fit for purpose and no tools fail.
- All maintenance is done off the critical path.
- The casing program is the minimum needed for well integrity.
- Each section of hole is drilled with one assembly and one bit.
- ROP is the on-bottom composite best in the field or region (including a 5-minute connection time per stand).
- Hole condition is perfect and does not require wiper trips or cleanout trips.

Some of the above might not be possible at the present moment in time (e.g., not all maintenance can be done off the critical path). However, the objective is to show what is possible and to change the thought process. It should now be a question of how can this be done safely and more efficiently rather than how did we do this last time.

TECHNICAL LIMIT—WELL PLANNING

Technical limit should not be considered as a discrete step in the well planning process. It should be considered as an integral part of the complete process.

It involves the following steps:

- **Downtime review.** By identifying the main areas of downtime from offset wells, a suitable improvement plan can be developed and implemented, to ensure that downtime is not repeated.
- **Peer reviews.** Peer reviews at various stages of the well design are held to ensure that opportunities are being identified and risks mitigated.
- **Technical limit workshops.** Technical limit workshops, using operator, drilling contractor, and service company personnel from onshore and offshore, are held to ensure:
  - the complete understanding of technical limit by all parties,
  - to challenge existing practices,
  - to identify additional opportunities, and
  - to set technical limit times for the individual steps that make up the total time of each phase.
- **Improvement plan.** All ideas and suggestions from the above steps are consolidated onto an improvement plan. The improvement plan details specific actions, the individual responsible for closing them out, and the deadline for closing them out (if required the improvement plan can form part of the lessons learned—see below).
- **Communication.** An important aspect of technical limit is the emphasis on communicating not only the improvement plan but the status of the actions on the improvement plan to all parties involved.

TECHNICAL LIMIT—OPERATIONS

Technical limit should not be considered as a discrete step in the drilling of the well. It should be considered as an integral part of the complete process.

It involves the following steps:

- **Communication.** An important aspect of technical limit is the emphasis on communicating. The objective is to get the involvement of all personnel, and supporting and encouraging them to challenge existing practice and identify new ways of doing things.
- **Prejob planning meeting.** Meeting held at least a day prior to the start of an operation (e.g., casing running, drilling a hole section, logging, etc.). Involves the senior supervisors from all relevant disciplines associated with the operation and any others as required. Objective is to review the work plan as laid out in the drilling program, review the equipment requirements, review the lessons learned from the previous time the operation was performed, challenge existing practices, and identify and record changes that can be made to the benefit of the operation.
- **Task planning meeting.** Meeting held the shift prior to a specific task being undertaken (e.g., run BOP, picking up DP, rigging up to run casing). Involves all personnel associated with the operation. Objective is to review the specific task plan, review the lessons learned from the previous time the task was performed, challenge existing practices, and identify and record any changes that can be made to the benefit of the operation.
- **Wash-up meeting.** Meeting held at the end of a job, or as required. Involves all personnel associated with the job. Objective is to review the job, capture the lessons learned, and identify future improvements.
- **Lessons learned.** Lessons learned are gathered all the time, not only from all of the above meetings, but also from one-to-one conversations, safety meetings, tool box talks, THINK drills, START cards, etc. The lessons learned will be entered onto the lessons learned register by the rig-site nominated technical limit champion. The lessons learned register is available at any time for review and is used in the planning process for future wells.
- **Monitoring and reporting.** The technical limit monitoring process revolves around comparing the actual with the planned or target times and displaying these graphically. These charts can then be displayed on the rig and in the office so that the performance becomes highly visible to everyone. Other performance indicators (e.g., casing, tripping and BOP running times, percent downtime) can be measured and displayed.